Natural and Anthropogenic Hazards in Karst Areas: Recognition, Analysis and Mitigation

Geological Society books refereeing procedures

The Society makes every effort to ensure that the scientific and production quality of its books matches that of its journals. Since 1997, all book proposals have been refereed by specialist reviewers as well as by the Society's Books Editorial Committee. If the referees identify weaknesses in the proposal, these must be addressed before the proposal is accepted.

Once the book is accepted, the Society Book Editors ensure that the volume editors follow strict guidelines on refereeing and quality control. We insist that individual papers can only be accepted after satisfactory review by two independent referees. The questions on the review forms are similar to those for *Journal of the Geological Society*. The referees' forms and comments must be available to the Society's Book Editors on request.

Although many of the books result from meetings, the editors are expected to commission papers that were not presented at the meeting to ensure that the book provides a balanced coverage of the subject. Being accepted for presentation at the meeting does not guarantee inclusion in the book.

More information about submitting a proposal and producing a book for the Society can be found on its web site: www.geolsoc.org.uk.

It is recommended that reference to all or part of this book should be made in one of the following ways:

PARISE, M. & GUNN, J. (eds) 2007. *Natural and Anthropogenic Hazards in Karst Areas: Recognition, Analysis and Mitigation.* Geological Society, London, Special Publications, **279**.

NISIO, S., CARAMANNA, G. & CIOTOLI, G. 2007. Sinkholes in Italy: first results on the inventory and analysis. *In*: PARISE, M. & GUNN, J. (eds) *Natural and Anthropogenic Hazards in Karst Areas: Recognition, Analysis and Mitigation.* Geological Society, London, Special Publications, **279**, 23–45.

GEOLOGICAL SOCIETY SPECIAL PUBLICATION NO. 279

Natural and Anthropogenic Hazards in Karst Areas: Recognition, Analysis and Mitigation

EDITED BY

M. PARISE
National Research Council, Research Institute for Hydrogeological Protection, Bari, Italy

and

J. GUNN
Limestone Research Group, University of Huddersfield, UK

2007
Published by
The Geological Society
London

THE GEOLOGICAL SOCIETY

The Geological Society of London (GSL) was founded in 1807. It is the oldest national geological society in the world and the largest in Europe. It was incorporated under Royal Charter in 1825 and is Registered Charity 210161.

The Society is the UK national learned and professional society for geology with a worldwide Fellowship (FGS) of over 9000. The Society has the power to confer Chartered status on suitably qualified Fellows, and about 2000 of the Fellowship carry the title (CGeol). Chartered Geologists may also obtain the equivalent European title, European Geologist (EurGeol). One fifth of the Society's fellowship resides outside the UK. To find out more about the Society, log on to www.geolsoc.org.uk.

The Geological Society Publishing House (Bath, UK) produces the Society's international journals and books, and acts as European distributor for selected publications of the American Association of Petroleum Geologists (AAPG), the Indonesian Petroleum Association (IPA), the Geological Society of America (GSA), the Society for Sedimentary Geology (SEPM) and the Geologists' Association (GA). Joint marketing agreements ensure that GSL Fellows may purchase these societies' publications at a discount. The Society's online bookshop (accessible from www.geolsoc.org.uk) offers secure book purchasing with your credit or debit card.

To find out about joining the Society and benefiting from substantial discounts on publications of GSL and other societies worldwide, consult www.geolsoc.org.uk, or contact the Fellowship Department at: The Geological Society, Burlington House, Piccadilly, London W1J 0BG: Tel. +44 (0)20 7434 9944; Fax +44 (0)20 7439 8975; E-mail: enquiries@geolsoc.org.uk.

For information about the Society's meetings, consult *Events* on www.geolsoc.org.uk. To find out more about the Society's Corporate Affiliates Scheme, write to enquiries@geolsoc.org.uk.

Published by The Geological Society from:
The Geological Society Publishing House, Unit 7, Brassmill Enterprise Centre, Brassmill Lane, Bath BA1 3JN, UK

(*Orders*: Tel. +44 (0)1225 445046, Fax +44 (0)1225 442836)
Online bookshop: www.geolsoc.org.uk/bookshop

British Library Cataloguing in Publication Data

A catalogue record for this book is available from the British Library.

ISBN 978-1-86239-224-3

Typeset by Techset Composition Ltd, Salisbury, UK

Printed by Antony Rowe, Chippenham UK

Distributors

North America
For trade and institutional orders:
The Geological Society, c/o AIDC, 82 Winter Sport Lane, Williston, VT 05495, USA
Orders: Tel. + 1 800-972-9892
Fax + 1 802-864-7626
E-mail: gsl.orders@aidcvt.com

For individual and corporate orders:
AAPG Bookstore, PO Box 979, Tulsa, OK 74101-0979, USA
Orders: Tel. + 1 918-584-2555
Fax + 1 918-560-2652
E-mail: bookstore@aapg.org
Website http://bookstore.aapg.org

India
Affiliated East-West Press Private Ltd, Marketing Division, G-1/16 Ansari Road, Darya Ganj, New Delhi 110 002, India
Orders: Tel. + 91 11 2327-9113/2326-4180
Fax + 91 11 2326-0538
E-mail affiliat@vsnl.com

Contents

Natural and anthropogenic hazards in karst areas: an introduction

M. PARISE[1] & J. GUNN[2]

[1]*National Research Council of Italy, IRPI, Bari, Italy (e-mail: m.parise@ba.irpi.cnr.it)*

[2]*Limestone Research Group, University of Huddersfield, Queensgate, Huddersfield HD1 3DH, UK (e-mail: j.gunn@hud.ac.uk)*

The distinctive hydrology and landforms of karst create a very special environment. Although several types of karst have been identified worldwide, a common thread is the dominantly subterranean drainage. The paucity of water flowing at the surface, a consequence of rapid infiltration underground through a network of discontinuities in the soluble rock mass, results in two important but contrasting points: the considerable value of karst water resources (representing about 25% of the drinkable supply in the world) is strongly counteracted by the ease with which human activities can negatively impact this precious resource. The same narrow discontinuities, and the larger dissolution conduits and karst caves, are the main pathways through which potential pollutants may travel swiftly to regional groundwater bodies, or directly to springs. Contaminants can be introduced by means of dispersed infiltration as well as from point sources and are frequently transmitted with minimal filtering. This example, just one of the many natural and/or anthropogenic hazards that may affect karst areas, illustrates the fragility of karst environments. Their high vulnerability is further expressed by a very simple concept that is true for many other environments but probably shows its best evidence in karst: it is very easy to damage or destroy natural resources but restoration to a pristine situation is an extremely difficult and commonly impossible, task. Where some degree of remediation is possible, the economic cost is commonly very high.

Since the early 1960s there has been a great deal of progress in understanding the processes and landforms of karst areas (e.g. Ford & Williams 1989; Gillieson 1996; Klimchouk *et al.* 2000; Gabrovsek 2002). There have also been several initiatives designed to translate theoretical knowledge into practical application, for example work by the International Geographical Unions Karst Commission (Williams 1993; Barany-Kevei & Gunn 2000), the IUCN (1996) and the World Bank (Vermeulen & Whitten 1999). However, projects continue to fail, and resources are lost or damaged owing to a failure in understanding the peculiarities of karst (for examples see Waltham *et al.* 2005). Hence, there remains a strong need to develop appropriate techniques to manage karst landscapes and protect karst resources.

There are many human activities that, intentionally or not, produce severe impacts in karst, often with irreparable damage. For example, in some regions land degradation has been intense with deforestation and overgrazing leading to soil erosion, destruction of the epikarst and rocky desertification. Rehabilitation of these desert lands is extremely difficult, although in some parts of the Mediterranean reduction of grazing has been followed by a return of shrubs and small trees. Legislation, and even more its practical enforcement, still appears to be extremely inadequate at facing these problems, even in well-developed countries. Lack of laws and acts specifically devoted to karst is very common, but even when legislation takes into account the complexity and peculiarity of karst the potential benefits deriving from these acts frequently remain on paper owing to a lack of enforcement and control by the authorities.

The best example is probably provided by the management of aquifers in carbonate rocks, as this requires an understanding of the character and type of water circulation within the rock mass. For many years the principles developed for non-carbonate aquifers, and the resulting methodologies created for groundwater vulnerability assessment, have also been applied to karst environments. However, it has long been known that the planning and management of carbonate aquifers poses different problems to those encountered in non-carbonate groundwater. Karst groundwater watersheds often do not coincide with topographic divides, to indicate just one of the main differences between karst hydrology and those in other environments. Only in recent years have some significant efforts been produced to give due weight to the peculiarity of karst hydrology and hydrogeology (for example, the COST 65 (1995) and subsequent COST (Co-Operation in Science and Technology) programmes in Europe and the report by Eckenfelder Inc. (1996) in the United States).

From: PARISE, M. & GUNN, J. (eds) *Natural and Anthropogenic Hazards in Karst Areas: Recognition, Analysis and Mitigation*. Geological Society, London, Special Publications, **279**, 1–3. DOI: 10.1144/SP279.1
0305-8719/07/$15.00

This Special Publication brings together papers presented at the Second General Assembly of the European Geosciences Union in Vienna, Austria, 24–29 April 2005, together with invited contributions. The authors discuss, and illustrate by means of case studies from a variety of different karst areas, some of the most common geohazards in karst, together with some appropriate mitigation procedures and operations. The book is divided into three sections dealing with collapse and subsidence hazards, hydrological hazards, and problems in managing karst. Collapses and subsidence phenomena are among the more subtle hazards in karst owing to the difficulty in predicting such events, and particularly the potential for rapid to very rapid velocity in their final stage, which often leads to catastrophic damage. The development of collapse dolines (sinkholes) provides a good illustration of the interconnections that exist in karst between the surface and the subsurface. Their genesis and later evolution, up to the eventual, potentially catastrophic, collapse, are related to the development of an underground cavity, with progressive upwards stoping leading to the formation of a collapse at the surface. The topic is treated through description of several examples from different parts of the world: **Brinkmann *et al.*** illustrate the distribution of sinkholes in a heavily urbanized county in Florida, USA, whilst **Mochales *et al.*** describe the use of geophysical prospections to identify dolines in the Ebro Basin in northern Spain, and **Ardau *et al.*** deal with cover-collapse sinkholes in the Sardinia Island of Italy. In addition to the case studies that illustrate doline development, distribution and evolution, the mechanisms of collapses over open natural caves are also discussed in the contribution by **Waltham & Lu**. This first section is further enriched by **Nisio *et al.*** with analysis of an inventory of dolines recently carried out in Italy, and by a contribution from **Santo *et al.*** that highlights the role of karst processes in the genesis and/or the evolution of instabilities other than dolines, namely rock falls and landslides.

Water is crucial for karst processes. It is one of the key factors, together with carbon dioxide, that initiate the dissolution of carbonate rocks and, later on, influence the type of karst. At the same time, groundwater contained in karst aquifers represents a remarkable resource that needs to be properly safeguarded. Water can also be destructive in karst, with natural hazards such as flash floods when the surface and subsurface karst systems do not have the capacity to transmit the huge amount of water deriving from intense or prolonged rainfalls. All these issues are examined in the second section of the book where the focus is on hydrological hazards.

Case studies, and the outcomes of research projects carried out to deepen our knowledge of the hydrogeological features of karst, the transport of water in karst areas and the possible contamination from pollutants are presented. **Cossu *et al.*** describe the flood that affected central-east Sardinia in 2004; **Gunn** presents an interesting attempt to define the protection zones of groundwater source in carbonate aquifers, aimed at mitigating the related hazard. Two contributions regard pollution of karst groundwater, through the example of the flow in the unsaturated zone of the Chalk in England presented by **Allshorn *et al.*** and a study of the Xingwen aquifer in Sichuan, China by **Bottrell**. This section is completed with the contributions by **Ducci** on the assessment of the vulnerability of karst groundwater in the Alburni Massif of southern Italy, one of the most important areas in the country as regards speleological research, and by **Delle Rose *et al.*** describing the impact on groundwater resources from extensive quarrying in a karst area of Apulia, again in southern Italy.

The complexity of karst, and the necessity for comprehensive studies requiring specialists from a great range of disciplines, pose many management problems in karst areas. In particular, every action carried out at the surface may have a rapid, and potentially dangerous, effect underground. It is extremely easy to unflict damage on the environment, but restoration of the original situation is a very difficult, sometimes impossible, task. To cover the issue of karst management, and the related natural and anthropogenic hazards, two examples are presented in the final section of the book. The first one, by **Day**, deals with tropical surface karst in Jamaica, and provide a overview of the many hazards affecting this country. In contrast, the second example comes from France, by **Chalmin *et al.***, and points to the need for protection of the heritage contained in natural caves, and specifically prehistoric paintings, that may be severely damaged by both natural processes and human impact. This last chapter of the book provides a link between the first presence of humans in the underground environment and the use (or, unfortunately, the frequent misuse) of natural karst resources by modern human society. It is probably correct to affirm that, notwithstanding human evolution and the development of technologies, humans have progressively lost their original ability to live ‘in a sustainable way’ in the natural environment. The concept of sustainability in management of human activity is based on attaining a balance between social, environmental and economic outcomes. Sustainable management of karst, and of its resources, must be a priority for all those working in and studying karst areas. To exploit without producing damage, to respect the environment and to offer to those populations living in karst the opportunity to develop and improve their style of life through the proper exploitation of karst, all of these tasks must be pursued. To this aim, it is necessary to attain a good knowledge of the intrinsic characters of karst systems, and of the effects produced by the human activities on it as well. As Editors, we hope that this book will provide a contribution to improve our understanding of the natural and anthropogenic hazards in karst.

References

BARANY-KEVEI, I. & GUNN, J. (eds) 2000. Essays in the Ecology and Conservation of Karst. *Acta Geographica Szegedensis*, **XXXVI**, (Special Issue).

COST 65. 1995. *Hydrogeological aspects of Groundwater Protection in Karstic Areas*. Final Report (COST Action 65). European Commission Directorate-General XII Science, Research and Development, Report EUR 16547 EN, Brussels.

ECKENFELDER INC. 1996. *Guidelines for Wellhead and Springhead Protection Area Delineation in Carbonate Rocks*. Prepared for US EPA Region 4.

FORD, D. C. & WILLIAMS, P. W. 1989. *Karst Geomorphology and Hydrology*. Unwin Hyman, London.

GABROVSEK, F. (ed.) 2002. *Evolution of Karst: From Prekarst to Cessation*. Postojna-Ljubljana: Zalozba ZRC.

GILLIESON, D. 1996. *Caves*. Blackwell, Oxford.

IUCN. 1996. *Guidelines for Cave and Karst Protection*. The World Conservation Union, Gland, Switzerland.

KLIMCHOUK, A., FORD, D., PALMER, A. & DREYBRODT, W. (eds) 2000. *Speleogenesis: Evolution of Karst Aquifers*. National Speleological Society, Huntsville, AL.

VERMEULEN, J. & WHITTEN, T. 1999. *Biodiversity and cultural heritage in the management of limestone resources: lessons from East Asia*. The World Bank, Washington, D.C.

WALTHAM, A. C., BELL, F. & CULSHAW, M. G. (eds) 2005. *Sinkholes and Subsidence: Karst and Cavernous Rocks in Engineering and Construction*. Springer, Berlin.

WILLIAMS, P. W. (ed.) 1993. *Karst Terrains: Environmental Changes and Human Impact*. Catena Supplement, **25**, Cetena, Cremlingen-Destedt, Germany.

Sinkhole distribution based on pre-development mapping in urbanized Pinellas County, Florida, USA

R. BRINKMANN[1], K. WILSON[1], N. ELKO[2], L. D. SEALE[1], L. FLOREA[1] & H. L. VACHER[1]

[1]*University of South Florida, Tampa, FL 33617, USA (e-mail: rbrinkmn@cas.usf.edu)*

[2]*Pinellas County Government, 513 S. Ft. Harrison Avenue, Clearwater, FL 33756, USA*

Abstract: Locating sinkholes in Pinellas County, Florida, is confounded by the presence of a cover of Quaternary sediments that mute the surface appearance of these sinkholes. As a first step in addressing the sinkhole hazard in the county, we analysed aerial photographs from 1926 and 1995 that covered the entire county. We digitized all identifiable sinkholes in each set of photographs in a GIS (Geographical Information System) using a set of criteria established to differentiate between karst depressions and depressions resulting from other geological processes. The 1926 photographs, although of low quality, helped to establish a baseline prior to urbanization. The 1995 photographs provided a post-urbanization distribution of natural sinkholes and man-made depression features (e.g. retention ponds). From these two data sets, we are able to assess natural and anthropogenic changes in the karst landscape of the study area. In particular, we discovered that 87% of the sinkhole features identified in the 1926 photographs are no longer present in the photographs from 1995. Many of the lost depressions have been incorporated into retention ponds.

Pinellas County, in west-central Florida (Fig. 1), in the USA, is located in one of the most active karst regions of the world. Karst, a collection of surface and subsurface landforms produced from the dissolution of limestone and other soluble rocks, is present in many regions throughout the world and is of distinct interest to individuals and organizations associated with these landscapes. This is due, in part, to unique engineering difficulties and contaminant transport properties of karst aquifers. Pinellas County is a peninsula and is underlain by distinct layers of Cenozoic limestone that are covered in most areas by Quaternary sands. The county is modified by coastal processes on the current shoreline, and marine terraces indicate the presence of higher sea levels. In the interior portions of the county, numerous karst depressions are present.

The most identifiable surface features in the karst landscape of west-central Florida are sinkholes. However, in contrast to classical karst landscapes of the Central Lowlands and Appalachian Mountains in the USA, the topographic expression of these depressions is partially masked by Quaternary sediments deposited by eolian processes or during higher sea levels. Within the past century, humans have filled many depressions and modified others for storm-water retention as part of the rapid and intense urbanization of this coastal area. Mapping and analysis of sinkholes was not completed prior to or after development to assess the distribution of karst features in Pinellas County or to document changes in topography as a result of urbanization. Pinellas County, like much of the Tampa Bay region and Florida in general, is rapidly urbanizing. Construction projects significantly impact the karst landscape and vice versa. Without a clear understanding of where karst features are located it is difficult to implement appropriate land-use strategies and management decisions that are suitable for the site geology and of benefit to the citizens of the county. As Pinellas County loses remaining green space, increases densities in existing developments and redevelops brownfields, it will be critical to understand the landscape that exists beneath the veneer of human development because the underlying geological processes that created the original surface features will continue to modify the soil, bedrock and topography.

In Florida, sinkholes occur in clusters or 'sinkhole regions' (Tihansky 1999). These clusters represent areas of past sinkhole formation and are locations where future sinkholes may develop, although the occurrence of topographic sinkholes does not predict the development of new cover collapses. Certainly, sinkhole regions may be hidden by geological deposition or human development, and may extend into or exist within urbanized regions. A useful method of identifying

From: PARISE, M. & GUNN, J. (eds) *Natural and Anthropogenic Hazards in Karst Areas: Recognition, Analysis and Mitigation*. Geological Society, London, Special Publications, **279**, 5–11. DOI: 10.1144/SP279.2

Fig. 1. Location of Pinellas County, Florida.

masked (obliterated) areas affected by subsidence is to map topographic depressions and wetland features that are probably the result of karst processes using historic air photographs and maps. Such an exercise will not only assist county planners, it will aid our scientific understanding of karst processes in Florida and the extent of sinkhole regions that existed in the pre-urbanized past. Such an exercise has rarely been conducted over such a large area.

Pinellas County is underlain by the unconfined Floridan Aquifer System and, as in most carbonate terrain, karst processes exert significant control on the geomorphology (White 1970, 1988; Schmidt & Scott 1984; Lane 1986; Miller 1986). Aggressive groundwater circulation and diagenetic alteration has dissolved limestone and increased porosity within the stratigraphic framework of the Floridan Aquifer. This solution porosity dominates groundwater flow. The complex nature of conduit systems in the unconfined Floridan Aquifer is becoming better understood (Miller 1997). However, the difficulty of locating and exploring submerged conduits and the complex porosity makes analysis and modelling difficult.

In Pinellas County, surficial expression of karst is limited to sinkholes, sinkhole-associated features and springs. Karst features provide unique problems for planners and environmental scientists. Land stability is perhaps the issue citizens are most familiar with and are the most concerned about, although contaminant transport and water quantity are also important issues.

Sinkholes form from a variety of processes (e.g. White 1988; Upchurch & Randazzo 1997). The bulk of sinkholes in Pinellas County are categorized as cover-collapse sinkholes that form when Pleistocene sands above limestone ravel into void space in bedrock (Tihansky 1999). The spatial distribution of these sinkholes appears linear on many Florida maps. Sinkholes form in linear trends because they are localized by joints and fractures that facilitate groundwater flow through the soluble rock.

Previous efforts have described and/or mapped karst features in Pinellas County. For example, Sinclair *et al.* (1985) analysed sinkholes in west-central Florida and described their formation based on cover and bedrock type. They also reviewed data on recently formed sinkholes in the region. However, they did not map them. Frank & Beck (1991) assessed the causes of subsidence damage in a small area of the town of Dunedin in northern Pinellas County. They examined the mechanisms behind the formation of subsidence in the area in the years 1990–1991 and found that almost all of these damaging subsidences were cover-subsidence sinkholes, although this could not be guaranteed for 100% of the damage.

In 1991, Beck & Sayad (1991) published a summary of the sinkhole hazards in Pinellas County. They utilized reported sinkhole events to assess the overall hazard from subsidence. Stewart *et al.* (1995) examined foundation failures in Pinellas County to assess the causes of failures. More recently, the Florida Geological Survey mapped closed depressions on United States Geological Survey (USGS) 1:24 000 topographic maps as part of their overall Florida Aquifer Vulnerability Assessment (FAVA) programme. A variety of other attempts have been made to map sinkholes in other parts of Florida, including nearby Pasco County (Currin & Barfus 1989).

Although all of the past research is useful, to date, there has been no effort to map all sinkholes in Pinellas County, including those lost to urbanization. Therefore, the focus of this project is to develop a county-wide map of sinkholes present on historic and recent aerial photographs in order to assess the number and spatial distribution of depressions lost through development, and in order to understand how pre-existing sinkholes may have been modified through urbanization.

One measure of the human impact on karst landforms is to assess the number of obliterated depressions and particularly those modified to create storm-water retention ponds. Storm-water retention ponds are used to manage storm-water drainage to prevent flooding during common high-intensity storm events in the rainy season and to allow time for polluted runoff to clear up before discharging into the aquifer or surface streams. If natural depressions like sinkholes are used as sites for storm-water pond creation, the degradation of groundwater quality may be aggravated because polluted surface water can quickly enter aquifer systems through sinkhole-ravelling zones. The unfortunate use of the sinkholes may also result in the reactivation of the subsidence features, with the consequent rise in the sinkhole frequency. Storm-water retention ponds became the best management tool for urban storm water in the 1970s when the combined effects of flooding and surface-water pollution caused environmental problems. Thus, many areas of Florida are dotted with these ponds.

Recent rules preclude the use of the sinkholes as sites for pond construction. Yet, many sinkholes were used prior to the approved establishment of these rules. Thus, we seek to understand the extent of storm-water construction in natural depressions in Pinellas County in order to assess the potential detrimental effects on the karst aquifer. This is particularly important because water-retention ponds are more frequently being used as sites of storm-water storage to prevent polluted water from entering nearby surface-water bodies like the important Tampa Bay estuary or many of the lakes and small streams that flow in Pinellas County.

Methods

Karst features were mapped on aerial photographs from 1926 and photographs from 1995 imported into a GIS (Geographical Information System). Four GIS layers were created using the 1926 black and white photographs: sinkholes, possible sinkholes, exclusion areas (areas that were obscured or unclear in the photographs) and developed areas (those areas that showed notable development such as clusters of roads and/or structures). The 1995 aerial photographs are full-colour high-resolution infrared images. Using these images, four GIS layers were constructed: sinkholes, possible sinkholes, water retention ponds and undeveloped and/or non-urbanized areas (mainly wetlands and forest).

Sinkholes were defined by vegetation changes or variation, presence or absence of water, soil moisture and shape. The criteria used to identify these features on both the 1926 and 1995 photographs were moisture content and circular and/or combined circular shapes. The features in the possible sinkhole layer on both the 1926 and 1995 air photographs were identified as having some moisture and circularity conditions, but lacking clear indications of one of those factors (moisture condition or circularity). Lakes that appeared to be depressional on the photographs were considered to be sinkholes. It must be noted that the 1926 air photographs were taken during the dry season in Florida so many of the sinks that would normally contain water were dry. So, vegetation patterns and coloration from soil moisture were used to identify the circular patterns. The 1926 air photographs are black and white, and of poor quality. Many of the areas of these photographs are completely unusable for photographic interpretation. Thus, an exclusion layer was created to remove these areas when carrying out density estimates and other analyses.

The features identified in the retention pond layer in the 1995 air photographs were those depressions considered to be man-made with at least one straight edge.

The sinkhole density in Pinellas County was calculated using a total land area of Pinellas County of 725 km^2 for 1926 and 1995. In addition, the total area and percentage of land area of sinkholes was calculated for 1926 and 1995. From this we were able to calculate the percentage loss of the features that were identified. Then, a map with a layer consisting of all sinks in 1926 and a layer of all sinks from 1995 was constructed. An intersect was then performed in ArcGIS and the results were the sinks that are still present in 1995. We also attempted to determine which sinkholes that were identified in 1926 are now storm-water retention ponds. To do this, we created a map that consisted of the intersect of the layer of total sinks identified in 1926 and the retention pond layer constructed in 1995.

Table 1. *Total area, density and per cent land area of sinkholes using the 1926 and 1995 air photographs*

	Total number of features	Total area (km^2)	Density (per km^2)	Per cent land area (%)
1926				
Likely sinkholes	1570	19.34	2.20	2.70
Possible sinkholes	1133	24.56	1.59	3.44
Total	2703	43.90	3.79	6.20
1995				
Likely sinkholes	261	1.60	0.36	0.22
Possible sinkholes	639	3.97	0.88	0.55
Total	900	5.57	1.24	0.77
Retention ponds	1646	13.25	2.27	1.83

Results

We identified 1570 depressions using the 1926 air photographs of Pinellas County. These probable, or likely, sinkholes account for nearly 20 km^2 (2.7%) of land (Table 1, Fig. 2). In addition, we identified 1133 depressions that are possible sinkholes. While the number of possible sinkholes is smaller than likely sinkholes, the area covered is larger at nearly 25 km^2 (3.4%). Collectively, this accounts for a density of 3.8 sinkholes km^{-2}. It must be noted that several areas of Pinellas County were already developed in 1926 (Fig. 3). These areas account for some of the earliest urbanized areas of the Florida peninsula in the vicinity of present-day downtown St Petersburg and the fishing community of Tarpon Springs. However, the developed portions account for only 14.5% of the total land area. Certainly, some sinkhole modification occurred in these areas, but the

Fig. 2. Distribution of depressions in 1926 in Pinellas County, Florida.

Fig. 3. Developed areas of Pinellas County, Florida.

vast majority of the county remained undeveloped at the time the air photographs were collected.

A total of 261 depressions were identified on the 1995 air photographs that are likely sinkholes (Table 1, Fig. 4). This number contrasts sharply with the number of sinkholes identified on the 1926 air photographs and accounts for only 0.2% of the modern land surface. Likewise, only 639 possible sinks were identified in 1995. While this number is greater than the number of likely sinkholes found in 1995, these depressions only cover 0.6% of the land surface. Collectively, the sinkhole density in 1995 was only 1.2 sinkholes km^{-2}. A great deal of urbanization has taken place over the 1926–1995 period. The undeveloped land portions account for only 2.7% of the county. It is evident that development significantly altered the karst landscape over the 72 year period of study. Most of the undeveloped land areas were lowlying and adjacent to Tampa Bay or in terrestrial wetlands in the NE portion of the county. A total of 87% of the sinkholes and possible sinkholes mapped in 1926 were buried or masked by 1995.

It is striking to compare similar land areas in 1926 and 1995 in order to see the loss of depressions and the extent of urbanization that has taken place. One of the most obvious examples can be found in central Pinellas County and clearly demonstrates the dramatic loss of depressions through urbanization (Fig. 5). The dark areas in the older air graphphoto in Figure 5 largely represent depressional features probably formed through karstification and subsidence

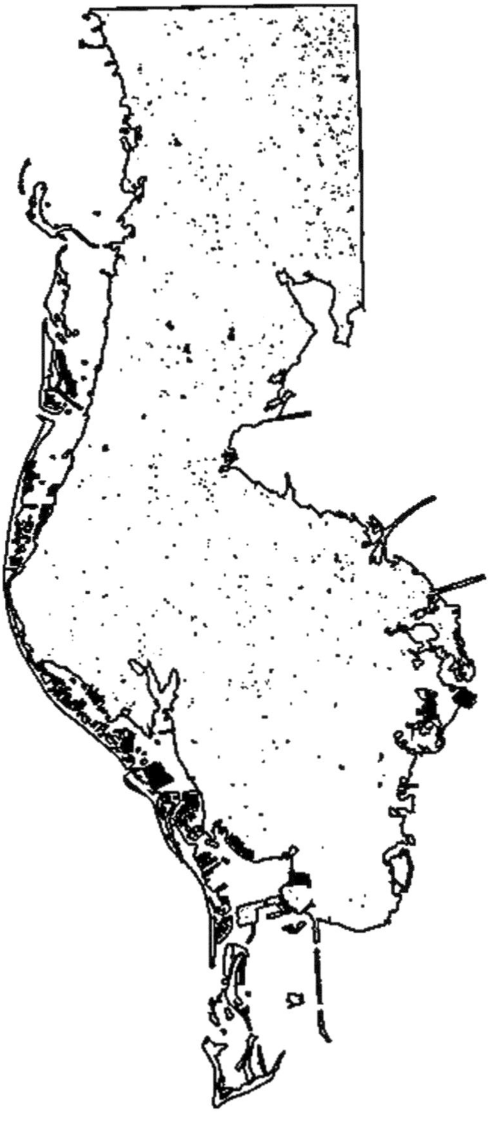

Fig. 4. Distribution of depressions in 1995 in Pinellas County, Florida.

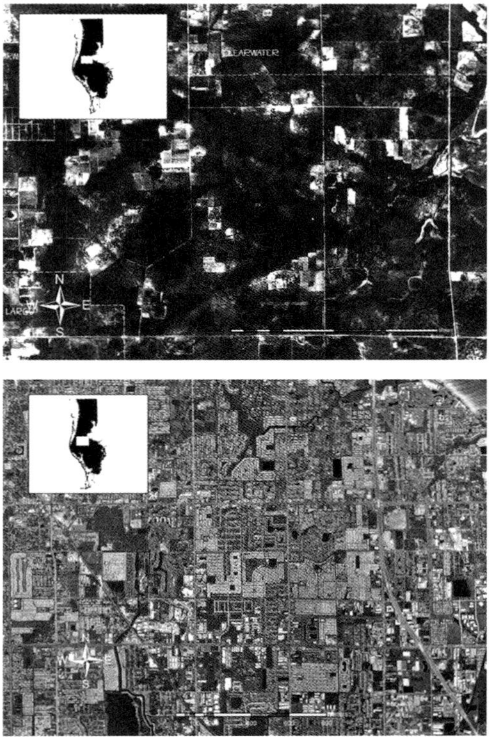

Fig. 5. Comparison of 1926 and 1995 air photographs. The photograph at the top is an image of an undeveloped portion of Pinellas County. The image below shows the same area in 1995. Note the loss of depressions throughout the area.

processes. Many of these features had gone by 1995, a time when urbanization and the associated regular grid system significantly modified the landscape, destroying many depressions. Sinkholes were most probably filled or otherwise modified as development progressed in the region.

Of particular interest is the presence of water retention ponds in Pinellas County present in the 2000 air photographs and how the distribution of these features coincides with depressions mapped on the 1926 air photographs. We identified 1646 water retention ponds on the 2000 air photographs. Approximately 21.8 km^2 of retention pond area overlapped with depressions mapped on the 1926 air photographs (Fig. 6). Of the 2703 depressions mapped in 1926, 499 of them have been converted or altered in some way into areas of water retention ponds. This figure, approximately 18% of all natural depressions in the county, does not necessarily mean that these water retention ponds pose a threat to groundwater. Indeed, we did not attempt to evaluate whether or not the retention ponds were lined or in some other way mitigated to prevent harmful effects of drainage through ravelling zones. However, it does demonstrate that there is a potential hazard to the subsurface system.

Although we did not intend to conduct any geomorphic description of sinkholes in the county, it is important to note that there are distinct differences in the karst landscape in the region (Fig. 2). The northern portion of the county had the most complex karst depressions. Many of these features were uvalas that certainly interconnected during high water flow. There may have been some fluvial modification of these features that occurred when subsurface drainage could not keep up with heavy rainfalls common during summer or during tropical storm events. Further south, the depressions are more rounded and exist in a less dense pattern than those in the north. It is not particularly surprising that the northern portion of Pinellas County is the last portion to undergo urbanization. Much of the area is in the form of low karst depressions and is thus difficult to develop.

Fig. 6. Distribution of water retention ponds in Pinellas County, Florida in 1995.

Conclusions and recommendations

The results of this study provides new information on the distribution of depressions in Pinellas County, Florida, as well on the effects of urbanization on Florida karst landscape. Specifically, a total of 2703 depressions were identified on the 1926 air photographs. These likely sinkholes accounted for 43.9 km^2 and 6.1% of the total land area. By 1995, only 900 of these depressions remained. These features totalled only 5.6 km^2 and 1% of the total land area of the Pinellas County. From these data we measured a loss of 87% of depressions between 1926 and 2000. It is likely that this loss is owing to the rapid urbanization that occurred since 1926. Many of the existing depressions were filled or modified into storm-water retention ponds.

We mapped 1646 storm-water retention ponds in Pinellas County. A total of 499 of these ponds are located in areas that were mapped as depressions in 1926. It is clear that there is a potential threat of subsurface pollution from leakage in the retention ponds owing to the overlap of these structures with ravelling zones of sinkholes. It must be noted that we did not evaluate individual water retention ponds to determine if they are lined or in some other way managed to prevent subsurface pollution. This would be a valuable future project.

The presence of so many buried or otherwise modified depressions in an urbanized area is rather striking. We are very interested in the impact of the modified depressions on overall ground stability, and surface-water and groundwater flow. Florida law mandates that insurance carriers cover damage to homes from subsidence caused by sinkhole activity (Eastman *et al.* 1995). There are dozens of claims filed each year in Pinellas County by property owners who have some form of structural damage to buildings caused by subsidence. While the focus of our research did not address the correlation between lost depressions and structural damage, we do believe that a follow-up study that addresses this issue is warranted.

References

BECK, B. F. & SAYED, S. 1991. *The Sinkhole Hazard in Pinellas County: A Geologic Summary for Planning Purposes.* Florida Sinkhole Research Institute. University of Central Florida, Orlando, FL.

CURRIN, J. L. & BARFUS, B. L. 1989. Sinkhole distribution and characteristics in Pasco County, Florida. *In*: BECK, B. (ed.) *3rd Multidisciplinary Conferences on Sinkholes.* A.A. Balkema, Rotterdam, 97–106.

EASTMAN, K. L., BUTLER, A. M. & LILLY, C. C. 1995. The effects of mandating sinkhole coverage in Florida homeowners insurance policies. *CPCU Journal*, September, **48**, 165–176.

FRANK, E. F. & BECK, B. F. 1991. *An Analysis of the Cause of Subsidence Damage in the Dunedin, Florida Area 1990/1991.* Florida Sinkhole Research Institute, University of Central Florida, Orlando, FL.

LANE, E. 1986. *Karst in Florida.* Florida Geological Survey, Special Publications, **29**.

MILLER, J. A. 1986. *Hydrogeologic Framework of the Floridan Aquifer System in Florida and Parts of Georgia, Alabama, and South Carolina.* USGS, Professional Paper, **1403-B**. 91 p.

MILLER, J. A. 1997. Hydrogeology of Florida. *In*: RANDAZZO, A. F. & JONES, D. S. (eds) *The Geology of Florida.* University of Florida Press, Gainesville, FL, 69–88.

SCHMIDT, W. & SCOTT, T. M. 1984. Florida karst – Its relationship to geologic structure and stratigraphy. *In*: BECK, B. F. (ed.) *Sinkholes: Their Geology, Engineering and Environmental Impact.* A.A. Balkema, Rotterdam, 11–16.

SINCLAIR, W. C., STEWART, J. W., KNUTILLA, R. L., GILBOY, A. E. & MILLER, R. L. 1985. *Types, Features, and Occurrence of Sinkholes in the Karst of West-central Florida.* US Geological Survey, Water Resources Investigations Report, **85-4126**.

STEWART, M., LEI, D., BRINKMANN, R., AANGEENBRUG, R. & DUNLAP, S. 1995. *Mapping of Geologic and Hydrologic Features Related to Subsidence-induced Foundation Failures, Pinellas County, Florida.* Report prepared for Pinellas County Government.

TIHANSKY, A. B. 1999. Sinkholes, West-central Florida. *In*: GALLOWAY, D., JONES, D. R. & INGEBRITSEN, S. E. (eds) *Land Subsidence in the United States.* USGS Circular, **1182**, 121–140.

UPCHURCH, S. B. & RANDAZZO, A. F. 1997. Environmental geology of Florida. *In*: RANDAZZO, A. F. & JONES, D. S. (eds) *The Geology of Florida.* University of Florida Press, Gainesville, FL, 217–250.

WHITE, W. A. 1970. *The Geomorphology of the Florida Peninsula.* State of Florida Department of Natural Resources Bureau of Geology, Geological Bulletin, **51**.

WHITE, W. A. 1988. *The Geomorphology and Hydrology of Karst Terrains.* Oxford University Press, Oxford.

Natural and anthropogenic rock collapse over open caves

T. WALTHAM[1] & Z. LU[2]

[1]*Civil Engineering Department, Nottingham Trent University, Nottingham NG1 4BU, UK (e-mail: tony@geophotos.co.uk)*

[2]*School of Civil Engineering, Nottingham University, Nottingham NG7 2RD, UK*

Abstract: Natural rock collapse that reaches the ground surface to form a collapse doline is relatively rare in limestone karst. The anthropogenic karst geohazard is posed by the possibility of rock collapse when additional loading is imposed by engineering works directly over a known or unknown cave. An intact rock-cover thickness that exceeds half the cave width appears to be safe in most karst terrains formed in strong limestone. Guidelines suggest that drilling or probing prior to construction should prove sound rock to depths ranging between 3 and 7 m in most of the various types of karst.

Conduits are a ubiquitous feature of karst (and pseudokarst) terrains, and many (but not all) reach dimensions whereby they are accessible by man and are then known as caves. The potential for gravitational collapse of rock and/or soil into them, either naturally or under induced load, therefore renders them a notable karst geohazard. As a real hazard to the construction industry, this primarily concerns the larger conduits (i.e. the caves), but soil failure into the smaller conduits (i.e. smaller than caves) can cause significant ground subsidence. The dominant hazard occurs on strong limestones, but assessment of cave instability can be extended to sites in chalk, gypsum, salt, basalt and some unconsolidated sediments (generally known as soils by civil engineers). Karst is distinguished by its formation in soluble rock terrains, so caves in basalt and soil are considered as features of pseudokarst.

Natural collapse and doline development

Rock that constitutes the roof span over a natural cave has an element of inherent instability. Some caves are cylindrical tunnels that have developed slowly over geological timescales, so that they have equilibrium roof arches in sound and massive rock, but even these may become unstable when surface lowering reduces their roof thickness. Far more caves are developed in fractured rock, whose rock mass integrity is locally variable, and progressive roof failure on small or large scales is a natural process within the evolution of a cave that was initiated by dissolutional processes (Fig. 1). Repeated failures in a cave roof constitute natural stoping and upward cavity migration. Where this propagates as far as the ground surface, it generates a collapse or caprock doline.

The surface depressions that are diagnostic of karst landscapes are known to geomorphologists as dolines, but in the American and engineering literature are generally known as sinkholes (Sowers 1996; Waltham *et al.* 2005). These may be classified into six main types, distinguished by their genetic processes and morphology (Fig. 2). Collapse dolines are formed by single or multiple collapses of cave roof spans; these are not common, and new events of rock collapse without imposed load are extremely rare. Where cave roof stoping migrates to the ground surface through a covering insoluble rock, a cap-rock doline is formed; these are similar to collapse dolines except that they form in an insoluble outcrop, and new events are equally rare.

The mechanisms and rates of cave roof failure, and ultimately surface collapse, are dependant largely on rock structure, most significantly on the bedding and jointing densities and attitudes within the rock mass. In roughly horizontal, and minimally disturbed, strong limestones, individual beds across a cave roof fail when the cave's unsupported span exceeds between 10 and 20 times the bed thickness (but such figures are major generalizations across greatly variable situations and processes; see Waltham *et al.* 2005, fig. 3.5). Joints across beds of rock clearly reduce their ability to survive in cantilever or in unconstrained beams across cave roofs, but the joint influence is greatly reduced where potential displacements are restrained within the zone of ground compression that naturally develops as an arch over a void. Consequently, many large caves have stable arched roof profiles that follow the shapes of their compression zones; these have developed by beds falling away from the tension zone beneath the compression arch. Although their geometry is not perfect, some cave roofs in heavily fractured rock can approach the stability of a voussoir arch in uncemented masonry (an arch formed of shaped blocks designed to be stable in compression). However, heavily shattered rock or intersecting sets of inclined fractures can

From: PARISE, M. & GUNN, J. (eds) *Natural and Anthropogenic Hazards in Karst Areas: Recognition, Analysis and Mitigation*. Geological Society, London, Special Publications, **279**, 13–21. DOI: 10.1144/SP279.3
0305-8719/07/$15.00

Fig. 1. Collapse of strong, well-bedded limestone in a chamber within the Sof Omar cave system in Ethiopia; the metre-thick bed above the man's head height has fallen away from the chamber roof, to form the breakdown floor, but the same bed still spans the narrower passage beyond the man.

produce serious instability, so that some caves are heavily collapsed when only a few metres wide.

Parameters for natural cave roof collapse

The great variability of structures within limestone means that cave stability is equally variable. Numerical modelling of cave roof failure under imposed load (see text below and Fig. 5 later) has provided peripheral data on unloaded cave failures. These suggested that a rock roof 2 m thick would be marginally stable across caves 10–25 m wide, depending on the quality of the limestone (indicated by rock mass strengths that are representative of cavernous karst ground). Similarly, rock 8 m thick would fail where caves reached widths of 20 m in weak rock or well over 50 m in strong rock. A different method of numerical modelling indicated that a cave 130 m wide in strong and massive limestone in Slovenia would fail when the roof was thinned to about 6 m (Kortnik 2002). Observations of cave chambers, both surviving and failed, confirm a huge range of values in stability parameters, dependant largely on the immediate patterns and densities of rock fractures.

Cave collapse events may be induced by increased water input. In the short term, enhanced rock dissolution is not normally significant, but greatly increased water flow could reduce stability by washing out fissure fills that had been contributing to the integrity of a compression arch within the roof rock. In the long term, the accelerated opening of fissures within the drainage zone beneath the floor of a solution doline is recognized as contributing to major rock failure over cave chambers to form the giant collapse dolines known as tiankengs (Zhu & Chen 2005). At a recent collapse event in Kentucky, rock failure appears to have been induced by impact loading created when soil arches collapsed over voids that had developed within the soil profile over fissures in the limestone (Kambesis & Brucker 2005), but this cave roof was already very thin (Fig. 3).

There is a shortage of data on cave roof stability in weaker rocks. Gypsum may be analogous to the weakest limestones, and roof spans of more than about 25 m do not appear to survive; the many larger collapse dolines in gypsum have evolved by multiple collapses. Salt is even weaker and is also subject to plastic flow at low stress levels; large caves in salt are rare and ephemeral. Both gypsum and salt are so rapidly soluble in flowing water that contemporary rock dissolution is a significant factor in ground stability over caves within them;

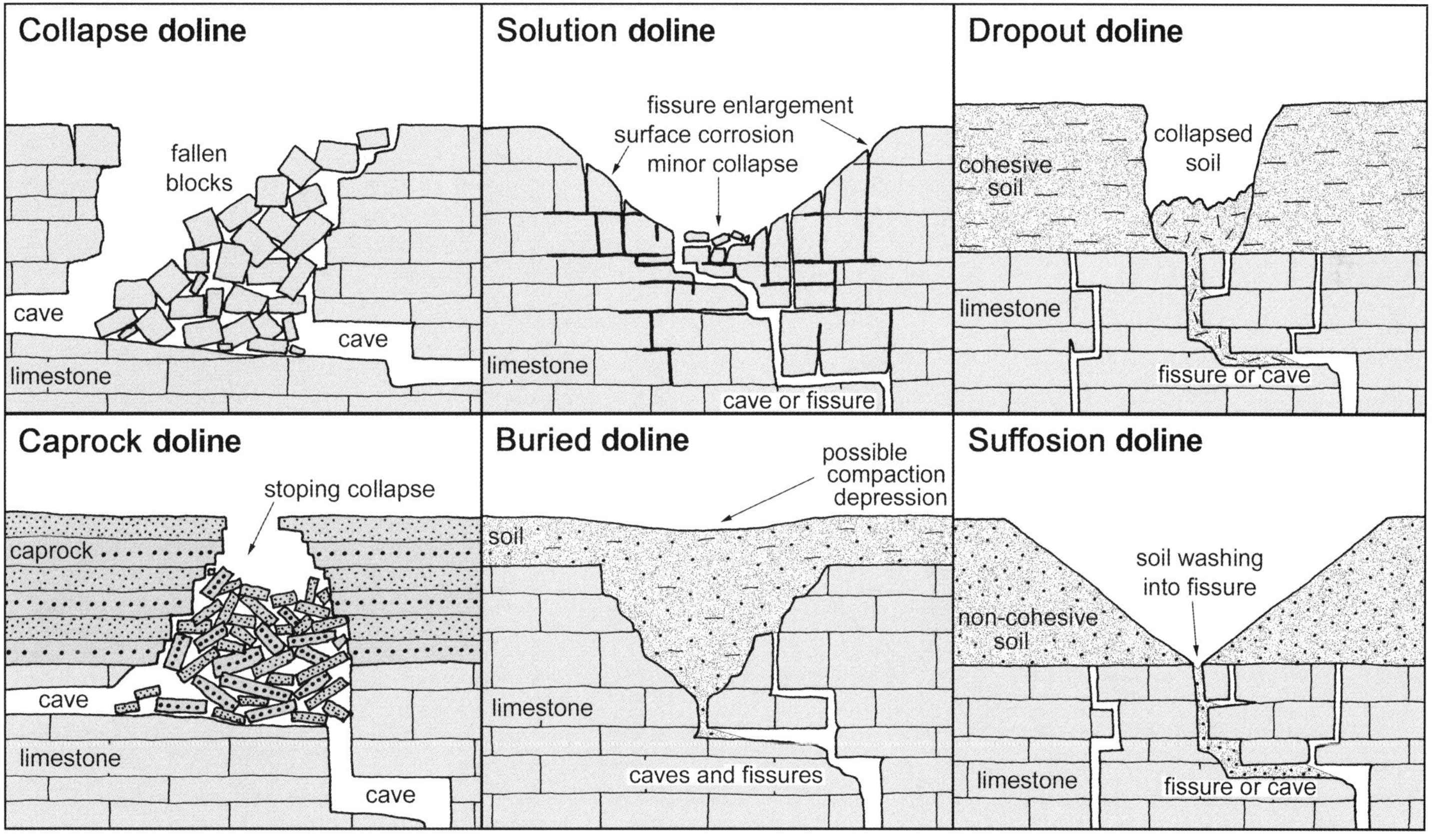

Fig. 2. The six main types of dolines; the two on the left are those involving rock collapse; the two on the right are together known as subsidence dolines, developed entirely by failure of the soil mantle. (After Waltham *et al.* 2005.)

Fig. 3. The edge of the collapse doline that destroyed a road in Kentucky in 2002, viewed when remediation was under way; the broken edge shows the wide cave chamber spanned by a very thin roof, of which the upper half was a zone of soil-filled fissures between rockhead pinnacles.

groundwater flows that are enhanced, either naturally, accidentally or by mismanaged engineering works, may modify cave profiles enough to induce collapse within engineering timescales but without any imposed loading. Chalk is only slowly soluble in water, but accelerated drainage can induce liquefaction of material previously weakened by Pleistocene frost shattering and its consequent loss into any underlying void; most recorded cases of liquefaction failure in Britain have been over old mines, and are therefore more allied to crown hole failures (surface holes that open up when progressive mine roof stoping failures reach rockhead), but comparable new collapse dolines are known in the chalk of France (Waltham *et al.* 2005).

The development of new cap-rock dolines is a function of the rock mass strength of the cap-rock. Gritstone capping the interstratal karst of Wales is analogous to the strongest limestone, whereas clays over some buried karst in Russia are so weak that collapse events mimic the mechanisms of subsidence dolines within soils. Large cavities in salt, formed rapidly by dissolution that is either natural or enhanced by brine drainage, have been known to migrate to the ground surface by stoping through hundreds of metres of weak cover rocks within periods of just months or years. These events have implications for ground stability over any soluble rock, but risks are reduced by their extreme rarity.

This brief review of the karst geohazard owing to natural, unloaded, rock collapse is more fully explored by Waltham *et al.* (2005). As a karst geohazard, rock collapse is totally overshadowed by the

Fig. 4. Houses in the town of Ogulin stand in complete safety atop a thick arch of massive limestone that spans a wide cave passage in the inland karst of Croatia.

hazard of dolines that form by downwashing of unstable soil into fissures within underlying, stable limestone; these are dropout dolines (formed rapidly, and widely known as cover-collapse dolines in North America) or suffosion dolines (evolving slowly), which are together known as subsidence dolines. The high rates of new doline appearances reported from some karst terrains (exceeding one per km^2 per year) all refer to new subsidence dolines that develop in the soil mantle over cavernous limestones, notably during periods of high rainfall.

Cave roof collapse under imposed load

Collapses that are induced by man's activities in karst terrains (thereby anthropogenic collapse) is a long-recognized hazard where increased loading is imposed in construction works. The great majority of caves lie at depths that are irrelevant to engineering loading on the surface because they lie beneath thicknesses of stable rock far greater than the cave widths. However, there is a potential hazard in caves at shallow depth, where the rock roof is thin in comparison to the underlying cave width.

A critical value, for the ratio of roof thickness to cave width needed to ensure stability under construction loading, has long been questioned, but rarely concluded. It is clearly a function of the thickness and rise profile of the compression arch within the ground that is needed both to span the cave and to support the load, ameliorated by an added thickness of rock that allows stress distribution above the notional arch. The concept that collapse is unlikely where the imposed load is less than about 10% of the existing overburden stress (Sowers 1996) relies on stress distribution within the bulb of pressure, but takes no account of stress concentration (and bulb distortion) in fissured karst over an open cave. It is more realistic to assess cave roof stability in terms of the cover ratio (t/w, where t is roof thickness and w is cave width) that is appropriate for any limestone (or other cavernous rock) of given rock mass quality. The popular 'rule of thumb', that a limestone cave is stable and can be ignored when its cover thickness exceeds its width, is very convenient, although it does appear to be rather conservative.

Some very thin rock spans have been inadvertently loaded and yet have survived. A large cave was found under the main runway at Palermo, Sicily, after it had been in use for many years; with a cover ratio of about 0.1, this was deemed unsafe and is now full of concrete (Jappelli & Liguori 1979). The survival of the open cave, and the runway, may have relied on stress distribution within the reinforced concrete of the runway. There are numerous sites in the populated karsts (notably in Croatia and China) where houses and villages stand directly over caves with rising or sinking rivers, but most of these rock arches are comfortably massive (Fig. 4).

In contrast, there are records of structural collapse into unseen caves inadvertently loaded by engineering works. A large column supporting a freeway, and heavily loaded during construction, dropped 5 m into a cave in Tampa, Florida, in 2004; the cave dimensions were not seen, but the rock below the column base had been probed for only 3 m, which was clearly inadequate for the properties and morphological conditions locally well known in the local karst limestone. An entire five-lane road collapsed into a cave in Bowling Green, Kentucky, in 2002 (Kambesis & Brucker 2005), and this cave was known and mapped before the road was built only a few years previously; the road lay over the widest part of the cave where the cover ratio was less than 0.2 (Fig. 3), so it was a collapse waiting to happen, although the failure process was probably complicated by collapsing soil voids between the rockhead and the roadbed.

Investigation of a safe cover ratio over caves

Clearly, some form of guidelines for safe cover ratios for engineering loading over caves would be useful in the construction industry, especially as the vagaries of cavernous karst morphology are so little understood by engineers normally more concerned with settlements on soils. Most of the worldwide geohazards databank on mine collapses is barely relevant to caves, as it is concerned with pillar failure or with crown hole development in mainly weak rock sequences.

The artificial sandstone caves under the city of Nottingham have been investigated with respect to the extensive new construction on top of them (Waltham & Swift 2004). This included the full-scale test loading of a cave roof, when failure was induced by a load of 340 kN on a small bearing pad on moderately weak sandstone with a cover ratio of 0.13 over a cave 4 m wide. Numerical modelling was calibrated with these test data, but was based on rock that was homogeneous except for a few defined fractures, and it could not model the fractured and fissured rock mass that pertains in karst limestone. The stability assessment of a karst cave roof depends on adequate evaluation of the rock fracturing. This could be based on direct observation so that it is very detailed but site-specific, but this does not help where the potential hazard posed by an unseen cave has to be assessed prior to engineering construction.

Numerical modelling has therefore been advanced by defining fractured rock masses in terms of their 'rock mass ratings' (RMR). The geomechanics system derives RMR by summing rating values ascribed on the basis of RQD (rock quality designation, based on fracture intersections in borehole core), mean fracture spacing, fracture conditions, fracture orientation, unconfined compressive strength of the intact rock and groundwater state (Bieniawski 1973). RMR values range from more than 80 for very good rock of rock mass class I to less than 20 for very poor rock of rock mass class V; they may be correlated with *Q* values derived from the alternative Norwegian classification scheme (Barton *et al.* 1974). For application in numerical modelling, packages of strength and deformation values have been created for each of various values of RMR; these have been developed through extensive research, and their realism has been confirmed by empirical correlation with various measurable situations (Asef *et al.* 2000).

By installing these definitions of rock mass properties, the second author has modelled the effects of loading over cave roofs in two dimensions, using the finite-difference code Fast Lagrangian Analysis of Continua (FLAC). Caves, 3–50 m wide with flat roof profiles at depths of 2–10 m, have been modelled under loads applied to pads of 1×1 m at the ground surface above the centre-line of the caves. Loads were increased until failures were defined by settlements of 25.4 mm, a value that indicates loss of integrity and is likely to precede total collapse, besides causing significant damage to built structures. The caves were modelled in materials with RMR values of 20–50, which are considered to encompass the ground conditions to be found in most types of cavernous karst in limestones.

Results from this numerical modelling effectively indicate ultimate bearing pressures in terms of cave width, roof thickness and rock mass strength (Fig. 5). For any design load, selected factor of safety, estimated RMR and known or inferred cave width, a safe roof thickness can therefore be estimated from these nomograms. Cave dimensions at failure loads of 5 MN can be extracted from these FLAC models to define safe conditions for the single case of 1 MN loading (at 1 MPa bearing pressure on a pad of 1 m^2) pad, with a factor of safety of 5 in any given rating of rock mass (Fig. 6). If RMR for typical cavernous karst in strong limestone is taken conservatively as between 30 and 40, a cover ratio of 0.5, where the roof thickness is at least half the cave width ($t = w/2$), appears to be adequate for most engineering practice where bearing pressures greater than 1 MPa are rarely invoked. In karst terrains on chalk and some other weak or thinly bedded limestones, RMR may be estimated as nearer 20, and a cover ratio of 1, where roof thickness equals cave width ($t = w$), may be required for safe construction. It is notable that these data are derived from purely two-dimensional modelling, and failure loads are likely to be higher where some roof support is provided in the third dimension; this would increase the factor of safety in any interpreted results.

Guidelines for construction over caves

Combined with an estimation of likely cave width in a given karst, the above results allow definition of the depth to which rock should be proof-drilled to eliminate the hazard of rock failure over caves with respect to the stability of engineered structures. The widths of potential caves that remain unseen beneath any given construction site can only be estimated as 'most likely values'. Such estimates are best made after perusal of local records of observed caves in that particular karst environment; alternatively, failing the existence of a useable cave database, estimates can be derived from the engineering classification of the karst, which may be assessed from broad visual inspection (Waltham & Fookes 2003). More conservative values for generalized safe cover ratios may then be taken as 0.7, to make due allowance for poor data on likely cave dimensions, even more local variation within the fissured karst bedrock and higher imposed loads from engineering works.

Values for the required safe cover thickness then become the guidelines for the minimum depths of exploratory drilling or probing that should prove sound rock beneath foundation levels in engineering works (Table 1). At most sites on strong cavernous limestone (typical of most major karst terrains), drilling is required to depths that varies between 3 and 7 m, depending on the karstic maturity and, hence, the cave size. The most common weaker cavernous rocks are chalk and gypsum; although they would both require higher values of safe cover ratios, they require roughly comparable depths of drilling to ascertain ground integrity under the lower loading stresses that building codes normally set as maxima on these weaker rocks (Table 1). Basalt is commonly very strong and also structurally massive in the types of lava flows that may contain tubes or caves (Waltham *et al.* 2005), and proof drilling within basalt may therefore be less, based on a safe value of 0.5 for the cover ratio.

A survey of available drilling guidelines, that have been applied in various ground investigations on karst (Waltham *et al.* 2005, table 7.2), reveals values that range from 1.5 to 5.0 m. Although these encompass considerable variety in limestone lithologies, karst morphologies and engineering requirements, some appear rather conservative and others may be open to question. Collapse of the Florida freeway viaduct, as referred to earlier, followed exploratory drilling that could not be regarded as appropriate. The freeway's support column stood on a large-diameter pile that reached 19 m below ground level

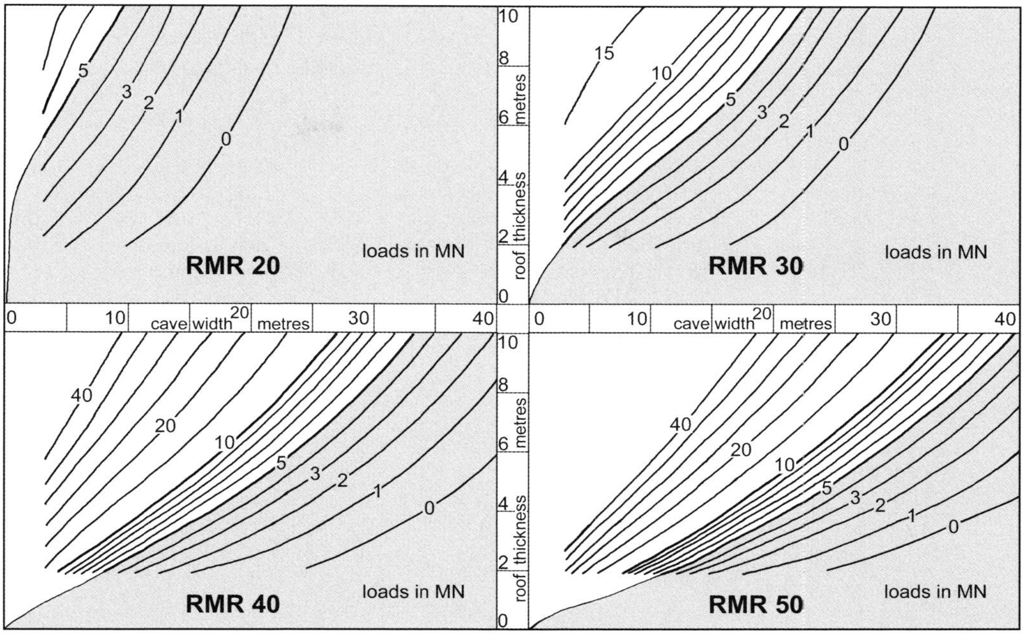

Fig. 5. Nomograms that relate failure loads to cave width and roof thickness in ground of various rock mass ratings. Numerical models were generated with incremental loads applied to foundation pads, 1 m^2, on the surface directly above the caves. The shaded areas represent situations with respect to cave width and roof thickness where a loading of 1 MN on the pad leaves a factor of safety of less than 5, and are therefore considered unsafe. (From Waltham *et al.* 2005.)

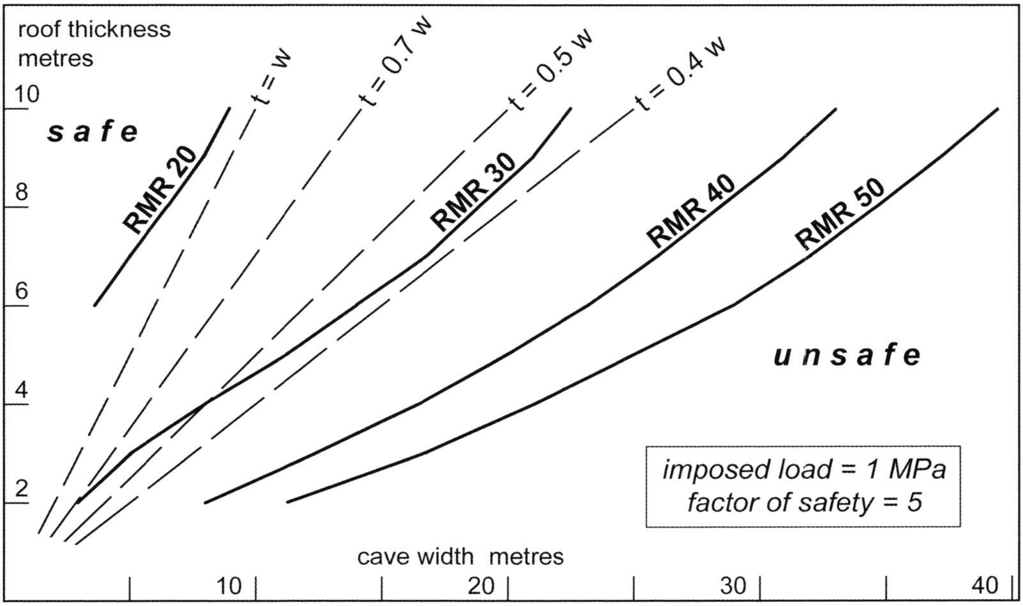

Fig. 6. Envelopes of acceptability, with respect to cave width and roof thickness, where engineering loads of 1 MN are applied, with a factor of safety of 5, to pads of 1 m^2 on the surface directly over the caves, in ground of various rock mass ratings; broken lines indicate various cover ratios, where t is roof thickness and w is cave width. (From Waltham *et al.* 2005.)

Table 1. *Safe roof thicknesses for various cave situations. The karst classes refer to the designations of Waltham & Fookes (2003)*

Rock	Imposed load (kPa)	Karst class	Cave width – likely maximum (m)	Safe roof thickness (m)
Strong karstic limestone	2000	kI–kIII	5	3
		kIV	5–10	5
		kV	>10	7
Weak limestone and chalk	750		5	5
Gypsum	500		5	5
Basalt lava	2000		5–10	3

(and therefore 8 m below rockhead), but prior drilling proved only 3 m of sound limestone below the heavily loaded base of the pile. This was because the engineers' guidelines had been based on a total drilling depth from the surface, in respect of the pile's conceptual skin friction within the profile of soil and karst limestone, while the pile's high end-loading caused the rock failure in this event.

While construction projects in typical karst on strong and massive limestone require guidelines that drilling or probing should prove between 3 and 7 m of sound rock, gypsum is a slightly weaker rock and, therefore, requires drilling to the greater depths. Further expansion of the guidelines may be required in active karst terrains on gypsum, where dissolution by flowing water may significantly

Fig. 7. A roadside cave in the limestone of Krk, Croatia, where the wider view (**a**) exposes a solid rock arch over the front of the cave, but a closer view (**b**) reveals an unstable pile of fallen blocks and debris extending upwards at the back of the cave.

enlarge or modify ground cavities within the lifetime of a built structure.

It is essential to note that the numerical modelling, with respect to cave roof integrity under imposed load (Fig. 5), is only based on generalized estimates of the strength parameters for rock masses of the various rating values. These estimates can only approximate the notoriously variable conditions in karst, where strong intact rock is broken by open or soil-filled fissures, in styles very different from those in insoluble rock masses. Whereas parts of some open caves may have structurally sound roofs within relatively intact rock, another site may have zones of broken and collapsed ground over ruckles of breakdown blocks inside what was once an open cave (Fig. 7). Such ground conditions in karst can be a nightmare to structural engineers, and do confirm that there is no substitute for careful examination of each individual site within a terrain of cavernous karst.

The authors thank Dr D. Reddish and his team at Nottingham University who made possible, and greatly assisted, the programme of numerical modelling.

References

ASEF, M. R., REDDISH, D. J. & LLOYD, P. W. 2000. Rock–support interaction analysis based on numerical modelling. *Geotechnical Geological Engineering*, **18**, 23–37.

BARTON, N., LIEN, R. & LUNDE, J. 1974. Engineering classification of rock masses for tunnel design. *Rock Mechanics*, **6**, 189–236.

BIENIAWSKI, Z. T. 1973. Engineering classification of jointed rock masses. *Transactions of the South African Institute of Civil Engineers*, **15**, 335–343.

JAPPELLI, R. & LIGUORI, V. 1979. An unusually complex underground cavity. *In: Proceedings of the International Symposium on Geotechnics of Structurally Complex Formations*, Volume **2**. Associazione Geotecnica Italiana, Rome, 79–90.

KAMBESIS, P. & BRUCKER, R. 2005. Collapse sinkhole at Dishman Lane, Kentucky. *In*: WALTHAM, T., BELL, F. & CULSHAW, M. (eds) *Sinkholes and Subsidence: Karst and Cavernous Rocks in Engineering Construction*. Springer, Berlin, 277–282.

KORTNIK, J. 2002. Stability appraisal of the Medvedova Konta pothole. *International Journal of Speleology*, **31**, 129–135.

SOWERS, G. F. 1996. *Building on sinkholes*. ASCE Press, New York.

WALTHAM, A. C. & FOOKES, P. G. 2003. Engineering classification of karst ground conditions. *Quarterly Journal of Engineering Geology and Hydrogeology*, **36**, 101–118.

WALTHAM, A. C. & SWIFT, G. M. 2004. Bearing capacity of rock over mined cavities in Nottingham. *Engineering Geology*, **75**, 15–31.

WALTHAM, T. BELL, F. & CULSHAW, M. 2005. *Sinkholes and Subsidence: Karst and Cavernous Rocks in Engineering and Construction*. Springer, Berlin.

ZHU, X. & CHEN, W. 2005. Tiankengs in the karst of China. *Cave & Karst Science*, **32**, 55–66.

Sinkholes in Italy: first results on the inventory and analysis

S. NISIO, G. CARAMANNA & G. CIOTOLI

APAT (Environmental Protection Agency and Technical Services), Via Curtatone, 3-00185 Roma, Italy (e-mail: stefania.nisio@apat.it)

Abstract: The Italian Geological Survey (APAT) carried out field surveys and analysis of collapse phenomena (sinkholes) in Italy. The main goal of the project is to collect geological, geomorphological, geochemical and hydrogeological data about the sinkhole-prone areas in Italy in order to develop a spatial database of the characteristics of each phenomenon. The preliminary results of this study provide information on the distribution, geological setting, and monitoring and remediation actions associated with these natural collapses in Italy. Many Italian regions are affected by these natural disasters. Some of them are caused by karst collapses or anthropic activity. However, some occur in areas characterized by buried carbonate bedrock (up to 190 m), as well as by peculiar geological–structural and geochemical scenarios. In these areas it is not reasonable to ascribe the formation mechanism to karst activity. Instead, these types of cavities quickly develop in terrains with a variable granulometry, often in connection with upwelling fluids. In this work some natural specific cases have been studied in order to define the relationships between the geology (regional tectonic elements, mineral spring waters and strong gas vents) and the genesis of the sinkholes. A first attempt of sinkhole classification is also presented.

Sinkhole definition

Collapse structures occur throughout the world and are particularly common in the USA (California and Florida), where they constitute a significant hazard owing to their unpredictability and the fact that it is difficult to avoid resultant damage (Newton 1984, 1986; Beck & Jenkins 1986; Sinclair & Steward 1985; Snyder *et al.* 1989). These natural processes are generally more important whereas anthropogenic causes are typically secondary, although it should be noted that the number of cases of collapse-structure formation has increased worldwide over the last 20 years, probably owing to urban expansion (Sinclair 1982; Newton 1984, 1986; Tihansky & Galloway 2000).

In the last few years the term *sinkhole* has been used often to indicate collapse phenomena of various natures; however, this has created confusion, particularly in the different usages of the term in the Italian literature v. English literature. Such a term defines a subcircular surface depression or collapse structure formed by the collapse of small subterranean karst cavities (Fairbridge 1968; Monroe 1970). This definition is synonymous of doline, which includes various genetic subtypes such as solution, collapse or subsidence sinkholes or dolines (Sweeting 1972; Jennings 1985; Castiglioni 1986). At present in the United States and Great Britain the term sinkhole is frequently used to define any subcircular cavity regardless of genetic origin (Beck 1984; Beck & Wilson 1987); the term is also used to indicate open cavities caused by anthropogenic activities that are not necessarily subcircular in form.

In Italy the term sinkhole is used to indicate a subcircular cavity that opens suddenly on the surface and is used as a synonym of collapse. A collapse is a rapidly formed depression of variable shape, even if typically subcircular, which occurs primarily in karst terrains, in plains and in areas where subterranean cavities already exist. Sudden collapses are also defined as those events that are not directly linked to karst phenomena, such as subterranean cavities formed by anthropogenic activities in urban areas, above mines, excavations and ancient catacombs.

Of the investigated sinkholes *senso stricto* 38% are filled with water, thereby forming small lakes or ponds. As this water can often be mineralized, it is not surprising that they can occur in correspondence with mineralized springs and are often aligned along tectonic lineaments that are stressed by the presence of water and gas chemical anomalies.

The formation of sinkholes can occur during a single event or with the slow and progressive collapse of the structure. In the first case the subsurface is affected by a single conduit and the cavity maintains a relatively constant dimension over time. In the second case the sinkhole develops with the slow washing away of the sediments from the base to the top of the cavity. It has been shown that the majority of the sinkholes are caused by upward erosion (Littlefield *et al.* 1984; Derbyshire & Mellors 1988; Derbyshire *et al.* 1991; Billiard *et al.* 1992; Faccenna *et al.* 1993; Muxart *et al.* 1994; Nisio 2003). *Deep piping* phenomena are probably originated by this process (Nisio 2003).

From: PARISE, M. & GUNN, J. (eds) *Natural and Anthropogenic Hazards in Karst Areas: Recognition, Analysis and Mitigation.* Geological Society, London, Special Publications, **279**, 23–45. DOI: 10.1144/SP279.4
0305-8719/07/$15.00

The deep piping process is a mechanical effect due to the upward flow of groundwater in granular rocks (from clay to gravel-sized particles) that occurs when the deposit is fractured or porous and when the water is under high pressure in a flow path in which high groundwater velocities (turbulence) are maintained. The flow of water results in the erosion of material and the formation of conduits (pipes) (Massari *et al.* 2001; Nisio 2003).

Many Italian regions are affected by these natural collapses, but some of these seem to be karst collapses and some are caused by anthropic activity. However, some sinkholes occur in areas characterized by a deep buried carbonate substratum (up to 170 m), as well as in areas where peculiar geological–structural, hydrogeological and geochemical settings facilitate sinkhole formation. In these areas it is not reasonable to ascribe the formation mechanism of these sinkholes to the mere gravity collapse of a karst cavity, but some other triggering and prone factors should be considered.

The Italian Geological Survey – APAT (National Agency for Environmental Protection and Technical Services) – carried out studies and field surveys on collapse phenomena occurring in sinkhole plains in Italy in the framework of the Sinkhole Project. In this work some natural specific cases were studied in order to define the relationships between the typical geological scenario (regional tectonic elements, mineral springs and strong gas vents) and the genesis of sinkholes. The main goal of the project is to collect geological, geomorphological, geochemical and hydrogeological data about the sinkhole-prone areas to construct a sinkhole hazard map in Italy. A Microsoft Access relational database (RDB) was designed to store, analyse and map the sinkhole-prone areas in Italy. The database (SH-RDB) includes geological, hydrogeological, geochemical and geotechnical data about known sinkhole collapses in Italy. The goal of the sinkhole database project is to provide easy access to sinkhole data via data query and reporting options. A geographical information system (GIS) was used to create thematic maps and to analyse the data.

This paper reports the first results of the Sinkhole Project on the distribution, classification and evolution of sinkholes in Italy. In addition, preliminary results of selected case studies help to reveal the geological setting.

Sinkhole classification

Classification of karst features can be based on genetic mechanisms; our aim is to clarify the genetic mechanism of collapse triggering and to present a general classification of the phenomena. This first attempt of classification considers the genesis of cavities that show similar morphological characteristics but that originate from very dissimilar genetic mechanisms.

Doline is a term derived from the Slav word 'dol' that means valley; these were the most common hollows in the landscape of the Dinaric karst. The term refers to natural enclosed depressions found in karst landscapes (Cramer 1941; Sweeting 1972; Castiglioni 1986; Ford & Williams 1989 and many others), principally owing to chemical dissolution of calcareous deposits. Dolines are also sometimes known as sinkholes.

In a strict sense, the term sinkhole in Italy and sometimes in Europe indicates a topographic cavity of various shapes, mainly subcircular, caused by a quick collapse of the soil surface. The origin of a sinkhole, again in the strict sense, is not necessarily related to karst phenomena. Other factors can contribute to the genesis. For this reason the terms doline and sinkhole are not synonymous. Indeed, the international terminology can be very confusing. Hence, in order to avoid the ambiguity that sometimes arises in the general use of these terms, further qualification is required such as 'solution sinkhole' or 'collapse sinkhole'. Table 1 lists the terms used by different authors and the extent to which genetic types are subdivided.

According to the main genetic processes, collapse phenomena can be subdivided into the following groups: (1) anthropogenic sinkhole; (2) karst phenomena; and (3) sinkhole *sensu stricto* or deep piping sinkhole (Nisio 2003; Nisio & Salvati 2004) (Fig. 1). It is widely recognized that enclosed depressions can be formed by four main mechanisms: dissolution, collapse, deep piping and subsidence.

Anthropogenic sinkhole

Anthropogenic sinkholes are caused by the collapse of the roof of an artificial underground cavity (i.e. mining, catacombs, etc.).

Karst phenomena

A genetic subdivision of the karst phenomena is based on the mechanisms that cause cavities in the shallow environment. Such mechanisms consist of: (a) collapse; (b) subsidence; and (c) dissolution mechanisms.

Collapse doline/sinkhole. These depressions formed mainly by mechanical processes. This fact may cause a considerable variation in nomenclature largely owing to the variety of materials and processes involved, and the tendency of some authors to group types and/or to subdivide types (Fig. 1, Table 2). Collapses refer to rapid downward

Table 1. *Sinkhole terminology**

Definition	Main genetic processes	Sedimentological setting	Morphology
Solution sinkhole/doline	Chemical dissolution of karst bedrock	Outcrops of carbonate/evaporitic rocks	Enclosed depression funnel-shaped with a flat bottom. Thin cover of red soil
Subsidence sinkhole, cover-subsidence sinkhole, cover sinkhole	Subsurficial dissolution or collapse of karst voids in the underlying rocks	Karst carbonatic bedrock with non-consolidated cover (i.e. sand, gravel).	Topographic depression with dimension of some tens of metres both in diameter and depth
Collapse/cave-collapse sinkhole, collapse doline	Collapse of the roof of caves	Karst cave overlain by lithoids deposits (i.e. tuff, limestone, travertine)	Deep sinkholes with steep walls and truncated cone vertical shape
Rock-subsidence sinkhole, subsidence doline	Collapse on cohesive, permeable and non-soluble rocks placed over soluble sediments	Karst cave overlain by cohesive deposits (i.e. clay, silt)	Funnel-shaped sinkholes
Piping sinkhole, cover-collapse sinkhole	Upwelling of water and gases with a piping process followed by the collapse of the soil	Sediments of different granulometry and less cohesive	Variable shape from smooth depression to steep sinkholes

*Terms used to describe the sinkholes, the main genetic processes, the sedimentological setting and the morphology (partly taken from Gunn 2004).

movement of the ground, whereas subsidence refers to gradual movement sometimes without even ripping the surface. These processes can occur in karst bedrock (collapse doline or collapse sinkhole: Tharp 1997, 1999; Capelli *et al.* 2001; Salvati *et al.* 2001). They can be subdivided into *cave (or cap-rock)-collapse doline or sinkhole* and, where unconsolidated sediments overlie the bedrock, into *cover-collapse doline or sinkhole*. In all cases, the collapse could be preceded by dissolution of the karst rock to form a void into which material can fall.

A cave-collapse doline or sinkhole is a well-shaped cavity in limestone and karst rocks. The karst cave must be overlain by rock (i.e. tuff, limestone or travertine) and the formation follows the collapse of the cave roof (Cramer 1941; Castiglioni 1986; Ford & Williams 1989). The trigger of the collapse is a progressive thinning of the roof of the cave followed by the failure of the overlying rock. This kind of catastrophic subsidence is often characterized by a large bell-shaped cavity and the absence of premonitory signs. The cover-collapse sinkhole or doline has a thin cover of permeable sediments (20–30 m of sand or gravel). It is promoted by the infiltration of surface water, and the cavity originates by the collapse of the roof of the underlain karst void. Sometimes a collapse extends from a cave below the water-table level; many sinkholes host small ponds.

Subsidence doline/sinkhole. Depressions caused by cover subsidence are called *subsidence dolines*. They can be subdivided into *rock-subsidence dolines/sinkholes* and *cover-subsidence dolines/sinkholes* (Fig. 1).

These landforms originate in unconsolidated deposits such as alluvium, glacial moraine, loess, sands, etc. The sedimentary cover slowly filters downward into the underlying karst voids through corrosionally enlarged fractures and results in gradual subsidence of the surface. The main process involves the gradual winnowing and down-washing (ravelling) of fine sediments by a combination of physical and chemical processes.

Rock-subsidence doline/sinkholes occur in soluble rocks with a thin cover of permeable sediments. The fracturated, soluble rock slowly settles without any surface evidence. The result is a bowl-shaped landform (Castiglioni 1986).

Cover-subsidence dolines/sinkholes are closed depressions in non-cohesive cover sediments (i.e. alluvial deposits), and are caused by the presence of karst voids in the underlying rocks (Castiglioni 1986). The triggering of this process is strictly correlated with the permeability of the materials and the thickness of the cover deposits (Newton 1984, 1986). The evolution of this type of collapse is typically progressive, with final dimensions of some tens of metres in diameter and shallow depth.

Solution doline/sinkhole. Shallow morphologies without collapse structures are called *solution dolines/sinkholes* and are caused by rock dissolution by infiltrating water (Cramer 1941; Castiglioni 1986). The bowl-shaped form of this landform

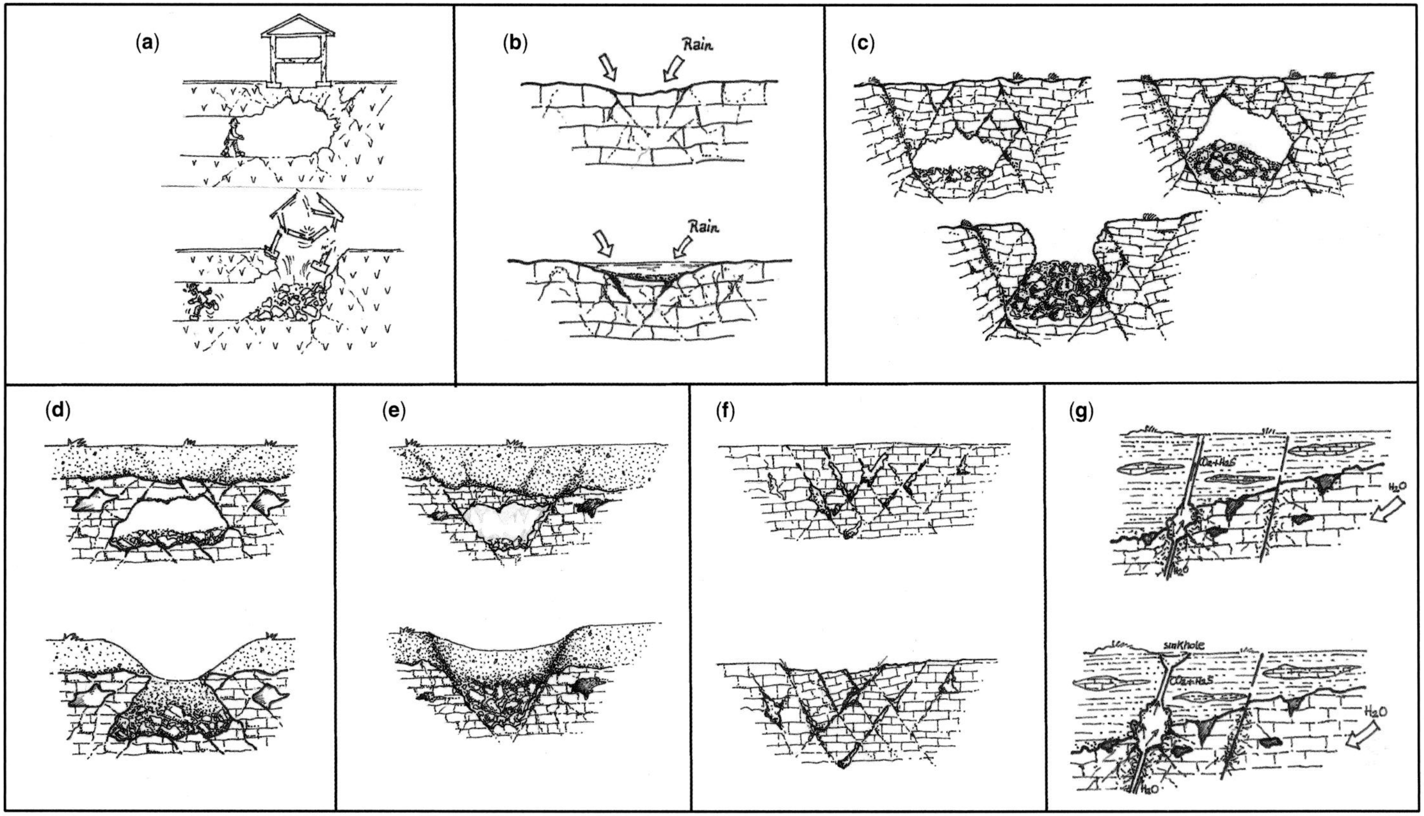

Fig. 1. Sinkhole classification according to literature information and field observations of Italian case studies. (**a**) Anthropogenic sinkhole; (**b**) solution doline/sinkhole; (**c**) cave collapse; (**d**) cover collapse; (**e**) cover subsidence; (**f**) rock subsidence; and (**g**) piping sinkhole.

Table 2. *Sinkhole genetic processes terminology**

Genetic process	Ford & Williams (1989)	White (1988)	Culshaw & Waltham (1987)	Beck & Sinclair (1986)	Jennings (1985)	Bogli (1980)	Sweeting (1972)	Other terms
Dissolution collapse	Solution collapse	Solution collapse	Solution collapse	Solution collapse	Solution collapse	Solution collapse (fast) Subsidence (slow)	Solution collapse	
Cap-rock collapse	Collapse	Collapse	Collapse	Collapse	Subjacent collapse	Collapse (fast) Subsidence (slow)	Collapse	Interstratal collapse
Dropout	Collapse	Cover collapse	Subsidence	Cover collapse	Subsidence	Alluvial	Alluvial	
Suffusion	Suffusion	Cover subsidence	Subsidence	Cover subsidence	Subsidence	Alluvial	Alluvial	Ravelled sinkhole
Burial			Buried					Filled sinkhole
Water filling								Palaeo-sinkhole Drowning

*Sinkhole terms used by various authors related to the genetic process (modified after Waltham & Fookes 2003).

indicates that more material has been removed from its centre than from around its margins. The amount of limestone that can be removed in solution depends upon two variables: the concentration of the solute; and the volume of the solvent (i.e. the amount of water draining through the doline). Variations in either or both of these quantities could be responsible of dissolution near the centre of the depression. The development of this type of doline depends on the ability of water to sink into and flow through karst rocks to outlet springs. The exposure of limestone by erosion provides a boundary for the infiltration of water. These landforms develop slowly without catastrophic subsidence, so they cannot be associated with the sinkhole phenomena.

Sinkhole sensu stricto *or deep piping sinkhole*

The main causes of the *sensu stricto* or deep piping sinkholes are complex and interacting. They can be genetically subdivided into the following categories (Faccenna *et al.* 1993; Nisio 2003).

- Soluble substrate (carbonates, evaporites or sulphate-rich rocks) affected by karst phenomena. The presence of small- or large-scale irregular bedrock morphology with karst macroforms and with karst cavities in the upper substratum (soil–rock interface) results in corrosive and dissolution processes.
- Impermeable or semi-permeable sediments that overlay the bedrock.
- Weak physical–mechanical characteristics of the materials in the upper cover.
- Presence of a network of faults or fractures that allow elevated groundwater circulation and mechanical erosion.
- High groundwater flow and recharge rates.
- The presence of acidic gases, such as CO_2 and H_2S, which dissolve into the groundwater, thereby increasing dissolution rates of the surrounding material.

Recognized triggering mechanisms for surface collapse include loading (such as construction), wetting (intense rainfalls, snowmelt), drying (droughts) and shaking (earthquakes; Snyder *et al.* 1989; Ferreli *et al.* 2004).

The terms sinkhole or deep piping sinkhole should be used with care owing to the great complexity of the phenomenon. Thus, we propose the following nomenclature.

- Deep piping sinkhole: This type of sinkhole occurs in areas characterized by mineralized groundwater with bubbling phenomena caused by the high concentrations of free and dissolved gases. The deep piping process is a mechanical effect caused by the flow of groundwater in fractured or porous granular rocks (from clay to gravel-sized particles) with abundant, high pressurized waters (Fig. 1). In this process a flow path with high groundwater velocities results in the erosion of material and the formation of channels. Furthermore, the presence of fault and fracture systems acts as route for the migration of deep acidic fluids. Mineral springs and/or gas vents enriched in CO_2 and H_2S, which dissolve and/or pass into the groundwater and make it more aggressive, have been recognized inside and/or close to the cavities. This phenomenon is also referred to as *deep suffusion* or *deep piping* (Nisio 2003; Graciotti *et al.* 2004; Nisio *et al.* 2004).

The geological model for a deep piping sinkhole requires the presence of a thick layer of cover deposits (up to 200 m) overlying a karst bedrock. Usually these sediments are alluvial deposits with vertical and horizontal granulometrical variations. The presence of a cover of clay deposits prevents deep deformations and the possibility that a cave collapse in the bedrock could create a sinkhole *sensu stricto* on the surface by ravelling processes. In the presented case studies, no signs of water flow from the surface to the underground occur. Instead, the water head can reach the soil level, and for this reason the presence of a spring in the cavity is possible (some authors define this type of sinkhole as a *spring sinkhole*). The presence of karst materials (i.e. travertine, limestone gravel) inside the cover deposits enhances the deep piping sinkhole formation.

Collapses suspected to be sinkholes *sensu stricto* have been widely reported from central Italy (from Tuscany to the Campania region). The majority of these cases seem to have a common genesis. They occur in groundwater discharge areas near the base of karst ridges in zones with major groundwater circulation. Thick overburden and proximity to a deep gas source and/or thermal upwelling is also present (Ciotoli *et al.* 2001; Annunziatellis *et al.* 2004). The occurrence of such sinkholes within a discharge zone is difficult to reconcile with the traditional conceptual model for two reasons. First, there is no downward moving water that could cause ravelling processes. Second, the upwelling acidic groundwaters have been in contact with limestone for long periods and, generally, these groundwaters are near $CaCO_3$ saturation. This limits the mechanism for creation of voids within the bedrock (Salvati *et al.* 2001; Salvati & Sasowsky 2002). In addition, caves occur within the cover sediments, as additional material is removed to create a substantial void in the cover material. Often the arch in the cover terrain will no longer support the overlying material, and the void propagates upwards until the shear stress becomes greater than the cohesion of the cover material. In these settings, the rising of pressurized and acidic fluids (gases and water) is the main factor of the deep piping sinkhole genesis. The upwelling of groundwaters is controlled by the presence of deep faults, which act as the main pathways for the deep fluids.

- Collapse piping sinkhole: This type of sinkhole differs from deep piping sinkholes for the following reasons: (1) It requires the presence of an unconsolidated cover sediment (i.e. pyroclastic deposits, gravel, sands, etc., with thicknesses of up to 100 m). (2) The carbonatic bedrock hosts a confined aquifer. In high recharge periods the aquifer should be pressurized with a strong increase of the water head that does not, however, reach the soil level. (3) Deep piping mechanisms act on the interface between the bedrock and the cover sediments. A cylindrical-shaped cavity forms along a fracture or fault in the lower sedimentary cover and propagates towards the surface. When the roof reaches a thickness that cannot support acting shear stresses, it collapses quickly. In the major part of the observed cases the cavities are dry. These sinkholes occur diffusely in volcanic areas of the Latium and Campania regions.

From the study of more than 500 natural sinkholes it is possible to suggest a specific classification for the Italian sinkholes. This classification is based on genetic mechanisms and on other factors affecting the sinkhole genesis: the lithology of the sedimentary cover and of the bedrock, the depth to the bedrock, and the geological and morphological setting of the sinkhole-prone areas. The sinkholes *sensu stricto* (or supposed to be sinkholes *sensu stricto*) are clustered into 10 different types. It should be noted that research is ongoing and that these are preliminary results.

Sinkhole evolution

Sinkholes are not 'static' features of the landscape. These phenomena are subjected to a sort of evolution with modification of the morphology, increasing or decreasing of the diameter, flooding and displacement. Figure 2 shows the main types of sinkhole evolution recognized in Italy.

Drowning

The drowning process affects sinkholes in areas where the table water is close to the soil surface.

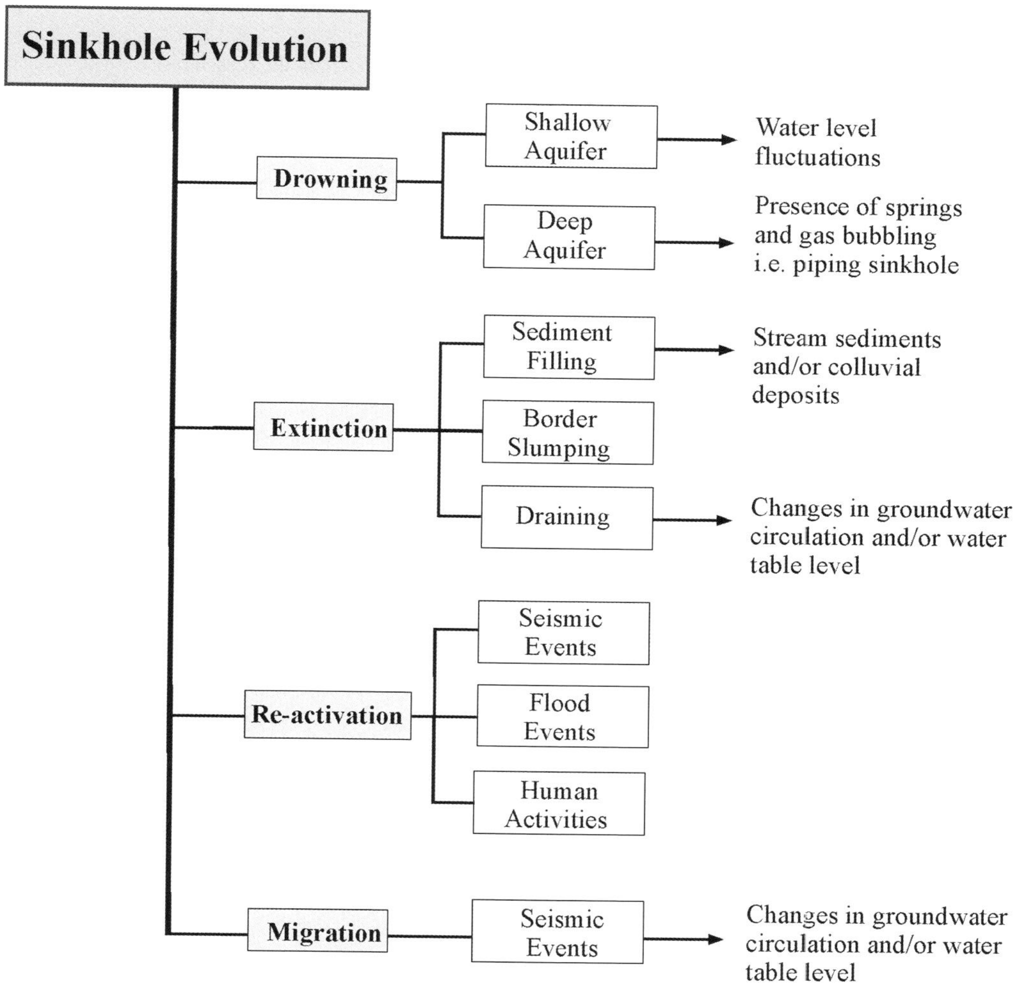

Fig. 2. Flow chart of sinkhole evolution.

In these zones, the formed subcircular depressions often contain small lakes or ponds. This phenomenon generally represents the first stage of sinkhole evolution (Caramanna *et al.* 2004, 2005; Nisio *et al.* 2004). Under such conditions, water levels may be affected by seasonal fluctuations caused by rain and/or by the presence of a water outlet. In some cases, fluctuations of the water surface are not recognized, but gas emissions and stream outflow from the pond may occur; this suggests the presence of a spring on the bottom of the lake fed by the deep aquifer.

These lakes are often characterized by bubbling phenomena caused by deep-origin gases (CO_2, H_2S, H_2, He) that rise along crustal discontinuities. The presence of acidic gases (CO_2 and H_2S), which dissolve into the groundwater, can cause dissolution and weakening of the supporting bedrock, travertine and other carbonate-rich sediments in particular. Carbonate dissolution combined with the siphoning off of fine particles results in a loss of structural integrity and subsequent collapse. The deep origin of the fluids has been verified by chemical analyses of the dissolved gases in the water filling the sinkholes (Ciotoli *et al.* 2001).

Extinction

Several years after formation, a sinkhole cavity could become dry and be filled by sediments until the depression is no longer visible on surface.

Some hypotheses may be advanced regarding the sinkhole extinction:

- filling with stream sediments and/or with colluvial material from range fronts (i.e. Lago Puzzo, Rome or Lago di Caira, Cassino);
- slumping of the sinkhole borders, which increases the diameter and simultaneously causes the filling of the cavity with the collapsed material;
- draining of the sinkhole caused by changes in groundwater circulation and/or water-table level, for example induced by seismic events.

Re-activation

Extinct sinkholes may be re-opened in the same location or undergo a sudden collapse that causes dimensional changes. Sinkhole re-activation may be caused by the conditions that caused the original collapse (seismic events, human activities, flood events, etc.).

The extinct sinkholes can be re-opened in the same position with the same shape or the morphology can change (i.e. variation of diameter and depth). For example, the Lago Puzzo re-opened in 1930 with a wider diameter following some phreatic explosions with gas emissions. The New Lake in the S. Vittorino Plain (Rieti, central Italy) formed in 1891 and underwent several cyclic re-activations from 1902 to 1930. The lake of Pra di Lama (Pieve Fosciana–Lucca, northern Italy) re-opened in 1828 after an earthquake in the same position of a thermal spring used for a health spa. On 15 August 1828 at 11 a.m. several bangs, along with gas and mud emissions, preceded a collapse. The resultant lake was 11 m deep with a diameter of 40 m. In 1842 it was almost totally dry and filled.

Migration

Sometimes extinguished sinkholes could re-open in a shifted position. This is called 'migration'. The causes of this displacement and of the possible preferential direction are still under study. For example, the collapses formed in the Fosso S. Martino, north of Rome, show a southern trend of migration according to a north–south-oriented tectonic discontinuity. In the Pontina Plain (central Latium) four sinkholes called 'Sprofondi' (Collapses) show a northward migration of the collapses along a SE–NW direction. In the S. Vittorino Plain a slow migration of sinkholes, showing southern and western trends from the carbonatic chain of Paterno Mt towards the western section of the plain, has been highlighted. Whether or not these cases indicate a real migration is still not clear (Centamore & Nisio 2002, 2003; Annunziatellis *et al.* 2004; Centamore *et al.* 2004; Nisio *et al.* 2004). Otherwise, the sinkholes' migration should be explained by the modification of the upwelling paths of fluids caused by seismic events and master fault re-activation (i.e. Fiamignano–Micciani fault in the S. Vittorino Plain) (Annunziatellis *et al.* 2004).

Data presentation

The goal of the Sinkhole Project is to map and study sinkhole *senso stricto* in Italy.

This census considered only collapses located in plain areas characterized by thick sedimentary covers and without bedrock cropping out. Some of investigated cases, of doubtful origins, were shown to be karst-related phenomena or anthropogenic collapses (see Fig. 5b late). Less than 50% of the analysed cases could be considered deep piping sinkholes. The first part of the project focused on the development of a bibliography of national and international literature on Italian sinkholes. In addition, historical data on particular sinkholes were collected to determine their formation date and their geostructural boundary conditions.

Actually more than 550 sinkholes, excluding those of anthropic and/or karst origin, have been investigated in the field in different geological scenarios of Italy (Fig. 3).

Multi-scaling and multi-temporal analysis of aerial photographs of the sinkhole-prone areas allowed the recognition of their morphotectonic features. This analysis integrated field data and facilitated the improvement of geological and geomorphological maps. Geological, structural and geomorphological field surveys were then conducted in order to verify the results obtained from photographic analysis, in particular to identify the thickness of the sedimentary cover above the bedrock, as well as the main tectonic elements that could trigger and affect the evolution of the collapses. Field hydrogeological and hydrogeochemical surveys were carried out to characterize hydrogeological boundary conditions (i.e. depth of water table, spring flow, etc.), as well as physicochemical and hydrochemical parameters of the fresh water and mineral springs, and of the water filling the collapsed areas (see Table 3). The depressions containing small lakes or ponds were investigated using a small inflatable boat equipped with a digital ecosounder in order to map the morphology of the bottom. The data were managed using a geographic information system (GIS). The Relational Database Systems Theory (RDST) provides rules for organizing representing phenomena as collections of attributes stored in tables so that a specific set of procedures can associate, transform and extract information from the

Fig. 3. Sinkhole distribution in Italy using the GIS database. ⊙ Sinkhole.

tables in a reliable way (Fig. 4). The storage of information in the SH-RDB is organized into rows and columns in a table, with a separate row for each entity and a column for each property stored about that set of entities.

Sinkholes are not isolated phenomena; usually these collapses are clustered or aligned along specific paths. The 555 studied sinkholes are found in 138 sinkhole-prone areas. Some areas are characterized by more than 30 collapses, in other areas only two or three sinkholes occur.

Excluding sinkholes located in northern Italy that are not yet part of this study, sinkholes *sensu stricto* are primarily concentrated along the Tyrrhenian margin near carbonate ridges in complex geological and hydrogeological scenarios.

Other collapses are located in Sicily, in Puglia and in northern Italy, in areas characterized by a widespread presence of evaporitic rocks. However, this census focused only on those cases that, due to the thickness of the cover sediments on the evaporitic bedrock, could be related to an upwards erosion. Collapses clearly related to karst phenomena have not been considered in this inventory.

The presence of regional faults with associated fractures, acting as preferential pathways for deep gas (i.e. CO_2 and H_2S) migration, has been recognized in many cases.

In this paper, the distribution of sinkholes in Italy with respect to the characteristics of the geological scenarios (i.e. fluvial, coastal and intermountain plains) has been evaluated. Cases with similar geological (substratum depth, sedimentary cover, presence of regional fault systems) and geochemical (mineral spring, presence of dissolved and free acidic gases, low pH, etc.) characteristics were identified, and clustered into five main recognized morphological scenarios. Figure 5a–d shows pie charts representing the distribution of the collapses in Italy, the collapse type, the morphological scenarios and the type of the sedimentary cover in which collapses occur. It is worth noting that about 58% of the sinkholes occur in central-southern Italy (Latium, 24.5%; Abruzzo, 13.2%; Campania 12.4%; Tuscany, 7.9%; Fig. 5). A high percentage of cases occurring in the Emilia Romagna and Friuli Venezia Giulia are based on historical information and/or reports. The piping and collapse piping sinkholes are the most frequently occurring types (37.1 and 33.3%, respectively) in alluvial and intermountain (seismically active) plains, at 28.1 and 32.6%, respectively. These plains (about 50% of the total database) are characterized by the presence of alluvial sediments, mainly clay and sandy-clay. These data have been registered in a data sheet subsequently used to construct the GIS database. A brief description of some selected case studies follows.

Sinkholes in fluvial plains

Aterno valley. The Aterno River valley, located in Abruzzo (central Italy), shows a general Apennine trend (NW–SE), with a straight-line path, that in the northern sector changes its direction to a NE–SW trend (Fig. 6). The valley is bordered by mountain slopes characterized by young tectonic activity (i.e. triangular facets) that highlights that this tectonic valley is bordered by a regional master fault with several reactivations up until the Pleistocene.

Five sinkhole-prone areas are located along the Apennine trending section (NW–SE) of the Aterno River within a distance of few tens of kilometres: (1) the northern area is around the village of Pizzoli; (2) the central section of the valley is the area around the S. Gregorio and Civita villages; (3) south of the central section is the third area (Demetrio nei Vestini); (4) south of this area is the Raiano–Prezze villages area; and (5) the fifth area is enclosed in the Sulmona Plain.

A total of 35 sinkholes have been individuated along the Aterno River: three close to Pizzoli; four in the S. Gregorio–Civita area; 11 in the Demetrio nei Vestini area; 10 near Raiano–Prezze; and seven in the Sulmona Plain.

The diameters of the sinkholes range from 30 to 40 m in the first area, with an enlargement trend in the central area of the river valley (up to several hundreds of metres). In the Raiano–Prezze and Sulmona Plain the diameters reach 50–100 m.

All phenomena are many centuries old. The Quaglia lake lies in the Raiano area. The origin of this pond seems to be correlated, from popular legends, with an earthquake in 1456 (in December of this year a strong earthquake affected central and southern Italy). The origin is probably correlated with the earthquake of the Maiella Mountain. In the same year other sinkholes probably also collapsed.

In the first area, on the left-hand side of the river, a NE–SW direct fault (Arischia–S. Pelino fault), SW dipping, occurs. The fault slope with 1.200 m of dip-slip, is highly tectonized. The right-hand side of the valley also shows structural evidence of active tectonics. The carbonatic bedrock, lowered by faults, lies more than 100 m below the cover sediments. These deposits are mainly composed of gravels and sands of various sizes.

The study area is located on the fluvial–alluvial deposits of the Aterno River near the village of San Vittorino (Fig. 6). In this zone two subcircular shaped ponds (Podere Giorgio 1 and Podere Giorgio 2) were recognized. A third irregularly shaped pond occurs close to the main road near the bridge 'Tre occhi'. The geological and structural setting of the area and the morphology of the ponds

Table 3. *Water chemistry**

ID	Ca^{2+} (mg l^{-1})	Mg^{2+} (mg l^{-1})	Na^{+} (mg l^{-1})	K^{+} (mg l^{-1})	HCO_3^- (mg l^{-1})	SO_4^{2-} (mg l^{-1})	Cl^- (mg l^{-1})	NO_3^- (mg l^{-1})	DO (mg l^{-1})	TDS (mg l^{-1})	Conductivity (μS cm^{-1})	pH	Eh (mV)	Temp. (°C)	County
1	420	98.8	12.58	10.61	228.2	1303	27	2.74		2112	2130	7.42	309	17	Tuscany
2	439.2	123.4	12.04	2.63	212.3	1378	18.76	2.68		2188	2090	7.57	244	21.5	Tuscany
3	432	113.4	12.65	2.35	196.5	1366	19.02	2.57		2144	2150	7.51	134	21.3	Tuscany
4	431.6	109	12.15	2.02	200.1	1272	18.97	2.49		2049	2160	7.55	241	16.7	Tuscany
5	457	100.8	12.8	2.29	228.4	1384	18.7	2.63		2206	2180	6.98	137	21.3	Tuscany
6	454	114.8	10.47	2.71	255.1	1344	16.05	3		2199	2200	6.75	231	22	Tuscany
7	71.35	8.24	6.41	1.33	298.5	6.12	8.82	6.25		342.2	394	8.17	237	14	Abruzzo
8	28.5	5.5	16.8	10.7	183	5.85	15.2	2.9		232	279	7.49		15.3	Campania
9	17.73	3.97	9.23	9.99	143.6	0.31	10.22	2.08		151	194	6.54	202	17	Campania
10	79.95	7.67	6.17	11.65	452.4	3.74	9.16	8.41		400	465	6.56	270	17.4	Abruzzo
11	82.1	13	3.17	1.5	279.5	30.5	4.79	3.8		418	479	7.1		7.5	Abruzzo
12	145	19	6	4.2	358.8	130	13.9	11.8		688	825	7.29		9.6	Abruzzo
13	111.9	32.3	15	5.7	407.6	43.5	34	0.3		655	757	7.37		9.1	Abruzzo
14	113	5.54	5.99	0.96	430	7.9	8.6	2		429	600	7.16			Abruzzo
15	46.88	7.7	7.61	1.73	234.7	6.15	10.3	3.63		245	316	6.73	213	18.3	Abruzzo
16	97.2	2.3	5.5	0.72	270.9	8.2	19	18.3		171	468	7.94		10.4	Abruzzo
17	21.63	12.47	2.24	0.57	134.2	4.54	3.04	3.05	6.96	299	586	7.46	192	18	Latium
18	25.42	14.64	21	8.56	195.3	5.26	14.94	0	3.48	284	361	8.46	114	28.3	Latium
19	160.8	74	325	15.15	453.6	172.8	546	1.29	3.07	1747	2300	7.91	125	26.2	Latium
20	23.3	12.96	2.06	0.54	122	4.54	2.91	2.8	5.86	171	603	7.14	216	17.8	Latium
21	65.55	1.3	1.75	8.89	211.1	0	5.94	5.1		299	351	7.12	253	17.6	Abruzzo
22	25.04	13.09	2.17	0.55	134.2	5.63	2.95	2.56	4.85	186	584	7.09	210	19.3	Latium
23	83.45	13.05	2.33	0.58	439.3	4.38	3.05	2.5	3.67	548	558	7.09	195	15	Latium
24	107.8	24.51	7.41	2.93	457.6	53.24	10.86	0	3.55	662	761	7.75	139	27.1	Latium
25	134.5	38.1	80.8	5.15	456.3	47.74	136.7	7.87	3.41	695	1162	8.06	128	27.3	Latium
26	38.09	12.51	2.19	0.71	197.8	4.5	3.09	3.06	6.99	210	596	7.08	210	18.4	Latium
27	33.08	9.73	16.63	7.16	804.6	2.26	18.39	51		235	355	5.56	247	22	Campania
28	132.7	37.8	117	11.15	530.9	63.95	207.4	0	3.66	1098	1432	7.69	163	27	Latium
29	87.3	26.9	64.1	4.42	366.1	28.3	117.7	4.6	2.41	697	907	7.32	148	26.2	Latium
30	41.6	15.7	11.58	5	241	1.25	18.51	0	3.68	695	400	8.32	154	28.5	Latium
31	46.28	27.88	22.83	6.24	264.8	11.19	33.14	0		333	498	7.03	218	20	Campania
32	35.79	9.36	15.22	4.9	243.4	4.81	13.3	0.5		332	268	6.82	218	18	Campania
33	19.84	34.78	4.48	0.12	231.9	9.7	6.42	8.67	4.94	269	543	7.06	188	17.5	Latium
34	126.2	21.5	152.3	6.72	122	218.7	230.4	11.6		934.6	1233	6.66	146	4	Tuscany
35	585	127	1230	5.16	281	1899	1824	0		5996	7910	6.66	−30	32.4	Tuscany
36	41.9	11.9	28.5	1.82	171	42.5	22.42	3.64		256	358	7.66	107	8.6	Tuscany

(Continued)

Table 3. *Continued*

ID	Ca^{2+} (mg l^{-1})	Mg^{2+} (mg l^{-1})	Na^{+} (mg l^{-1})	K^{+} (mg l^{-1})	HCO_3^- (mg l^{-1})	SO_4^{2-} (mg l^{-1})	Cl^- (mg l^{-1})	NO_3^- (mg l^{-1})	DO (mg l^{-1})	TDS (mg l^{-1})	Conductivity (μS cm^{-1})	pH	Eh (mV)	Temp. (°C)	County
37	112.3	23.5	28.9	5.64	153	227.2	53.11	10.9		522	730	7.2	107	10.4	Tuscany
38	60.1	13.3	54.4	2.7	195	45.3	79.5	15		424.7	594	7.29	121	10.9	Tuscany
39	123.9	256	5.25	2.6	366	6.76	8.7	12.2		407.6	570	6.93	122	11	Abruzzo
40	144.8	103.5	477.1	24.8	301.4	484	727	1.9		2880	3800	9.2		18	Sicily
41	261.5	237	1178	106	368.6	1219	1873			6163	8130	9.2		18	Sicily
42	346.5	43.67	10.85	1.18	1003	83.2	14.98	0.18	5.62	1868	1190	6.3		18	Latium
43	448	70.5	20.4	3.63	2043	240.8	30.19		0.47	2257	2400	6.13	−268	14.4	Latium
44	427	65	17.85	3.36	1537	207.5	25.29	0.07	0.29	2287	2270	6.47	117	14	Latium
45	339.5	50.67	13.73	2.09	1220	140.2	19.15		1.82	1788	1670	6.17	−212	12.9	Latium
46	162.3	24.2	3.65	0.96	634.4	38.36	5.02	1.82	6.36	870.9	936	6.97	262	13.3	Latium
47	353.8	54.5	11.44	2.16	1293	123.8	14.59	0.27	0.43	1857	1867	6.16	−159	13.8	Latium
48	410	59.5	14.69	2.68	1513	156.1	20.7	0.15	0.35	2180	2150	6.08	−1	13.9	Latium
49	551	88.5	26.52	5.26	2062	226	36.53	0.08	0.34	3001	2820	6.23	241	14.3	Latium
50	357.5	48.5	5.11	1.17	1293	102.3	6.29	0.88	0.11	1817	1878	6.31	163	12.9	Latium
51	333.8	37.3	7	2	1220	83	8.17	0.07	0.33	1693	1713	6.08	−385	14	Latium
52	353.8	79	22.8	3.9	1623	233.2	34.05		2.9	2464	2120	6.24	63	13.2	Latium
53	227	30	3.82	1.12	878.4	28.16	4.71		5.9	1174	1208	6.76	293	12.1	Latium
54	238	23.8	3.82	1.68	927.2	18.41	6.05	15.66	4.02	1235	1313	6.69	−50	15.1	Latium
55	272.5	39.2	8.09	1.96	951.6	77.72	9.67	0.24	6.92	1362	1478	6.55	219	14.1	Latium
56	89.2	20.6	4.56	1.27	378.2	33.55	6.01		10.37	533.6	605	7.97	306	12.6	Latium
57	118.5	20.4	6.58	3.59	463.6	36.76	7.41	0.74	13.22	657.3	748	7.75	325	16.1	Latium
58	104.1	17.34	4.9	3.2	402.6	23.96	6.39	0.46	9.35	562.8	656	7.83	324	15.1	Latium
59	179.7	25	5.26	1.12	683.2	66.08	7	1.49	7.07	969.3	1037	6.67	293	11.8	Latium
60					439.2				6.79	438.5	648	7.65	230	20.9	Latium
61	190	9	12	5	29	740	7			534	937	7.2			Latium
62	514	83	97	25	740	1417	102			1830	3210	6.25			Latium
63	517	80	88	26	754	1417	109			1824	3200	6.3			Latium
64	169	57	60	8	201	1389	106			901	1580	7.3			Latium

*Main ion concentration and physicochemical parameters of some studied sinkhole waters.

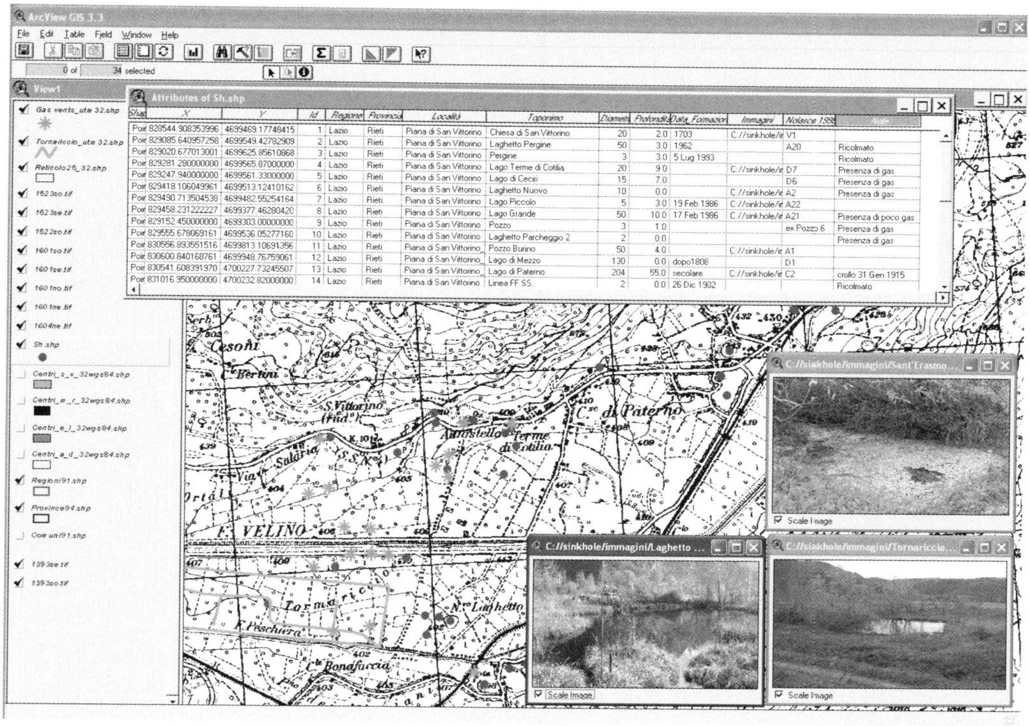

Fig. 4. Example of the Sinkhole Relational Database constructed in the Microsoft Access environment and used to store information of each case (row) and for each property of the single case (column). The archival Access format of the information regarding the descriptive characteristics, as well as georeferenced and numerical data of the known Italian sinkholes, provides a dynamic and flexible structure in which information can be extracted by using a simple standard database query language (SQL) in order to obtain tables to join with geometric information according to the operator's wishes. The ESRI ArcGis program is used to organize, manage and visualize in thematic 'layers' all collected information.

indicate that these collapses can be considered as deep piping sinkholes. The diameter of the lakes varies from 10 to 40 m, with shallow depths ranging from 1.5 to 3 m. The lakes, Giorgio 1 and Giorgio 2, are surrounded by clay and silt and are joined by an artificial channel. The water table is around 2 m below the soil surface. There are three springs on the bottom of the lake Giorgio 1 (unpredictable flow). The water level of Giorgio 2 is quite a bit higher than in Giorgio 1, so there is a flow towards Giorgio 1. The water is used for agricultural purposes and the surface level in both lakes is quite stable.

The date of the formation is unknown but the lakes, as reported by the landowner, have existed at least since 1960. Giorgio 2 is not reported on the IGM map so it probably formed in the second half of the 1900s. The lake of 'Tre occhi' bridge probably did not form from a sinkhole collapse. It is a depression in the soil fed by a close spring.

Sinkholes in intermountain basins

S. Vittorino Plain (eastern Latium). The S. Vittorino Plain, crossed by the Velino River, is a tectonic valley bordered by direct and transtensive faults (Fig. 7). These displacements affect the morphology of the plain, creating a triangular-shaped plain with the apex facing south (Centamore & Nisio 2003; Centamore *et al.* 2004).

The plain is filled by fluvial and lacustrine deposits of the upper Pleistocene–Holocene; gravel lenses and travertine (mostly sandy) are found locally (Fig. 7). The thickness of the cover increases toward the middle of the plain, and reaches 130–170 m over a faulted and displaced calcareous sequence (Nisio 2003). The S. Vittorino Plain is the drainage basin of the water from the surrounding recharge areas (Boni *et al.* 1995; Capelli *et al.* 2000). Springs are aligned along the northern and southern boundaries of the plain. Some springs, with a flow rate around $0.1-2\,l\,s^{-1}$, feed some

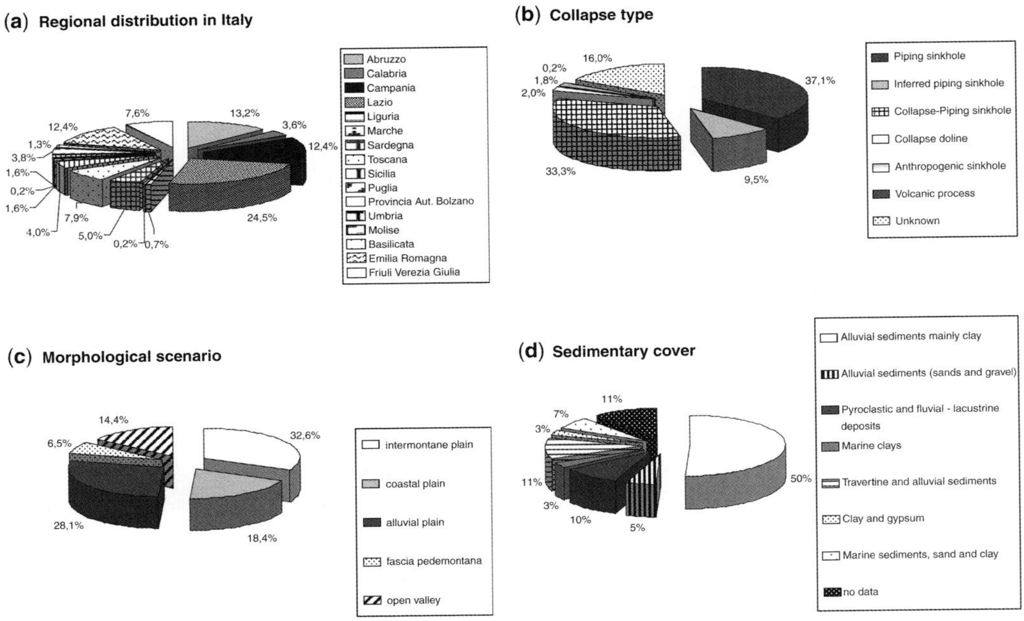

Fig. 5. Pie charts from the data elaboration of more than 500 studied cases: (**a**) regional distribution of the sinkholes; (**b**) genetic mechanism; (**c**) morphology of the area; and (**d**) typology of the sedimentary cover.

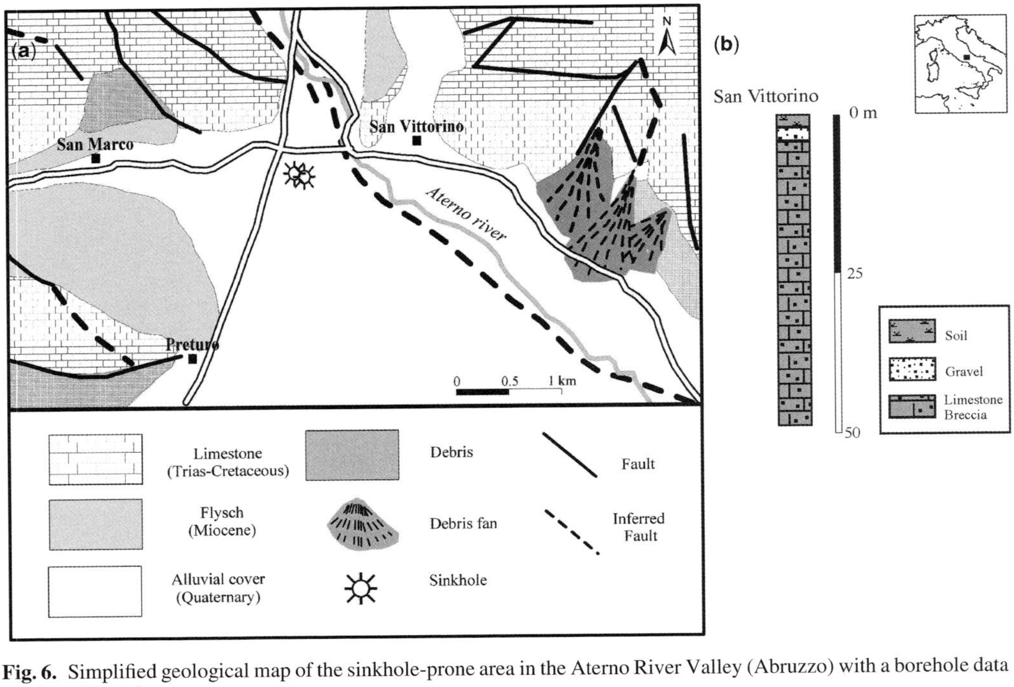

Fig. 6. Simplified geological map of the sinkhole-prone area in the Aterno River Valley (Abruzzo) with a borehole data stratigraphic column.

sinkhole ponds inside the plain. The S. Vittorino Plain is characterized by the presence of collapses (Faccenna *et al.* 1993; Capelli *et al.* 2000; Nisio 2003). Some collapses affect the bordering calcareous structures. We have recognized at least 35 subcircular sinkholes with diameters varying from a few metres to 100 m. Some collapses are naturally or artificially filled. Sinkholes are mainly located in the northern sector of the plain and are aligned along the NE–SE path of the Fiamignano–Micciani fault. Sinkholes host small lakes or spring ponds with mineralized water and gases (free and dissolved). The most recently formed sinkhole lies in the northern sector of the plain close to the Cotilia thermal bath. It is a small circular pond with springs and gas vents. Mineralized springs discharge mainly sulphuric water and ferric water, and are aligned along tectonic displacements (Faccenna *et al.* 1993; Ciotoli *et al.* 2001; Nisio 2003).

Fucino Plain. The Fucino Plain is a rhomboidal intermountain tectonic depression located in the middle of Apennines (central Italy) (Fig. 8). A Cenozoic thrust and fold belt has been undergoing uplift and extension since the Late Pliocene (Blumetti *et al.* 1988, 1993). During the Quaternary, the basin was deprived of a natural outlet and thus hosted a wide lake (artificially drained in the second half of the last century) that collected a thick sequence of Pleistocene and Holocene silty lacustrine deposits interbedded with coarse alluvial fan deposits at its borders. In 1915 the plain was hit by the strong Avezzano earthquake, which testifies to the activity of the faults bordering and crossing the basin (Blumetti *et al.* 1988; Michetti *et al.* 1996). The ENE-trending, SE-dipping Tre Monti (TMF) and Avezzano–Celano (ACF) faults; the NW-trending San Benedetto–Gioia dei Marsi fault (SBGMF) reactivated during the 1915 Avezzano earthquake; and the NW-trending, SW-dipping Ortucchio (OF) and Vallelonga–Trasacco–Avezzano (TF) faults exist in the basin (Blumetti *et al.* 1993). Data from subsurface investigations, as well as geophysical surveys, demonstrate that a network of capable faults affects the very recent deposits of the basin floor and outline two sub-basins that are separated by a near-surface horst located along the NW extension of the Vallelonga–Trasacco ridge. The two sub-basins appear to be half-grabens with the main normal faults (SBGMF and TF) dipping SW without any significant listric component. Faults either bordering or cutting the basin show evidence of Holocene reactivation (Blumetti *et al.* 1988; Michetti *et al.* 1996).

Moreover, palaeoseismic studies performed at several sites along the San Benedetto–Gioia fault (SBGMF) have demonstrated that coseismic

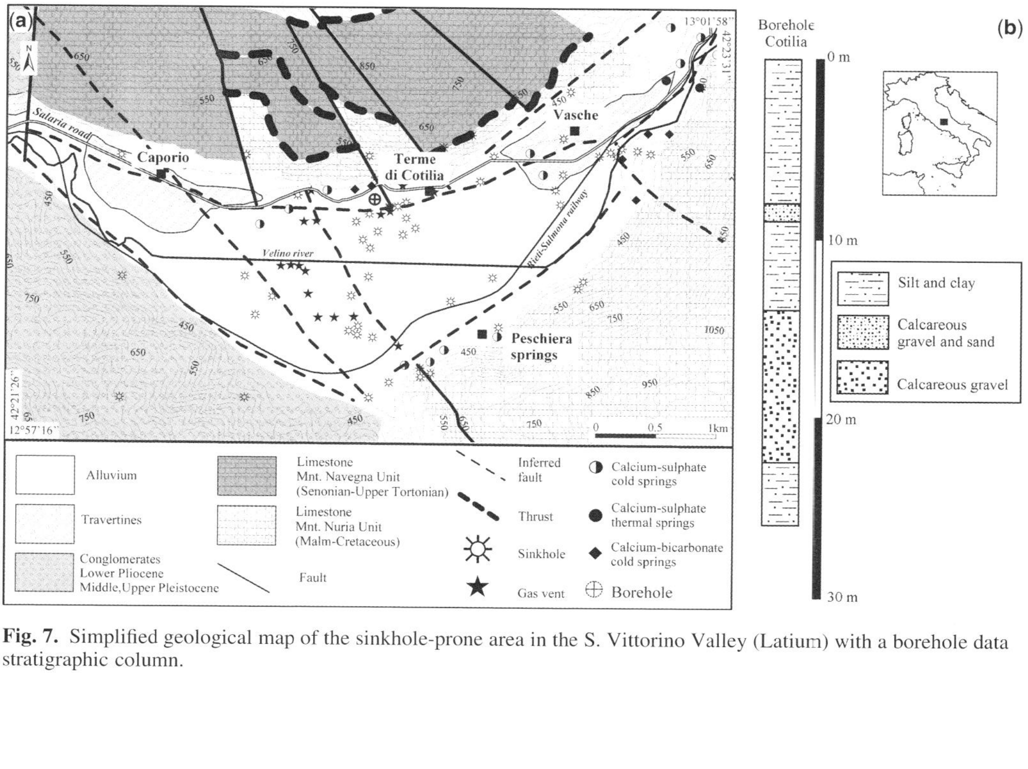

Fig. 7. Simplified geological map of the sinkhole-prone area in the S. Vittorino Valley (Latium) with a borehole data stratigraphic column.

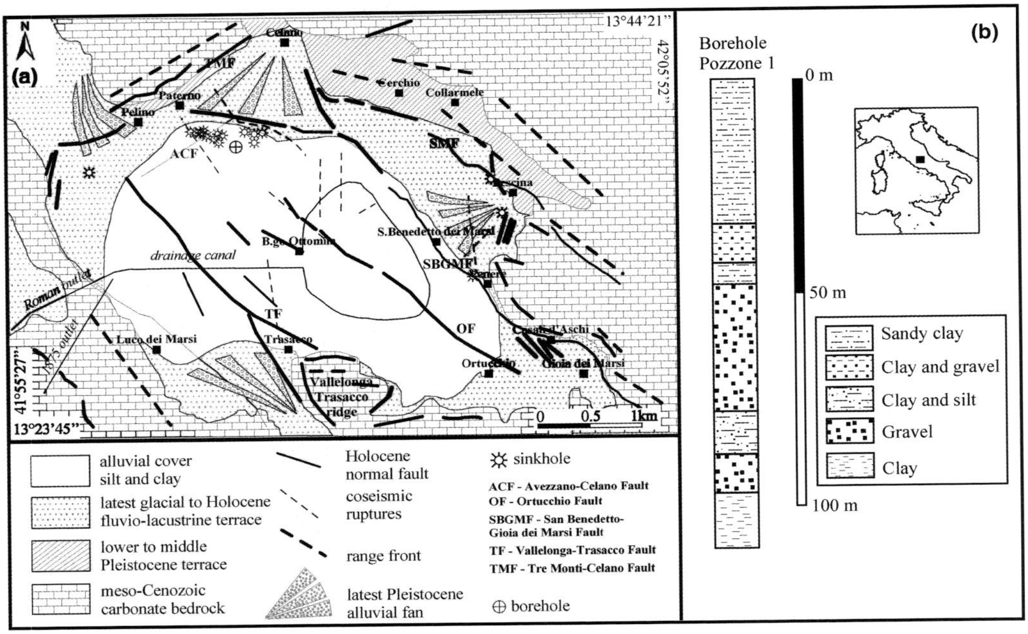

Fig. 8. Simplified geological map of the sinkhole-prone area in the Fucino Plain (Abruzzo) with a borehole data stratigraphic column.

displacement of a magnitude comparable to that of 1915 occurred at least once, and possibly twice, during the Middle Ages (Michetti *et al.* 1996). During the 1915 earthquake, typical seismo-geological effects, such as liquefaction, spring anomalies, and gas and water emissions, were observed at several sites inside the plain (Oddone 1915).

A total of 30 subcircular-shaped cavities (Fig. 8), with diameters ranging from 5 to 30 m and a few metres of depth, have been identified in the plain. The sinkholes occur mainly in the NW sector of the Fucino Plain with an east–west trend similar to the direction of the ACF. In this area, high soil gas concentrations (CO_2, He and Rn) of deep origin have been measured indicating fluid upwelling from a deep reservoir along buried tectonic discontinuities (Ciotoli *et al.* 1998). Some cavities actually host small ponds. Others are buried. Some of these sinkholes are multiple or twin sinkholes. The 1915 earthquake triggered at least three sinkholes. Borehole data show the presence of a thick cover (mainly clay and few gravel levels), of at least 110 m, overlying calcareous bedrock. A pressurized aquifer is hosted in the bedrock; some perched aquifers are in the gravel levels. The collapses have been identified as deep piping sinkholes.

Sinkholes in the coastal plain

Pontina Plain (central Latium). The Pontina Plain is on the border of the carbonatic ridge of the Lepini–Ausoni Mounts (Fig. 9). Since the Pliocene, the area has been subsiding. It has an average altitude ranging from few metres (Sezze and Migliara zone) to 35–50 m above sea level. Quaternary fluvial–lacustrine sediments, lagoon sediments, aeolic deposits and pyroclastic products from the Albani Hills volcano, and from some local emission points, crop out in the area. Travertine is found at various depths under the plain deposits. Furthermore, borehole data and some outcrops also indicate the presence of Pliocene marine sediments. The carbonatic bedrock is characterized by direct faults with a graben structure.

The Lepini–Ausoni Mounts structure, bordered on the western slope by a master regional fault and affected by a widespread karst erosion, hosts an imposing aquifer. Several springs are fed by this aquifer along the contact between the carbonatic ridge and the alluvial cover of the plain. Below the plain there is a confined aquifer that recharges in the Lepini–Ausoni Mounts. Locally, there are some small perched aquifers fed by direct rainfall or from the carbonatic ridge through some faults and fractures. A natural fluvial and artificial

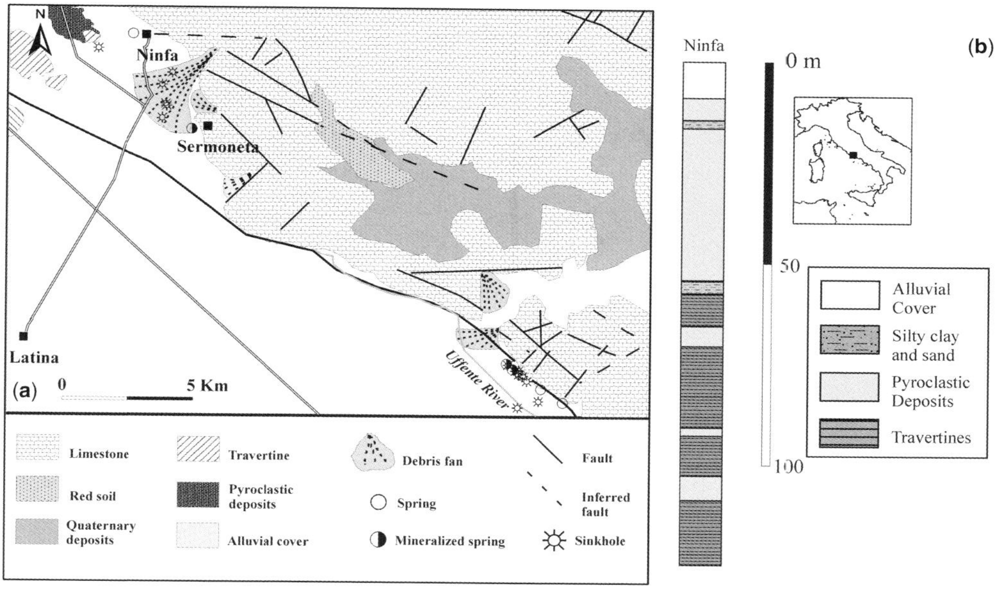

Fig. 9. Simplified geological map of the sinkhole-prone area in the Pontina Plain (Vescovo and Sprofondi lakes sector – Latium) with a borehole data stratigraphic column.

drainage network drains toward the Tyrrhenian shore. Furthermore, a thermal groundwater circulation with two end members (bicarbonatic–calcic and sulphuric waters) occurs (Colombi *et al.* 2001).

The Pontina Plain is considered a sinkhole-prone area with many subcircular ponds. Morphological, geological and hydrochemical studies have been conducted on some of these flooded sinkholes: Vescovo Lakes group, Sprofondi and Doganella di Ninfa sinkhole.

The Vescovo lakes group (from north to south: S. Carlo Lake, Vescovo lakes and Mazzocchio Lake) are aligned along a NW–SE regional fault where some mineralized springs also occur.

The S. Carlo Lake is circular and is located about 200 m from the Lepini–Ausoni carbonatic ridge. The lake is fed by some underwater springs and shows bubbling phenomena that indicate gas rising in the water. The origin of the lake from collapse, according to popular legend, happened some centuries ago. The lake has been plotted on geographical maps since 1777.

The Vescovo lakes are located in the area south of the village of Sezze, close to the carbonatic ridge. There are five ponds (White Lake, Green Lake, Black Lake, Bagno pond and one more without a name) with mineralized and/or opalescent waters showing high levels of dissolved and free CO_2 and H_2S. The Black Lake is the least mineralized of all, with no presence of sulphurs or sulphate, no opalescent water and a pH of greater than 7. On the shores are outcrops of travertine that should be of hydrothermal origin, proving the ancient presence of mineralized warm fluids in the pond. The Green Lake is characterized by acidic ($6.3 < pH < 6.6$) water with a decreasing pH and increasing electric conductivity from the surface to the bottom. Borehole data show the presence of an alluvial cover (clay and silt) with a thickness of more than 80 m, with some travertine levels at around 50–60 m. The bedrock is at a depth of 170 m.

All the lakes seem to be formed by more than one collapse. In particular, the results obtained by means of a bathymetric survey show that the Black Lake presents an 8-shaped feature characterized by two collapses of 9–10 m depth joined together and separated by a shallow bank occurring at a depth of 2 m.

On the map of 1777 all the lakes were represented with the exception of Bagni pond and the unamed pond. The same map shows the presence of another lake, Manello Lake, that now is buried.

The Mazzocchio Lake is located at about 500 m from the Lepini–Ausoni carbonatic ridge. It is fed by a channel of a pumping station. The outlet is an artificial channel linked to the close Uffente River. This lake has been present for more than three centuries.

The Sprofondi group. These collapses are located in the municipality of Sermoneta (southern Latium). There are five subcircular-shaped aligned sinkholes along the Apennine direction (NW–SE). Two are now buried. One is close to the Sermoneta–Bassiano railway station and the second is a few hundred metres to the SW. The former is mapped on an official IGM map of 1938 as a depression that hosts a small pond of surficial water. Buried after the Second World War, the depression is now not recognizable. Two more depressions with small lakes are on private proprieties. The water surface is about 10 m below the soil level. The regional water table lies 13 m below the soil level. Borehole data show an alluvial cover 50 m thick with pyroclastic deposits. The presence of deep-water supply wells, deeper than 100 m with pressurized water, suggests that the bedrock lies more than 100 m below the alluvial cover.

The last of the Sprofondi, which is known as Sprofondo Lake in the official IGM map, is locally Blue Lake. It is a circular pond. The origin seems to be a catastrophic collapse of the soil on St Jake's day (25 July) of an unknown year, probably 1786. Chemical analyses of the water revealed a bicarbonate–calcic water with pH > 7 that decreases from the surface to the bottom with concomitant increases of conductivity. The water temperature is always less than the average temperature in the area and this should be proof of cold underground water feeding the lake. The lake is used as a fishing facility. The other cavities formed in the 1800s, the authors suggest either 1818 or 1870.

Doganella di Ninfa sinkhole. This sinkhole is close to the small village of Doganella in the NE area of the Pontina Plain (Fig. 9), a few kilometres from the city of Latina. On 22 August 1989, after seismic activity earlier in the month and during a dry period associated with lowering of the aquifer owing to water overpumping, there was a collapse in a field utilized as a kiwi plantation. In just 3 days the diameter of the depression increased from 1 to 3 m; after 3 years it reached 31–32 m. Now this sinkhole hosts a subelliptical lake 30 m deep with a bell-shaped vertical cross-section. The water chemistry is similar to the chemistry of Sprofondo Lake.

The carbonatic bedrock lies at a 124 m depth, as shown from a borehole in the area. Volcanic and volcanoclastics deposits crop out in the area, reaching a thickness of 54 m. Volcanic deposits and travertine are present to a 100 m depth. Below these deposits, travertine and sand beds extend to 124 m. At this depth the borehole ends because of the interception of a confined and pressurized aquifer.

Other authors (Bono 1995; Di Filippo *et al.* 2002) correlate the presence of a travertine bed at around 30–40 m depth with a cave that, after the collapse of the roof, created the sinkhole. Direct investigations of the lake by scuba divers revealed that the sediments are pyroclastic units down to the bottom (35 m from the soil level) with no sign of travertine or underwater springs.

Close to the sinkhole there is the Ninfa Spring with more than $2.000\,l\,s^{-1}$ flow and a typical bicarbonate–calcic low mineralized water.

From the collected data is possible to hypothesize the genetic mechanism of the Pontina Plain sinkholes. There is a thick alluvial cover (more than 100 m) overlying a deep carbonatic bedrock. The bedrock hosts a confined pressurized aquifer. In some cases (Vescovo lakes) high mineralized springs with a free gas phase (mainly with H_2S and CO_2) lie on the bottom of the flooded sinkholes. The water and gas chemistry indicates a deep origin of the fluids. There is a clear correlation between the sinkhole distribution and the fault systems. In particular, it is possible to identify an alignment of the sinkholes NW–SE for the Vescovo Lake and north–south for the Sprofondi group. Subsurface explorations did not reveal any void that could be correlated with the genesis of the collapses. This geological scenario strongly supports the hypothesis of the presence of conduits in the cover that link the bedrock with the surface. The enlargement of the cylindrical conduits owing to the upward erosion causes the collapse of the soil surface. The collapses should be defined as deep piping sinkholes.

Discussion

In Italy, sinkholes have been recognized since the Roman age. In fact, the presence of catastrophic collapses has been widely reported throughout Italian history. More than 50% of the studied sinkholes are of unknown age. Nevertheless, since 1960 an increase of the phenomena has been noted.

The collapses occur in different geological scenarios and in the same areas with gaps lasting centuries, during which time they could fill through natural or artificial means. Sinkhole extinction, coupled with the long gap between events, causes a lack of attention to this hazard.

In central Italy, sinkholes are mainly located in the Latium, Tuscany and Campania regions along the Tyrrhenian tectonic plains and/or thermal areas. Some collapses have been identified in other regions, but the periadriatic zone, with the exception of a few cases in the Puglia region, does not seem to be affected by sinkhole hazard.

The overall scenario obtained by the analysis of more than 550 cases suggests a specific geological setting that causes a high probability of sinkhole hazard. Field investigations provide a conceptual model for sinkhole genesis. This model could be extended to many other similar cases occurring in other intermountain basins (more than 40% of studied cases) in Tuscany (in the plains of Camaiore and Massa Marittima, and Capalbio towns), in Abruzzo (in the Fucino and Sulmona plains) and in Latium (Pontinia Plain and near Tivoli village).

The case studies demonstrate that sinkholes also occur in alluvial and coastal plains, along pedemountains zones, in small valleys in the immediate vicinity of carbonate ridges, and in complex geological–structural and hydrogeological settings. The observed sinkholes are mainly single collapse features (90%), with diameters ranging from 30 to 100 m, and depths ranging from 5 to 10 m. The following common features have been recognized in these areas.

- A deep carbonate bedrock (5% of cases between 30 and 50 m, 6% between 50 and 100 m, 30% deeper than 100 m and 59% have no data available) overlain by a thick sequence of unconsolidated sediments (predominantly silt and clay). Travertine lenses may be present. In some locations, the geological setting is characterized by the presence of permeable (i.e. sands) or impermeable (i.e. marine clays) covers overlying evaporitic sediments (i.e. gypsum).
- The presence of active faults of local and regional importance. The main direction of the faults are NW–SE and north–south. These faults act as paths for mineralized and/or geothermal fluids enriched in acidic gases (i.e. CO_2 and H_2S) that dissolve and/or pass into the groundwater and cause consumption of the carbonate matrix of shallow sediments and weakening of the supporting geological structure. In this context, the uprising of deep fluids seems to play an important role in the sinkhole formation mechanism. Flooded sinkholes account for 30% of the total; 5% of which are affected by gas emission (38% are dry or buried).
- The presence of confined or unconfined aquifers with high underground water flows and mineralized springs. The small lakes hosted in the sinkholes are usually fed by underwater springs (i.e. S. Vittorino Plain) or represent the outcropping of the water table (i.e. Pontina Plain). In some cases the presence of inlet streams causes a mixing between the low mineralized surficial water and the water inside the sinkholes with a reduction of total dissolved solids concentration.

From a chemical point of view the sinkhole waters may be divided into three main groups (Fig. 10).

- The first group (19% of studied cases) is represented by very shallow aquifers or surficial waters and is distinguished by low values of total dissolved solids (TDS around 300 mg l^{-1}) and pH around 7 or less for the influence of acidic soils (i.e. peat layers). Little increments of TDS are due to the presence of mineral-rich sediment covers (i.e. sinkholes hosted in pyroclastic deposits).
- The second group (19% of studied cases) clusters karst waters characterized by a medium TDS (500 mg l^{-1}), and pH values that are neutral or basic. The main ion concentration is HCO_3^{2-} and Ca^{2+} for the dissolution of the calcareous bedrock by the circulating water.
- The third group (28% of studied cases) collects mineralized waters derived from the mixing of karst waters with fluids (CO_2 and H_2S) of deep origin raised through faults and tectonic displacements. The waters could become very acidic owing to the high level of CO_2 as dissolved and free gas enhancing the chemical dissolution of the carbonatic fraction of the sediments and rocks causing a quick sinkhole development and enlargement. In these last cases mineralized springs are usually located close to the sinkholes and are aligned along tectonic displacements. High values of TDS have also been measured in the ponds hosted in evaporitic sediments. These values are not related to deep fluids mixing but only to the evaporitic rocks dissolution (i.e. some gypsum collapses in Sicily).

There are both natural and artificial triggering processes. Seismic activity, flooding or anthropic activities may cause catastrophic collapses. Earthquakes are associated with some collapses (24%); specific investigations concerning the correlation between earthquake and sinkholes have been conducted on 104 cases. The average distance between the collapses and the epicentre varies from 30 to 50 km. The maximum distance is 130 km. The gap between the seismic event and the sinkhole collapse is usually from 1 to 10 days. The maximum registered gap is 23 days.

Strong oscillation in the water-table level is another triggering effect. Human activity, like strong vibrations caused by heavy working machines, could also cause the sinkhole formation. All of the abovementioned triggering factors act like external forces increasing the stress on the structures.

In some cases, results obtained from structural field surveys indicate the presence of minor extensional fault systems causing the formation of a block mosaic characterized by different uplifting rate (e.g. San Vittorino Plain). The extensive movement of these blocks could be responsible for rejuvenating and reactivating older sinkholes and/or forming other new ones.

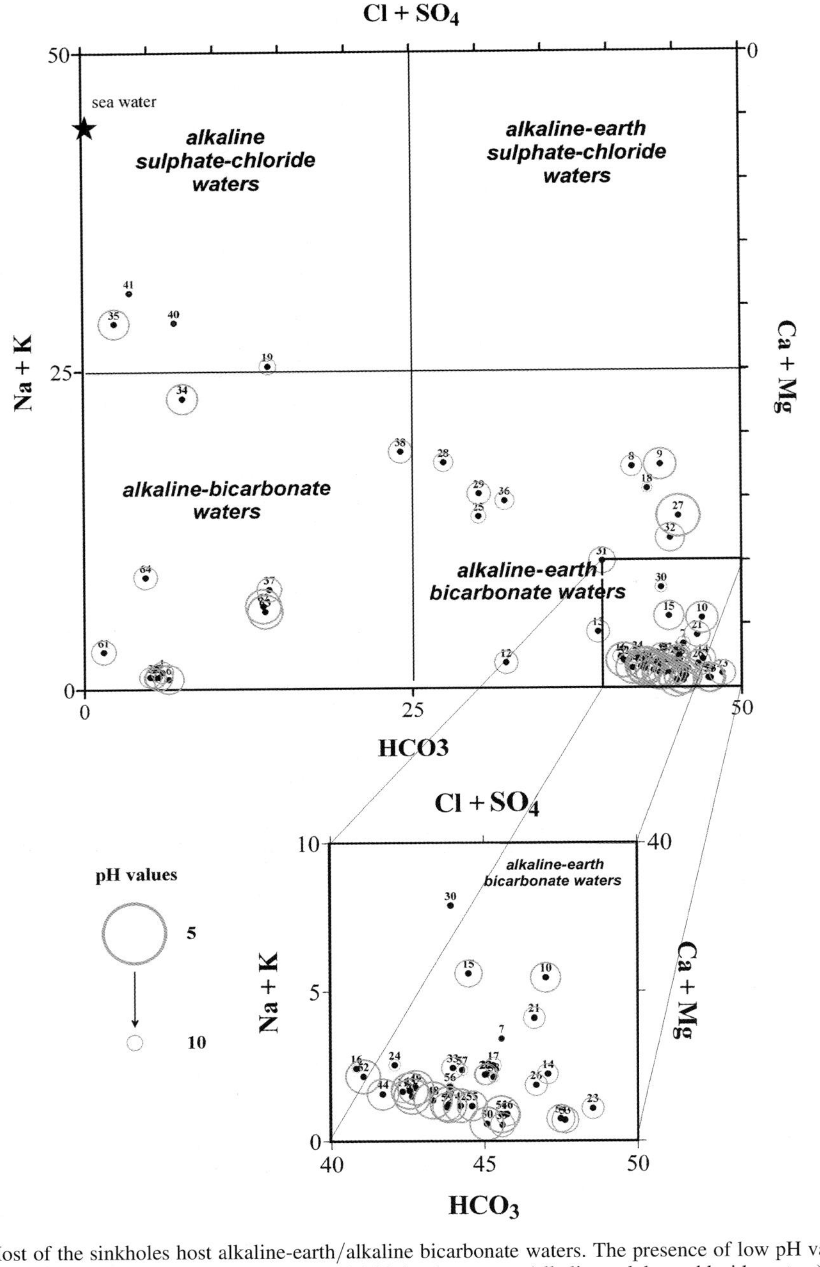

Fig. 10. Most of the sinkholes host alkaline-earth/alkaline bicarbonate waters. The presence of low pH values is related to acidic gas rising in faulted areas. Waters with high salt contents (alkaline-sulphate-chloride waters) are due to the presence of evaporitic bedrock (samples 40–41).

Conclusions

Sinkholes *sensu stricto* in Italy are mainly concentrated in alluvial and intermountain plains, close to carbonate ridges and/or in complex geological–structural and hydrogeological settings. In central-southern Italy, sinkhole-prone areas are characterized by the presence of a deep calcareous bedrock overlain by a thick sequence (100–200 m) of unconsolidated sediments (predominantly silt and clay) with weak physical and/or mechanical characteristics. Pressurized confined aquifers within the bedrock are often recognized.

The widespread occurrence of various faults or fracture systems has been observed in many of the studied areas; faults could act as paths for acidic gases (i.e. CO_2 and H_2S) migration. These gases increase the chemical aggressiveness of the groundwater on the soluble bedrock (i.e. limestone, travertine, gypsum, etc.). Some mineralized water springs with dissolved and free gas are aligned along the main faults close to the sinkholes (i.e. S. Vittorino Plain, Pontina Plain). The water chemistry of the ponds hosted by the sinkholes is not always affected by mineralized fluids. This, in many cases, could be explained by the migration of the springs. Travertine deposits, both as outcrops and in boreholes close to the cavities (i.e. Black Lake in the Pontina Plain), provide clues to the presence of warm mineralized waters.

The morphology of the investigated cavities is usually represented by steep walls and flat bottoms in a bowl or subcylindrical shape. In the flooded sinkholes, echo-sounder surveys show the presence of thick soft silt and clay strata on the bottom; this is also confirmed by direct investigations (i.e. Doganella di Ninfa sinkhole). The cavity is supposed to enlarge by step collapses of the rim. If the debris is not removed by an outlet stream, the depression will fill with sediment and become an 'extinct' (buried) sinkhole.

The main factor associated with the collapse seems to be a deep piping effect. The upwelling of groundwater results in the erosion of sediments and the formation of conduits within the unconsolidated sediments cover; the following loss of strength causes the collapse of the overlaying terrain.

Sinkhole formation could be triggered by various natural causes (seismicity, drought, flood, etc.) and human activity (water over pumping, mines, quarries, drilling operations). In several of the studied cases seismic activity was detected days before the collapse. This suggests that earthquakes could be responsible, in part, for sinkhole genesis.

Preliminary statistic data elaborations on some parameters have been conducted on over 500 cases; other data are still being analysed. The census is still ongoing with more than new 200 cases of possible sinkholes.

The construction of an updatable relational database that includes topographical, geological, hydrogeological, geochemical and other environmental data of the studied sinkhole cases in Italy, coupled with the GIS capabilities, allows analysis of the collected data to define genesis of features in sinkhole-prone areas. This may help the national and regional authorities in the assessment of the sinkhole hazard.

References

ANNUNZIATELLIS, A., BEAUBIEN, S. E., CIOTOLI, G., LOMBARDI, S., NISIO, S. & NOLASCO, F. 2004. Studio dei parametri geologici e geochimici per la comprensione dei meccanismi genetici degli sprofondamenti nella piana di S. Vittorino. *In*: *Proceedings of the Workshop 'State of the art on the Study of Sinkholes, and the Role of National and Local Authorities in the Management of the Territory'*, *Rome*, 20–21 May 2004, 63–82.

BECK, B. F. (ed.) 1984. *Sinkholes: Their Geology, Engineering and Environmental Impact: Proceedings of the First Multidisciplinary Conference on Sinkholes, Orlando*, FL, A.A. Balkema, Rotterdam, The Netherlands.

BECK, B. F. & JENKINS, D. T. (eds) 1986. *Geotechnical Considerations of Sinkhole Development in Florida. International Symposium of Environmental Geotechnology*, April 21–23, 1986, Allentown, PA.

BECK, B. F. & WILSON, W. L. (eds) 1987. *Karst Hydrogeology: Engineering and Environmental Applications. Proceedings of the Second Multidisciplinary Conference on Sinkholes and the Environmental Impacts of Karst, Orlando*, FL, A.A. Balkema. Rotterdam, The Netherlands.

BILLIARD, A., MUXART, T., DERBYSHIRE, E., WANG, J. T. & DIJKSTRA, T. A. 1992. Les glissements de terrain induits par les loess de la province de Gansou, Chine. *Annales de Gèographie*, **566**, 495–515.

BLUMETTI, A. M., DRAMIS, F. & MICHETTI, A. M. 1993. Fault-generated mountain fronts in the Central Apennines (Central Italy): geomorphological features and seismotectonic implications. *Earth Surface Processes and Landforms*, **18**, 203–223.

BLUMETTI, A. M., MICHETTI, A. M. & SERVA, L. 1988. The ground effects of the Fucino earthquake of Jan. 13th, 1915; an attempt for the understanding of recent geological evolution of some tectonic structures. *In*: MARGOTTINI, C. & SERVA, L. (eds) *Historical Seismicity of Central-eastern Mediterranean Region, Proceedings of the 1987 ENEA–IAEA International Workshop*. Ente per le Nuove Tecnol., l'Energia, e l'Ambiente, Rome, 297–319.

BONI, C., CAPPELLI, G. & PETITTA, M. 1995. *Carta idrogeologica dell'alta e media valle del F. Velino*. System cart, Roma.

BONO, P 1995. The sinkhole of Doganella (Pontina, Plain, Central Italy). *Environmental Geology*, **26**, 48–52.

CAPELLI, G., COLOMBI, A. & SALVATI, R. 2001. Catastrophic subsidence risk assessment. A conceptual matrix for sinkhole genesis. *In*: BECK, B. F. & GAYLE HERRING, J. (eds) *Geotechnical and Environmental Applications of Karst Geology and Hydrology*. Balkema, Rotterdam, The Netherlands.

CAPELLI, G., PETITTA, M. & SALVATI, R. 2000. Relationships between catastrophic subsidence hazards and groundwater in the Velino Valley (Central Italy). *In*: *Proceedings of the Sixth International Symposium on Land Subsidence SISOLS 2000*, Ravenna, Italy, Volume 1, 123–136.

CARAMANNA, G., CIOTOLI, G. & NISIO, S. 2005. A review of natural sink phenomena in Italian plain areas. *In*: *Sixth International Conference of*

Geomorphology, 7–11 September 2005, Zaragoza, Spain. Zaragoza University.

CARAMANNA, G., NISIO, S. & VITA, L. 2004. Fenomeni di annegamento dei sinkholes: casi di studio su alcuni laghetti di origine incerta. *In*: *Proceedings of the Workshop 'State of the art on the Study of Sinkholes, and the role of National and Local Authorities in the Management of the Territory', Rome*, 20–21 May 2004, 229–248. APAT, Rome.

CASTIGLIONI, G. B. 1986. *Geomorfologia.* Opere UTET di geografia e discipline affini.

CENTAMORE, E. & NISIO, S. 2002. Quaternary Morphodynamic between the Velino and Salto Valleys. *Studi Geologici Camerti*, Special Volume. Selca Camerino Macerata, **1999**, 37–44.

CENTAMORE, E. & NISIO, S. 2003. The effects of uplift and tilting in the Central Apennine. *Quaternary International*, **101–102**, 93–101.

CENTAMORE, E., NISIO, S. & ROSSI, D. 2004. Aspetti geologico-strutturali in relazione alla formazione della 'sinkhole plain' di S. Vittorino. *In*: *Proceedings of the Workshop 'State of the art on the Study of sinkholes, and Role of National and Local Authorities in the Management of the Territory', Rome*, 20–21 May 2004, 285–298.

CIOTOLI, G., DI FILIPPO, M., NISIO, S. & ROMAGNOLI, C. 2001. La Piana di S. Vittorino: dati preliminari sugli studi geologici, strutturali, geomorfologici, geofisici e geochimici. *Memorie della Società Geologica Italiana*, **56**, 297–308.

CIOTOLI, G., GUERRA, M., LOMBARDI, S. & VITTORI, E. 1998. Soil gas survey for tracing seismogenic faults: a case study in the Fucino Basin, Central Italy. *Journal of Geophysical Research*, **103**, 23,781–23,794.

COLOMBI, A., SALVATI, R. & CAPELLI, G. 2001. Sinkhole in the Latium region (Central Italy). Purposes of the main project. *In*: BECK, B. F. & GAYLE HERRING, J. (eds) *Geotechnical and Environmental Applications of Karst Geology and Hydrology.* Balkema, Rotterdam, The Netherlands.

CRAMER, H. 1941. Die Systematik der karstdolinen. *Neues Jahrbuch für Mineralogie Geologie und Paläontologie*, **85**, 293–382.

DERBYSHIRE, E. & MELLORS, T. W. 1988. Geological and Geotechnical characteristic of some loess and loessic soil from China and Britain: a comparison. *Engineering Geology*, **25**, 135–175.

DERBYSHIRE, E., WANG, J. *ET AL.* 1991. Landslide in the Gansu loess of China. *Catena Supplement*, **20**, 119–145.

DI FILIPPO, M., PALMIERI, M. & TORO, B. 2002. Studio gravimetrico del sinkhole di Doganella di Ninfa (Latina). *In*: *Le voragini catastrofiche, un nuovo problema per la Toscana. Proceedings of the Conference* 31 Marzo 2000, Grosseto R. Regione Toscana, 62–70.

FACCENNA, C., FLORINDO, F., FUNICELLO, R. & LOMBARDI, S. 1993. Tectonic setting and Sinkhole Features: case histories from western Central Italy. *Quaternary Proceedings*, **3**, 47–56.

FAIRBRIDGE, R. W. 1968. *The Encyclopedia of Geomorphology.* Reinhold, New York.

FERRELI, L., GUERRIERI, L., NISIO, S., VITA, L. & VITTORI, E. 2004. Relations among seismogenic structures, earthquakes and sinkhole phenomena: a methodological approach in the Apennines (Italy). *In*: 32nd International Geological Congress, Firenze, 20–28 August 2004, Volume Abstract, Part 1, 669.

FORD, D. & WILLIAMS, P. W. 1989. *Karst Geomorphology and Hydrology.* Unwin Hyman, London.

GRACIOTTI, R., NISIO, S. & VITA, L. 2004. Sinkholes in Italy: inventory of natural phenomena and some study cases. *In*: 32nd International Geological Congress, Firenze, 20–28 August 2004, Volume Abstract, Part 1, 670.

GUNN, J. 2004, *Encyclopedia of Caves and Karst Science.* Fitzroy Dearborn, New York.

JENNINGS, J. N. 1985. *Karst Geomorphology.* Kateprint, Oxford.

LITTLEFIELD, J. R., CULBRETH, M. A., UPCHURCH, S. B. & STEWART, M. T. 1984. Relationship of modern sinkhole development to large scale-pholinear features. *In*: BECK, B. F. (ed.) *Sinkholes: Their Geology, Engineering & Environmental Impact.* Balkema, Rotterdam, The Netherlands.

MASSARI, F., GHIBAUDO, G., D'ALESSANDRO, A. & DAVAUD, E. 2001. Water-upwelling pipes and soft-sedimentary deformation structures in lower Pleistocene calcarenites (Salento, southern Italy). *Geological Society of America Bulletin*, **113**, 545–560.

MICHETTI, A. M., BRUNAMONTE, F., SERVA, L. & VITTORI, E. 1996. Trench investigations of the 1915 Fucino earthquake fault scarps (Abruzzo, Central Italy): Geological evidence of large historical events. *Journal of Geophysical Research*, **101**, 5921–5936.

MONROE, W. H. 1970. *A Glossary of Karst Terminology.* US Geological Survey Water Supply Paper. GV Print, Washington.

MUXART, T., BILLARD, A., DERBYSHIRE, E. & WANG, J. 1994. Variation in runoff on steep unstable loess slopes near lanzhou, China: Initial results using rainfall simulation. *In*: KIRBY, M. J. (ed.) *Process Models and Theoretical Geomorphology*, 337–355. Wiley, Chichester.

NEWTON, J. C. 1984. Review of induced sinkhole development. *In*: BECK, B. F. (ed.) *Sinkholes: Their Geology, Engineering & Environmental Impact.* Balkema, Rotterdam, The Netherlands.

NEWTON, J. G. 1986. *Natural and Induced Sinkhole Development in the Eastern United States.* International Association of Hydrogeological Sciences, Publication, **151**.

NISIO, S. 2003. I fenomeni di sprofondamento: stato delle conoscenze ed alcuni esempi in Italia Centrale. *Il Quaternario*, **16**, 121–132.

NISIO, S. & SALVATI, R. 2004. Fenomeni di sprofondamento catastrofico. Proposta di classificazione applicata alla casistica italiana. *In*: *Proceedings of the Workshop 'State of the art on the Study of Sinkholes, and the Role of National and Local Authorities in the Management of the Territory', Rome*, 20–21 May 2004, 573–584. APAT, Rome.

NISIO, S., GRACIOTTI, R. & VITA, L. 2004. I fenomeni di sinkhole in Italia: terminologia, meccanismi genetici e problematiche aperte. *In*: *Proceedings of the Workshop 'State of the art on the Study of Sinkholes, and the Role of National and Local Authorities in the Management of the Territory', Rome*, 20–21 May 2004, 557–572. APAT, Rome.

ODDONE, G. 1915. Gli elementi fisici del grande terremoto marsicano-fucense del 13 gennaio 1915. *Bollettino della Società Sismologica Italiana*, **19**, 71–215.

SALVATI, R. & SASOWSKY, I. D. 2002. Development of collapse sinkholes in areas of groundwater discharge. *Journal of Hydrology*, **264**, 1–11.

SALVATI, R., THARP, T. & CAPELLI, G. 2001. Conceptual model for geotechnical evaluation of sinkhole risk in the Latium Region. *In*: BECK, B. F. & GAYLE HERRING, J. (eds) *Geotechnical and Environmental Applications of Karst Geology and Hydrology*. Balkema, Rotterdam, The Netherlands.

SINCLAIR, W. C. 1982. *Sinkhole Development Resulting From Ground-water, Development in the Tampa area, Florida*. US Geological Survey, Water Resources Investigation Report, 81–50.

SINCLAIR, W. C. & STEWART, J. W. 1985. *Sinkhole Type, Development and Distribution in Florida*. US Geological Survey, Map Series 110, Plate 1.

SNYDER, S. W., EVANS, M. W., HINES, A. C. & COMPTON, J. S. 1989. Seismic expression of solution collapse features from the Florida Platform. *In*: *Engineering and Environmental Impacts of Sinkholes and Karst. Proceedings of the Third Multidisciplinary Conference on Sinkholes and the Engineering and Environmental Impacts of Karst, St Petersburg Beach, Florida*, October 2–4, 1989.

SWEETING, M. M. 1972. *Karst Landform*. Macmillan, London.

THARP, T. M. 1997. Mechanism of formation of cover collapse sinkhole. *In*: *Proceedings of the 6th Multidisciplinary Conference of Sinkhole and the Engineering and Environmental Impact of Karst*. Balkema, Rotterdam, The Netherlands, 29–36.

THARP, T. M. 1999. Mechanism of upward propagation of cover collapse sinkhole. *Engineering Geology*, **52**, 23–33.

TIHANSKY, A. B. & GALLOWAY, D. L. 2000. Land and water-resource development activities increase sinkhole frequency in the mantled karst region of Florida. USA. *In*: *Proceedings of the Sixth International Symposium on Land Subsidence SISOLS* 2000, Ravenna, Italy, Volume 1, 77–90.

WALTHAM, A. C. & FOOKES, P. G. 2003. Engineering classification of karst ground conditions. *Quarterly Journal of Engineering Geology and Hydrogeology*, **36**, 101–118.

WHITE, W. B. 1988. *Geomorphology and Hydrology of Carbonate Terrains*. University Press, Oxford.

Assessment of cover-collapse sinkholes in SW Sardinia (Italy)

F. ARDAU[1], R. BALIA[1], M. BIANCO[1] & J. DE WAELE[2]

[1]*Dipartimento di Ingegneria del Territorio, University of Cagliari, Cagliari, Italy*

[2]*Dipartimento di Scienze della Terra e Geologico-Ambientali, Università di Bologna, Via Zamboni 67, 40126 Bologna, Italy (e-mail: dewaele@geomin.unibo.it)*

Abstract: The SW part of Sardinia has been afflicted, in recent years, by several cover-collapse sinkholes mostly occurring in low-density population areas. The study area, that lies in the Iglesiente–Sulcis region, is characterized by the cropping out of the Palaeozoic basement related to the South European Hercynian chain, covered with Tertiary–Quaternary sediments. The main rock types that crop out are Palaeozoic metasandstones, metadolostones, metalimestones, shales and metaconglomerates, and Tertiary–Quaternary fluvial–lacustrine continental sediments. The combined application of several geophysical techniques, integrated with boreholes and geotechnical as well as hydrogeological measurements, proved to be very useful and promising in defining in detail the geological context in which each sinkhole has formed. Moreover, the gravity method, even when used alone, proved to be very effective in detecting the regional geological structures to which sinkholes are related.

Eventually, the historical analysis of phenomena, the geological knowledge of the Iglesiente–Sulcis area and the results of properly designed geophysical surveys allows the most probable areas for cover-collapse sinkholes to occur in the future to be determined. In fact, this research pointed out that the depth of the sediment-covered Palaeozoic bedrock is one of the major constraints in delimiting hazardous areas, leading to the construction of a preliminary hazard map. This map shows a belt of high risk, and also suggests the areas in which further geophysical and geotechnical investigations should be carried out to estimate the depth of the bedrock.

Cover-collapse sinkholes forming on karst terrains covered with more or less loose sediments have been described from many areas of the world (Beck 1986; Newton 1987; Twidale 1987; Soriano & Simon 1995; Kaufmann & Quinif 1999; Salvati & Sasowsky 2002; Waltham *et al.* 2005). These phenomena are often triggered by human intervention, especially by water extraction and modifications in surface drainage patterns, and are further enhanced by the increasing occurrence of extreme meteorological events.

Similar hazards have been occurring in several areas of SW Sardinia (Italy) since the early 1990s with increasing frequency (Balia *et al.* 2001; De Waele & Muntoni 2001). One of the most interesting and impressive sinkholes formed in October 1998 in the Cixerri Valley, at Guardia Su Merti, only 50 m away from the main railway that connects Cagliari to Iglesias (Balia *et al.* 2001). Other major sinkholes occurred 1 year later some kilometres south, in an area named Planu Francau close to the main road from Villamassargia to Carbonia (Fadda & Fais 2002; De Waele *et al.* 2003). The increase in the formation of these sinkholes has led local and regional stakeholders and decision makers to consider studying these phenomena that could seriously endanger local infrastructures and human lives in the near future.

Geological setting

Cover-collapse sinkholes have been reported from several covered karst areas in SW Sardinia, near the villages of Carbonia, Iglesias, Narcao, Sant'Anna Arresi and Villamassargia (Balia *et al.* 2001; De Waele & Muntoni 2001; Fadda & Fais 2002; De Waele *et al.* 2003). The study area (Fig. 1) is characterized by the outcrops of Palaeozoic metasediments ranging in age from Early Cambrian to Upper Carboniferous, covered with Tertiary–Quaternary sediments and volcanic rocks. From bottom to top the Palaeozoic sequence is composed of metasandstones, shales, oolithic limestones, calcareous sandstones, dolostones, intensely karstified limestones and shales (Pillola 1989; Bechstadt & Boni 1996; Carmignani *et al.* 2001). Post- Palaeozoic sedimentation starts only during the Lower Tertiary with marly limestones, marls and mudstones with coal seams of Eocene age (Murru & Matteucci 2002). Sedimentation continues with alluvial and lacustrine deposits composed of conglomerates and sandstones of the Cixerri Formation of Middle Eocene–Upper Oligocene age (Carmignani *et al.* 2004). Immediately south of Villamassargia these sediments are overlain with andesites and basaltic andesites of Oligocene–Lower Miocene age (Beccaluva *et al.* 1985).

From: PARISE, M. & GUNN, J. (eds) *Natural and Anthropogenic Hazards in Karst Areas: Recognition, Analysis and Mitigation*. Geological Society, London, Special Publications, **279**, 47–57. DOI: 10.1144/SP279.5
0305-8719/07/$15.00

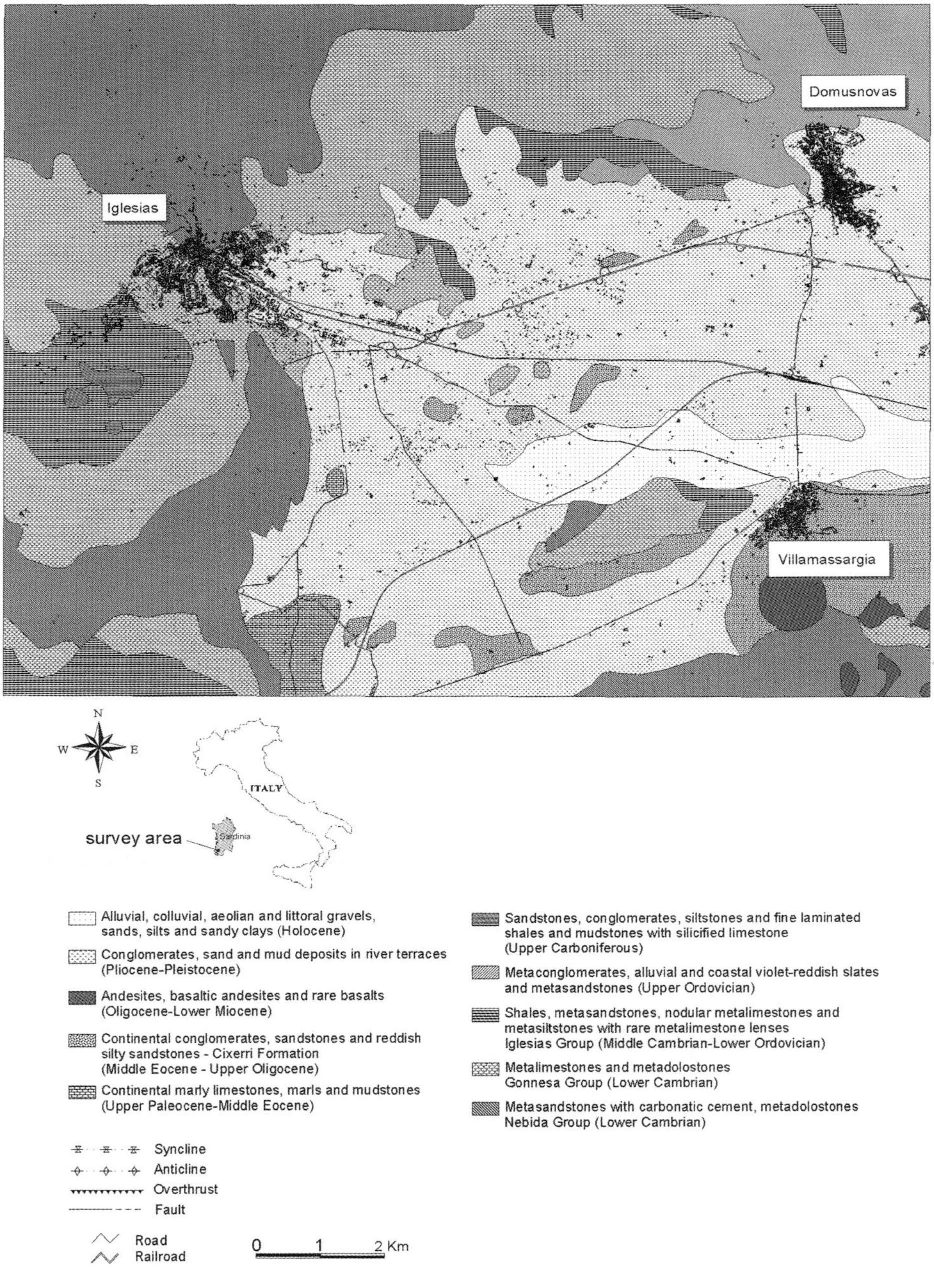

Fig. 1. Geological map of the study area and surrounding region.

The geological succession ends with Plio-Quaternary deposits mainly composed of alluvial sediments and landslide deposits.

Hydrogeology

Rocks with different permeability crop out in the study area, and the main aquifer is the one hosted by the more or less fractured and karstified Cambrian limestones (Civita *et al.* 1983). Other aquifers are the superficial Quaternary alluvium, river terraces and fractured volcanic rocks, the latter being extremely variable in importance and very localized.

The Palaeozoic karst aquifer has one main outlet in the study area, the Caput Aquas spring, located at an altitude of 116 m above sea level (a.s.l.), and giving water to the cities of Villamassargia, Carbonia and Iglesias. Until the late 1980s this spring had a flow rate ranging between 70 and 200 l s^{-1}, but in June 1990 it completely dried out (Bianco & De Waele 1992). The local water authorities had started pumping in several boreholes drilled close to the spring, thus lowering the water table for several tens of metres. In fact, water is pumped out from a depth of almost 100 m.

Another borehole operated until the beginning of the 1990s in the small Cambrian limestone hill of Guardia Su Merti, but it is now completely dried out.

Up to some tens of years ago water was also extracted from the superficial alluvial aquifers, but after a long period of drought during the 1990s most farmers and local residents have drilled boreholes directly in the Palaeozoic bedrock, sometimes at considerable depth (>50 m).

These pumping sites have caused a permanent lowering of the water table since the early 1990s, negatively impacting on the whole hydrogeological system.

The entire Cixerri sedimentary basin can be considered as a recharge area of the underlying carbonate aquifer. The Cixerri Formation sandstones are known to be aquitards; nevertheless, surface-water infiltration can occur locally, especially where carbonate bedrock is not deeply buried. It is in these localized recharge areas, or at their borders, that cover-collapse sinkhole formation is more likely to occur because of enhanced infiltration causing underground ravelling of sediments in draining fissures or karst conduits in the bedrock.

Mechanisms and causes for cover-collapse formation

Cover-collapse sinkholes, even though an apparently sudden phenomena, are the result of complex processes that occur through a succession of steps in terrains characterized by a soluble (karst) bedrock covered with a more or less thick and loose sedimentary cover, normally ranging in thickness between a few metres and approximately 30 m (White & White 1995) (Fig. 2). They normally form in recharge areas of covered karst terrains (Beck 1986), but can also develop, under certain circumstances, in areas of groundwater recharge (Salvati & Sasowsky 2002). Under-ground drainage through the epikarstic zone towards deeper lying karst conduits or along enlarged joints and mantled karren shafts (wide enough to enable turbulent flow) connected to the overlying sediments allows downward and lateral transportation of material by gravity or by seeping water. This downward erosion of covering sediments into the epikarst is called 'ravelling' (Beck 1988). In many cases this is not a purely natural phenomenon, but appears most often to be accelerated by anthropogenic activities such as pumping and constructing, as clearly demonstrated by numerous case studies published in the proceedings of the multidisciplinary conferences on sinkholes (Beck 1984, 1989, 1993; Beck & Wilson 1987; Beck & Pearson 1995).

The downward removal of sediment creates a void in the overlying cover material that can increase rapidly in size forming a bowl- or arch-shaped cavity. This is possible especially where sediment can be transported laterally (or vertically) away from the cavity. The arch-like roof of this void progressively grows towards the land surface eventually causing sudden collapse, often triggered by climatic (rainfall, frost etc.) or mechanical causes (nearby blasting, passage of heavy equipment, etc.).

The process of ravelling can cause collapse to occur very quickly (in the order of a couple of days), but usually takes several weeks or even months, depending on lithology, meteorology and thickness of cover.

Formation of sinkholes is often related to the lowering of the water table because of a loss of buoyant support, an increase in groundwater velocity because of a steeper hydraulic gradient, and induced water-level variations during and after rain. After prolonged drought even modest rainfall can trigger sinkhole formation. Sinkholes also form in places where surface runoff has been concentrated, such as along road ditches and close to other surface-water collectors (Newton 1987; Tharp 1999; Waltham *et al.* 2005).

Sinkhole description and distribution

Our field survey has concerned the Cixerri Basin, an east–west-oriented open syncline of Oligocene age (Carmignani *et al.* 2004). The main cover-collapse sinkholes have formed in several places, from north to south, as shown in Figure 3: Riu

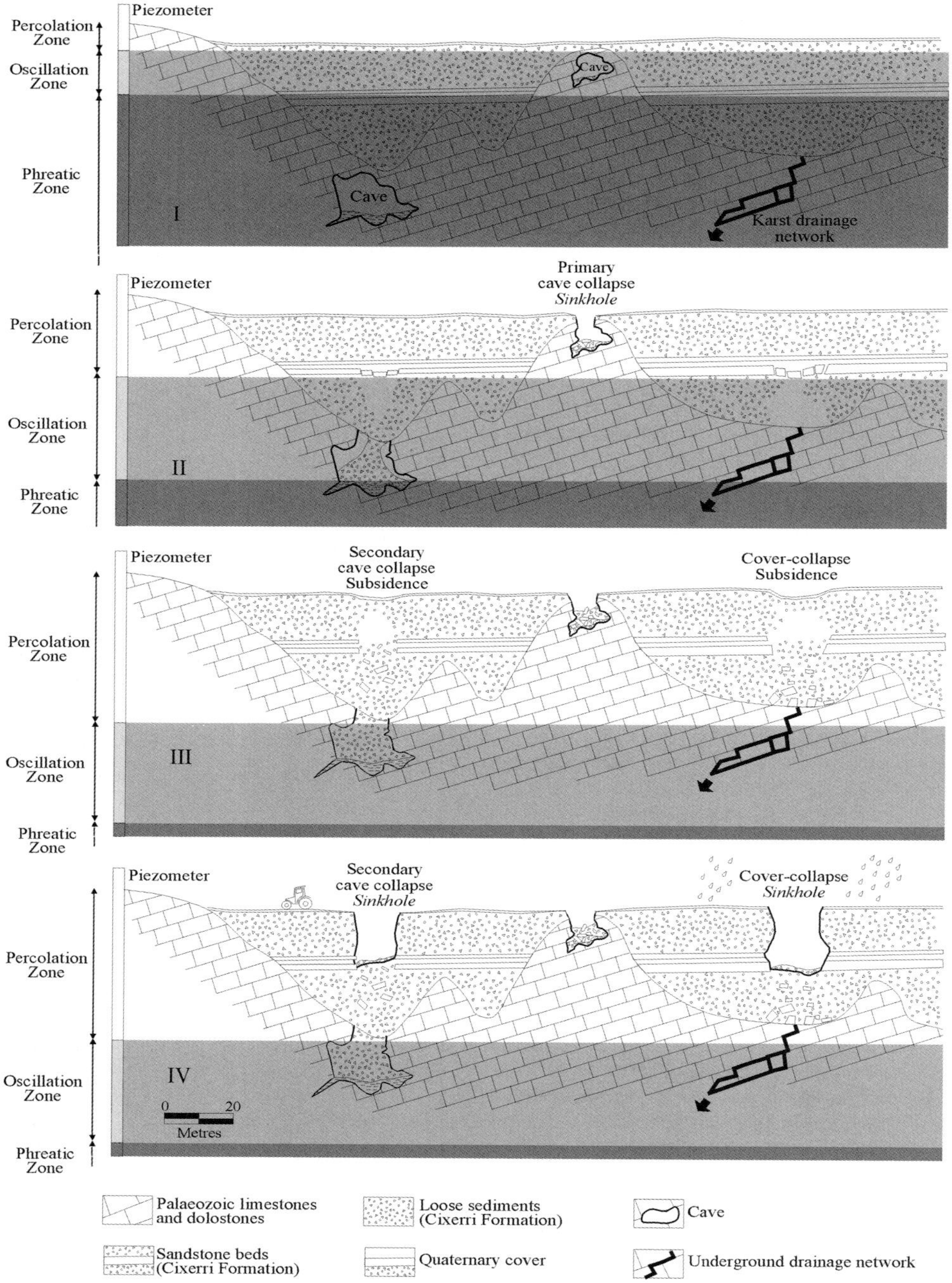

Fig. 2. Evolutionary stages in the formation of cover-collapse sinkholes in the Cixerri sedimentary basin: (**I**) natural equilibrium stage with normal water table. Carbonates are affected by more or less developed karst voids. (**II**) Anthropogenic lowering of the water table triggers ravelling in an underground collapsed cave (left) or in an underground drainage system (right). The collapse of a superficial cave causes a primary cave collapse sinkhole (centre). (**III**) Persistant lowering of the water table causes seasonal ravelling with rupture of the sandstone beds and formation of premonitory subsidence on the surface. (**IV**) Cover-collapse sinkhole formation triggered by vibrations (e.g. heavy equipment), rainfall or other catalysing situations.

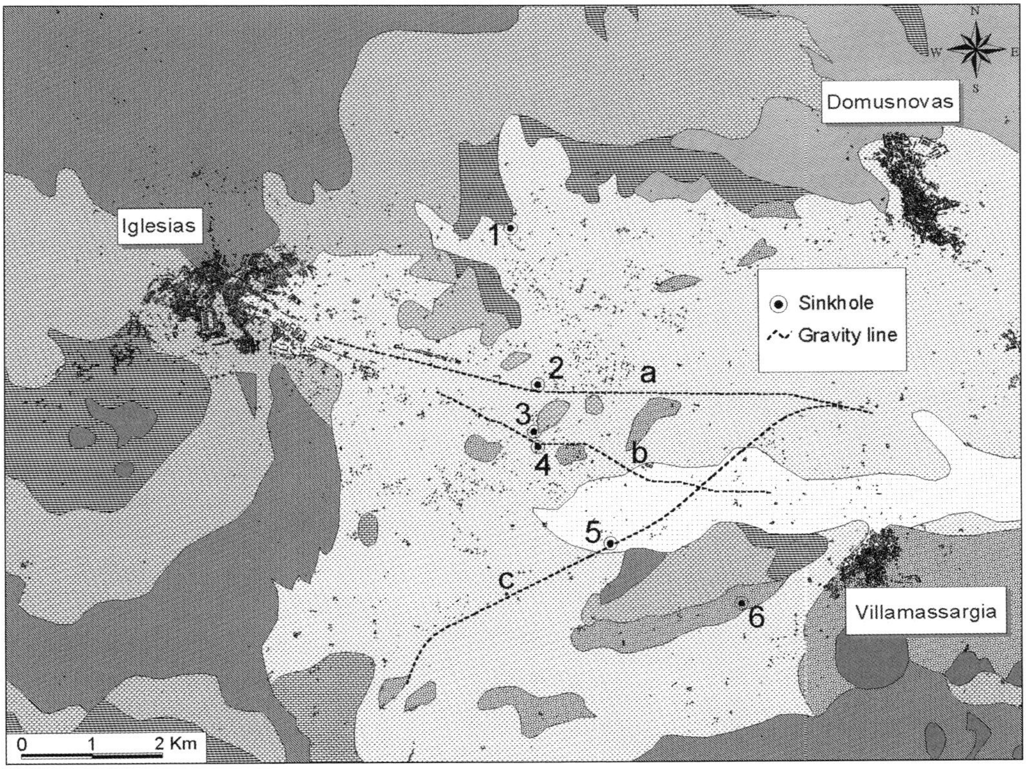

Fig. 3. Position of main cover-collapse sinkholes. Gravity lines a, b and c are marked with a dashed line.

Corongiu–Cuccuru Tiria, Guardia Su Merti, Caput Acquas and Planu Francau (Balia *et al.* 2001; De Waele & Muntoni 2001; De Waele *et al.* 2003).

In the alluvial plain of Riu Corongiu–Cuccuru Tiria some small sinkholes formed after heavy rains in the early 1990s (number 1 in Fig. 3). These were all filled with sediments by the local farmers and did not cause serious problems.

The sinkhole of Guardia Su Merti (number 2 in Fig. 3) formed in October 1998 only 50 m away from the railway that connects Cagliari to Iglesias (Balia *et al.* 2001). It is completely formed within Cixerri Formation sediments, characterized by brittle sandstones, and exposing, at a depth of 5–7 m, two massive beds of sandstone. Its diameter reaches almost 20 m for a depth of 15 m. Less than 100 m west of this major sinkhole, another small collapse exposes Palaeozoic limestones at a depth of only 2 m, proving that carbonate rocks are present under a relatively thin sedimentary cover.

Some kilometres south, two other sinkholes have formed in 1999, one just in the River Cixerri bed, the other one close to a tarmac road (numbers 3 and 4 in Fig. 3, respectively). These sinkholes have been filled with some natural clastic material and are no longer visible today. Another sinkhole, formed some kilometres SE of the karst spring of Caput Acquas, has also been filled with sediments (sinkhole number 5 in Fig. 3).

South of Monte Ollastu, at Planu Francau, a series of cover-collapse sinkholes formed in the late 1990s, and their evolution is still in progress (number 6 in Fig. 3). The largest of these has a diameter of more than 20 m and a depth of 10 m; some others are still developing and show clear evidence of subsidence (De Waele *et al.* 2003) (Fig. 4).

The role of geophysical techniques in the assessment of cover-collapse sinkholes

Several geophysical surveys with different techniques have been carried out in the Cixerri Basin since 1998, with the specific aim of studying sinkhole phenomena; the results of these surveys show that they are suitable for both detailed and large-scale investigation.

Fig. 4. A very large sinkhole, still in evolution, at Planu Francau (number 6 in Fig. 3).

The first survey, dating back to the period October 1998–February 1999, concerned a small area including the regional railway, situated in a place named 'Su Merti', and was begun immediately after the sudden appearance of a large cover-collapse sinkhole (number 2 in Fig. 3) shown in Figure 5. At that time electrical resistivity, gravity and seismic refraction techniques were employed, and some boreholes were drilled. One year later a seismic reflection survey along a profile parallel to the railway was added to previous surveys (Balia *et al.* 2001). While electrical data were very noisy and substantially useless owing to abundant clay and moisture in the near surface, the combined interpretation of gravity, seismic and borehole data allowed a fine-detail geological section to be drawn up, shown in Figure 6 along with the stratigraphy of borehole Bh1. In this section, strongly dependent on geophysical data and corresponding to a profile adjacent to the sinkhole, the geological structure to which the latter is related is clearly depicted, as well as the lowering of Palaeozoic limestones and dolostones on the central-eastern side of the section. Beyond this, the gravity anomaly map indicated unequivocally that the structure at the origin of the sinkhole crosses the railway, thus indicating a potential impending danger (Balia *et al.* 2001).

The above-cited case study, and some more detailed investigations carried out in other areas of the Cixerri Basin where the same kind of phenomenon has taken place (Fadda & Fais 2002), represent a typical example of how the combined application of different geophysical techniques can be very effective in the detailed assessment of sinkholes.

However, this kind of information is normally acquired – and thus be used for safety, geotechnical and remediation purposes – only *after* the occurrence of a collapse involving the ground surface; however, in terms of prevention and land-use planning one would prefer to know *in advance* where these events could most probably occur. In other words, although the knowledge of local conditions was (and is) very important, the benefit gained from highly detailed investigations carried out over limited areas was rather small with respect to the general understanding of sinkhole occurrence in the entire region of the Cixerri Basin. Thus, we recognized that finding possible correlations between sinkholes and the regional geological

Fig. 5. The sinkhole at Su Merti (number 2 in Fig. 3), the first place where geophysical methods were applied.

structure of the basin was strictly necessary. Consequently, in 2003–2005, thanks to grants received from the RFI – the Italian Railways Department – three fine-detail gravity profiles were acquired along two railway lines and along a road. On the whole, gravity measurements were taken at 1500 station points at 10 m spacing, and at 158 station points at 20 m spacing. The positions of the gravity profiles reported in Figure 7 are shown in Figure 3, together with the location of known sinkholes.

As can be seen in Figure 7, especially in profiles a and b, the gravity anomaly decreases very rapidly east of the sinkholes and witnesses the presence of a regional structure that deepens the Palaeozoic basement, in agreement with recent geological investigations (Carmignani *et al.* 2004). The modelling of the gravity anomaly for profiles a and b in terms of two-dimensional (2D) mass distribution is shown in Figure 8. As can be appreciated, the slope of the bedrock ranges between 10 and 20%.

It is clear that the cover-collapse sinkholes are strongly connected to the above-said regional structure surveyed by means of gravity. The same structure also represents the boundary between the region where Palaeozoic carbonates lie very close to the surface (western side) and the region where they are covered by conspicuous thicknesses of Caenozoic and Quaternary rocks. Obviously, the probability of cover-collapse sinkhole occurrence decreases as the depth to limestones and dolostones increases, and thus as the thickness of the overlaying Caenozoic sandstones and conglomerates increases. On the basis of the abovementioned considerations it is possible to define a 'dangerous' belt of land where restrictions should be carefully considered. This belt is qualitatively shown in Figure 9: its path, which is largely interpolated, should be better defined essentially by means of further gravity investigations – that

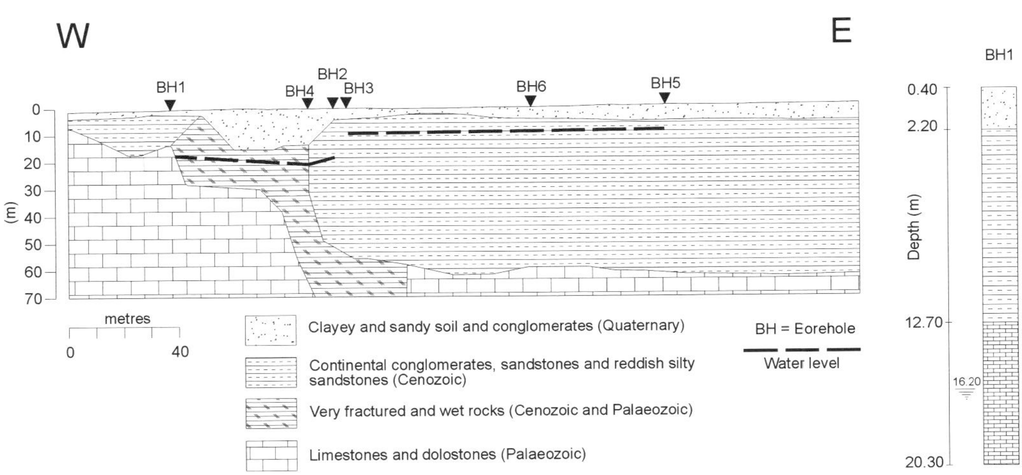

Fig. 6. Geological section based on geophysical and well data at Su Merti (after Balia *et al.* 2001, modified); the section is 40 m south of the sinkhole number 2 in Figure 3.

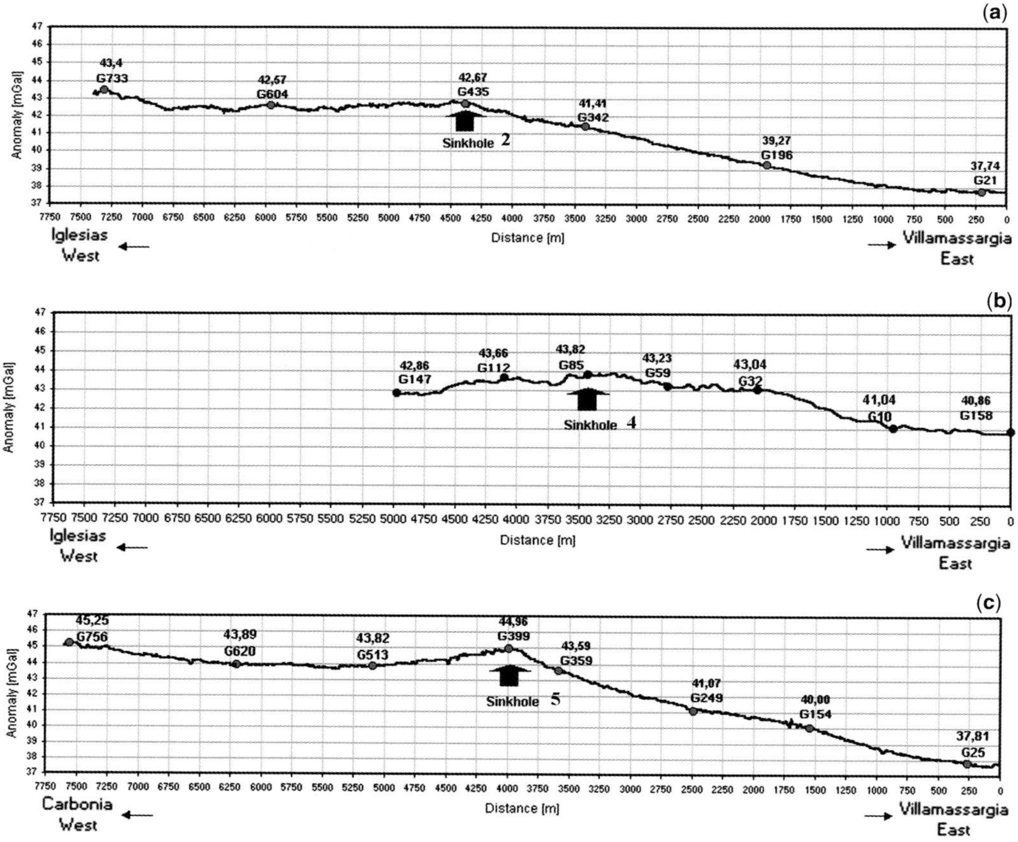

Fig. 7. Gravity anomaly profiles corresponding to lines **a**, **b** and **c** in Figure 3; the station interval was 10 m for lines a and c; 20 m for line b. The lines include, respectively, sinkhole numbers 2, 4 and 5 (see Fig. 3).

proved very effective – while its width should also be estimated on the basis of geotechnical and rock mechanics computations that are beyond the scope of this paper.

Towards a geohazard map

In the framework of regional planning, and in particular of the Piano di Assetto Idrogeologico (PAI) (Landslide and Flood Setting Plan) (cfr. Italian National Laws 183/1989 and 180/1998), natural hazards such as landslides and flooding have to be taken into consideration. Sinkhole risk is mentioned in these regional guidelines for the territory of Iglesias and appropriate measures should be taken to prevent risk in such vulnerable areas, including putting in place danger signs, enclosures and fences, raising public awareness on the problem and compiling a geohazard map.

Mapping of cover-collapse sinkholes in the Cixerri Basin, together with detailed geological interpretation of existing and field data, have allowed a first distribution of these phenomena to be outlined in a regional geological framework. The geophysical surveys have contributed to a better understanding of the covered bedrock topography along roughly east–west-striking profiles and enabling geological reconstructions.

On the basis of these data it has been possible to draw a preliminary hazard map of the Cixerri sedimentary basin that distinguishes three classes of hazard, from high to low (Fig. 9). The 'high' hazard class is related to the alignment of the existing sinkholes where the Palaeozoic limestone is covered by Tertiary and Quaternary insoluble sediments, ranging in thickness from 0 to approximately 50 m. These areas are also affected by artificial lowering of the water table by overpumping, triggering the cover-collapse phenomena.

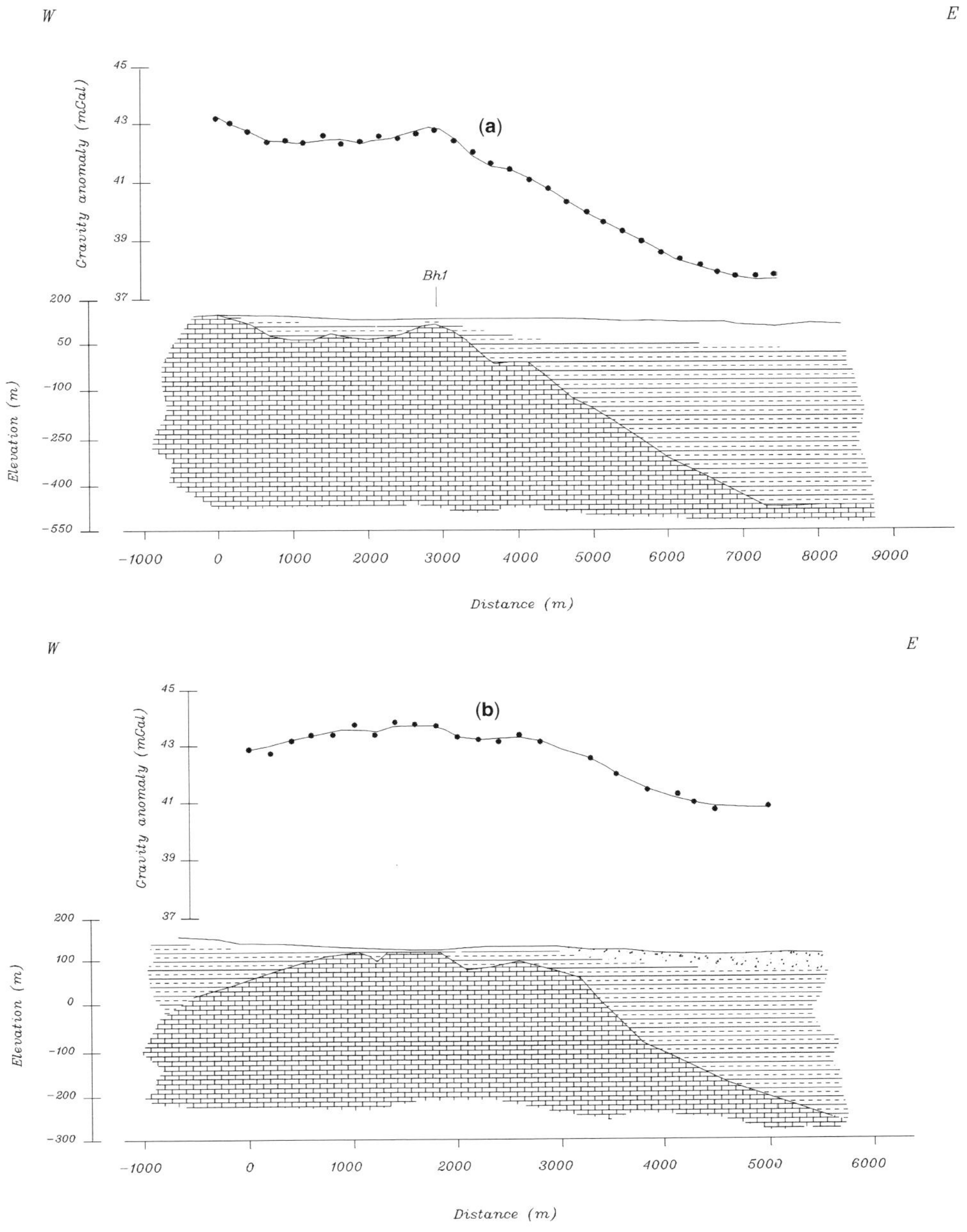

Fig. 8. Two-dimensional interpretation of the gravity anomaly along lines a and b. In the upper part of each interpretation dots represent experimental anomaly values, and the line is the computed anomaly corresponding to the model. For the geological legend refer to Figure 6.

East of this line the degree of hazard is 'low' owing to the fact that the bedrock deepens gradually, in agreement with the experimental results of our geophysical surveys and with the presence of the Cixerri syncline.

West of the line the situation is not yet clear, and the level of hazard can vary, depending on depth of Palaeozoic bedrock and on local geo-environmental conditions. Further surveys should be carried out to understand what is occuring where covered karst

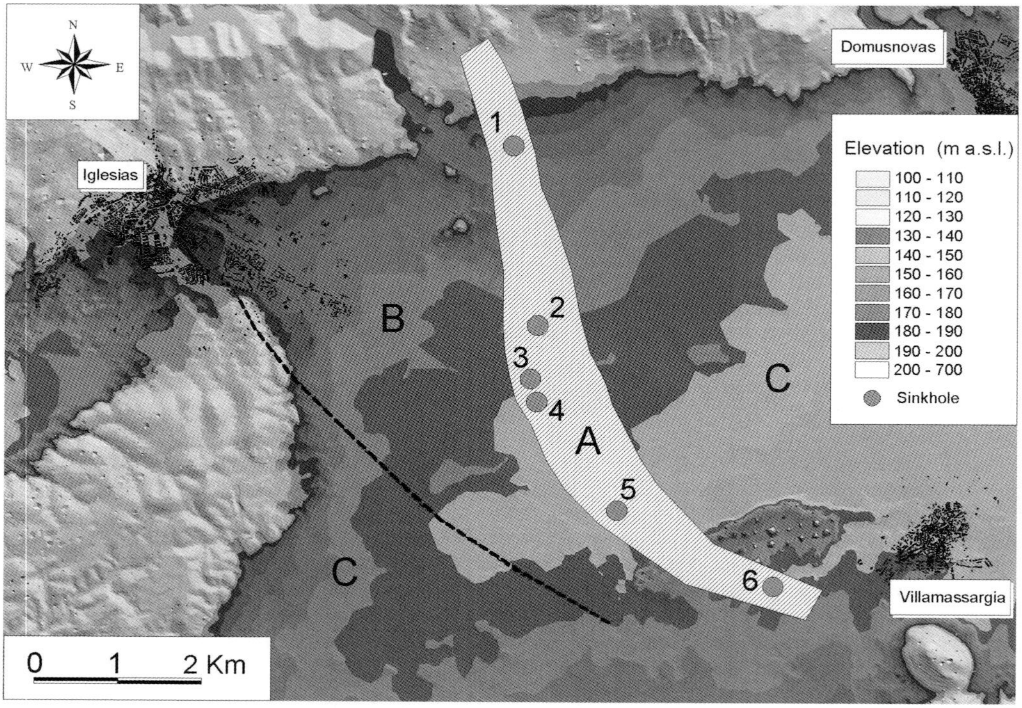

Fig. 9. Tentative hazard map of the study area, where A indicates the high hazard belt including known sinkholes, B indicates medium hazard areas and C low hazard areas. The dashed line corresponds to the presumed borderline between shallow Palaeozoic limestones and dolostones (east) and Palaeozoic sandstones (west).

topography persists. In particular, testing sites should be situated around Iglesias where extensive exploitation of the aquifer could result in some subsidence phenomena.

Conclusions

Cover-collapse sinkhole formation has been occurring in scarcely inhabited areas of SW Sardinia over the past 15 years, sometimes very close to transport routes (roads and railways). Local authorities, stakeholders and landowners have increasingly felt the importance of proposing prevention programmes of such phenomena that could result in serious causes of risk in the near future. The formation of these sinkholes, although representing a natural land-forming process, is enhanced by anthropogenic factors such as exploitation of aquifers, construction of roads, channels and buildings, and concentrated surface runoff.

Geological and geophysical surveys both on a local and regional scale have allowed an area (zone A in Fig. 9) to be recognized in which Palaeozoic limestones, buried beneath the unconsolidated Tertiary alluvial clastic sediments, rise to within 50 m of the ground surface. This zone, which forms a belt first running in a NNW–SSE direction and then from NW to SE, indicates the areas of hazard, but also suggests the areas in which further investigations should be carried out to better comprehend the characteristics of the sediment-covered Palaeozoic bedrock, whose depth is one of the major constraints in delimiting hazardous areas.

The authors would like to express their thanks to Dr A. Funedda for useful suggestions on the geological background of this paper, and to T. Waltham and D. Calcaterra for their comments that improved the quality of the paper. Thanks are also due to the geophysical field crew members G. Casti, A. Lai, L. Noli, M. Serci and G. Uda. Most of the funds for the geophysical work were granted by RFI – Rete Ferroviaria Italiana.

References

Balia, R., Gavaudo, E. & Ghiglieri, G. 2001. Geophysical survey of a karst area – a case study from Sardinia, Italy. *European Journal of Environmental and Engineering Geophysics*, **6**, 167–180.

BECCALUVA, L., CIVETTA, L., MACCIOTTA, G. P. & RICCI, C. A. 1985. Geochronology in Sardinia; results and problems. *Rendiconti della Società Italiana di Mineralogia e Petrografia*, **40**, 57–72.

BECHSTADT, T. & BONI, M. 1996. Sedimentological, stratigraphical and ore deposits field guide of the autochtonous Cambro-Ordovician of Southwestern Sardinia. *Memorie descrittive della Carta Geologica d'Italia*, **48**, 1–390.

BECK, B. F. 1984. *Sinkholes: their geology, engineering and environmental impact*. Proceedings of the First Multidisciplinary Conference on Sinkholes. A.A. Balkema, Rotterdam.

BECK, B. F. 1986. A generalized genetic framework for the development of sinkholes and karst in Florida, U.S.A. *Environmental Geology and Water Sciences*, **8**, 5–18.

BECK, B. F. 1988. Environmental and engineering effects of sinkholes – the process behind the problems. *Environmental Geology and Water Sciences*, **12**, 71–78.

BECK, B. F. 1989. *Environmental and engineering impacts of sinkholes and karst*. Proceedings of the Third Multidisciplinary Conference on Sinkholes and the Engineering and Environmental Impacts of Karst. A.A. Balkema, Rotterdam.

BECK, B. F. 1993. *Applied karst geology*. Proceedings of the Fourth Multidisciplinary Conference on Sinkholes and the Engineering and Environmental Impacts of Karst. A.A. Balkema, Rotterdam.

BECK, B. F. & PEARSON, F. M. 1995. *Karst geohazards*. Proceedings of the Fifth Multidisciplinary Conference on Sinkholes and the Engineering and Environmental Impacts of Karst. A.A. Balkema, Rotterdam.

BECK, B. F. & WILSON, W. L. 1987. *Karst hydrogeology: engineering and environmental applications*. Proceedings of the Second Multidisciplinary Conference on Sinkholes and the Environmental Impacts of Karst. A.A. Balkema, Rotterdam.

BIANCO, L. & DE WAELE, J. 1992. La sorgente di Caput Acquas. *Sardegna Speleologica*, **2**, 24–25.

CARMIGNANI, L., FUNEDDA, A., OGGIANO, G. & PASCI, S. 2004. Tectono-sedimentary evolution of southwest Sardinia in the Paleogene: Pyrenaic or Apenninic dynamic? *Geodinamica Acta*, **17**, 275–287.

CARMIGNANI, L. & OGGIANO, G. *ET AL*. 2001. Geologia della Sardegna. Note illustrative della Carta Geologica della Sardegna a scala 1:200,000. *In*: *Memorie Descrittive della Carta Geologica d'Italia 60*. Istituto Poligrafico e Zecca dello Stato, Roma.

CIVITA, M., COCOZZA, T., FORTI, P., PERNA, G. & TURI, B. 1983. Idrogeologia del bacino minerario dell'Iglesiente (Sardegna Sud Occidentale). *Memorie dell'Istituto Italiano di Speleologia*, **2**, 1–139.

DE WAELE, J. & MUNTONI, A. 2001. Cover collapse sinkhole risk in the Cagliari province (Sardinia, Italy): location, genesis and forecasting. *In*: *Proceedings of the XIIIth International Congress of Speleology*, Sociedad Brasileira de Espeleologia, Campinas, Sao Paulo, Brazil, paper 82-S1.

DE WAELE, J., MUNTONI, A. & VILLANI, M. 2003. I 'Sinkhole' della provincia di Cagliari: alcuni esempi. *Sardegna Speleologica*, **20**, 33–41.

FADDA, A. F. & FAIS, S. 2002. *Studio sulla natura ed estensione delle voragini apertesi in località Monte Ollastus nel Comune di Villamassargia*. Consorzio di Bonifica del Cixerri, Iglesias.

KAUFMANN, O. & QUINIF, Y. 1999. Cover-collapse sinkholes in the 'Tournaisis' area, southern Belgium. *Engineering Geology*, **52**, 15–22.

MURRU, M. & MATTEUCCI, R. 2002. Early Tertiary of Sardinia, Italy. *Rendiconti della Società Paleontologica Italiana*, **1**, 269–273.

NEWTON, J. G. 1987. *Development of Sinkholes Resulting From Man's Activities in the eastern United States*. US Geological Survey Circular, **968**.

PILLOLA, G. L. 1989. Trilobites du Cambrien inférieur du SW de la Sardaigne, Italie. *Paleontographica Italica*, **78**, 1–174.

SALVATI, R. & SASOWSKY, I. D. 2002. Development of collapse sinkholes in areas of groundwater discharge. *Journal of Hydrology*, **264**, 1–11.

SORIANO, M. A. & SIMON, J. L. 1995. Alluvial dolines in the central Ebro basin, Spain: a spatial and developmental hazard analysis. *Geomorphology*, **11**, 295–309.

THARP, T. M. 1999. Mechanics of upward propagation of cover-collapse sinkholes. *Engineering Geology*, **52**, 23–33.

TWIDALE, C. R. 1987. Sinkholes (dolines) in lateritised sediments, Western Sturt Plateau, Northern Territory, Australia. *Geomorphology*, **1**, 33–52.

WALTHAM, T., BELL, F. & CULSHAW, M. 2005. *Sinkholes and Subsidence: Karst and Cavernous Rock in Engineering and Construction*. Springer, Berlin.

WHITE, W. B. & WHITE, E. L. 1995. Tresholds for soil transport and the long-term stability of sinkholes. *In*: BECK, B. F. & PEARSON, F. M. (eds) *Karst Geohazards: Engineering and Environmental Problems in Karst Terrane*. Balkema, Rotterdam, 73–78.

Karst processes and slope instability: some investigations in the carbonate Apennine of Campania (southern Italy)

A. SANTO, S. DEL PRETE, G. DI CRESCENZO & M. ROTELLA

Section of Applied Geology, Department of Geotechnical Engineering, Federico II University of Naples, Italy (e-mail: santo@unina.it)

Abstract: Some investigations carried out in the Campania Region (southern Italy) are shown concerning instability phenomena, the development of which is strongly influenced by karst. The widespread presence of carbonate massifs close to important urban centres with dense road networks creates high-risk situations in many settings of this region.

Such phenomena can have very different dimensions, origin and geomorphological development, and can be traced back to the action of hypogean and epigean karst and to complex interactions with other erosional processes.

In particular, among the hypogean forms, we have analysed collapse sinkholes that have developed on carbonate slopes, especially along fault lines where there are aquifers and ascent of mineralized fluids, and which are sometimes connected to strong seismic events.

Among the forms connected to epikarst processes, the origin of pinnacles has been investigated. They are isolated rock pillars, whose origin depends on a particular interaction between the geostructural characteristics of the masses and the process of karstic dissolution.

Moreover, a wide variety of morphologies exist that are related to the interaction between epigean and hypogean karst and other typologies of erosional processes. Among these one group is represented by caves on carbonate slopes developed in cataclastic zones, where a slow karstic process leads to the formation of upwards caves, with dimensions of some decametres, and consequently to the high production of debris downhill. Similarly, this process has been observed along slopes set on talus. Finally, the complex combination of the karstic phenomenon with the erosional wave action forms both caves and natural rock arcs along the coasts.

Carbonate massifs in southern Italy, especially in Campania, are affected by karstic phenomena, both hypogean and epigean (Fig. 1). The numerous available studies on these areas have mainly a geomorphological and hydrogeological approach, while only a few deal with the relationships between karstic processes and slope stability, although several phenomena of instability can be found whose origin is strictly linked to karst.

Surveys started in some areas of the region have shown that karstic morphologies that appear to affect slope stability can be quickly grouped into three different categories. They can be mainly the result of hypogean and epigean karst or of interactions between karstic processes and other erosional agents (Fig. 2). In some cases such phenomena have damaged built-up areas and main roads, offering at the same time interesting scientific cues to their origin.

On the basis of these investigations, this paper is aimed at defining the various geological contexts affected by these problems.

In the first part we will focus on the engineering – geological aspects concerning the most significant karstic phenomenologies (sinkholes, pinnacles), pointing out the dangerous interaction often set up within built-up areas. Then, other less frequent phenomenologies are described; although they are little documented, in some particular settings they can cause serious damage to populated areas and road networks.

Karstic sinkholes and slope instability

It is generally acknowledged that a possible collapse of a sinkhole represents a serious problem for territory planning (Ford & Williams 1989), mainly because of the difficulty in predicting and localizing the event. This phenomenon has become so frequent in some areas of Florida (Sinclair & Stewart 1985) that the United States Geological Survey (USGS) carry out ongoing information campaigns for the inhabitants, who are trained to report, including through the Internet, any small amounts of soil deformation (www.sinkhole.org).

Recently, in Italy, the scientific community and public administration and local management agencies have become more aware of the frequent development of sinkholes (AA.VV. 2000, 2004). The sinkholes in Campania (Del Prete *et al.* 2004) can be distinguished (Nisio 2003; Sauro 2003; Waltham *et al.* 2005) into:

- *collapse sinkholes* of karstic origin, opening onto carbonate slopes;

From: PARISE, M. & GUNN, J. (eds) *Natural and Anthropogenic Hazards in Karst Areas: Recognition, Analysis and Mitigation*. Geological Society, London, Special Publications, **279**, 59–72. DOI: 10.1144/SP279.6
0305-8719/07/$15.00

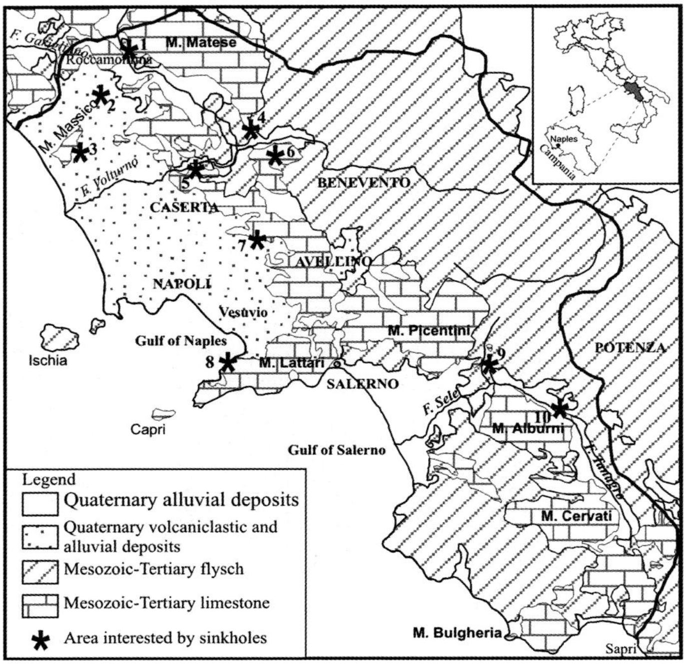

Fig. 1. Geological map of the Campania Region and the distribution of the main sinkhole areas: 1, Mastrati; 2, Vairano Lake; 3, Agro Falerno; 4, Telese; 5, Castelmorrone; 6, Solopaca; 7, Cancello; 8, Jala; 9, Contursi; and 10, Pertosa (after Del Prete *et al.* 2004, modified).

- *cover sinkholes*, wholly developing within detritical covers, lying at the base of carbonate massifs;
- *piping sinkholes*, developing in gravel, sand and silt deposits in the areas of alluvial plains.

In this work we will focus only on collapse sinkholes, which develop mainly along fault lines where there are aquifers or upward movement of mineralized fluids together with hyperkarst phenomena. In some cases there is evidence of their formation in conjunction with strong seismic events. Some examples can be seen on the mountains of Avella, along the coastline between Vico Equense and Castellammare (Nota D'Elogio 1979), and near the thermal springs of Telese and Contursi. The presence of mineralized aquifers in these areas seems to confirm, as suggested by various authors (Corniello & De Riso 1986; Forti & Perna 1986; Forti 1991, 2002; Corniello *et al.* 1999), the existence of a close connection between the origin of sinkholes, hyperkarst phenomena and, most probably, recent tectonic activity. It has been observed that sinkholes can develop in two different contexts: (a) deeply karstic areas with wide underground caves (collapse of cave); and (b) much fractured limestone and dolomite bedrock.

Collapse of cave

In the areas where carbonate massifs are characterized by slightly fractured formations with wide caves very close to the surface, collapse sinkholes are more likely to develop owing to the collapse of the cave vaults (Fig. 3). Basically, as also observed by other authors (Lolcama *et al.* 2002; Beccarisi *et al.* 2003; Delle Rose *et al.* 2004), the process starts with the formation of an embryonic cave connected to a karstic base level. Subsequently, the evolving dissolution process leads to a constant widening of the cave, also fostered by partial collapses of the vault and the walls. Thus, the cave tends to widen upwards and, if an underground stream is present, also at the base owing to erosion and dissolution processes. Such a constant widening of the cave also affects the geomechanic characteristics of the bedrock. The thinning of the vault near the surface and the increasing tangential stress affecting the vault rock at the same time can cause a sudden collapse with consequent formation of a collapse sinkhole.

When these phenomena take place near urban centres and main road networks they evidently represent a serious risk factor.

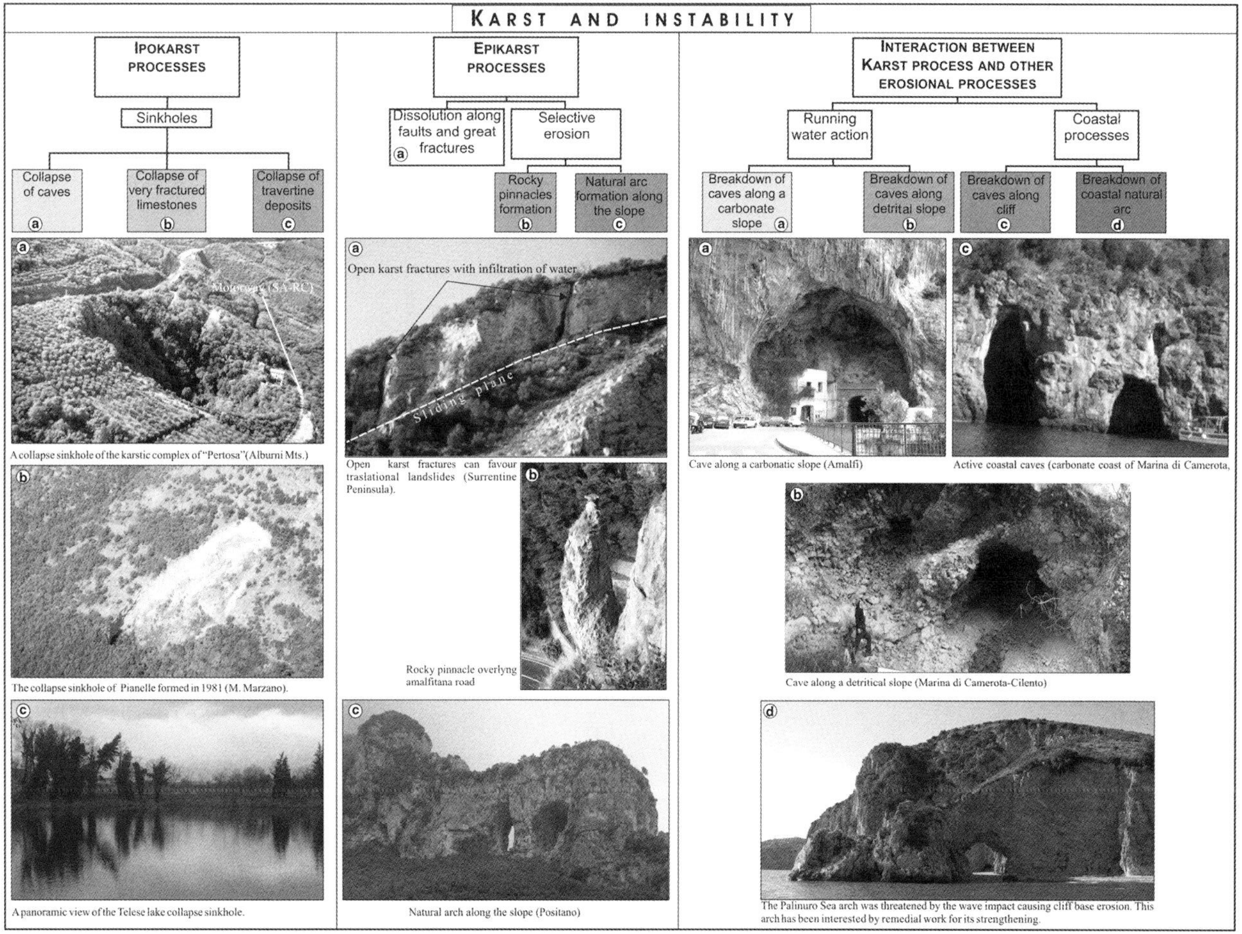

Fig. 2. The different types of karstic morphologies that can affect slope stability.

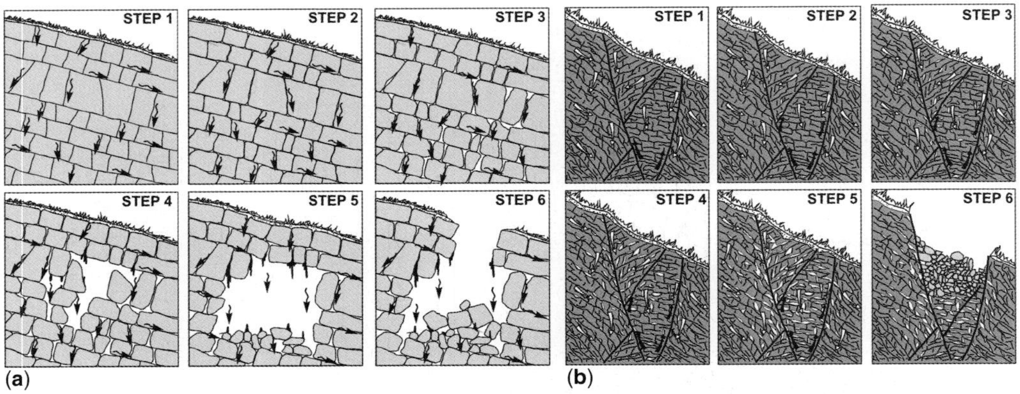

Fig. 3. Schematic representation of the various periods of a collapse sinkhole formation. (**a**) Collapse of a cave and (**b**) highly fractured limestones.

An example occurs in the historical centre of the village of Maddaloni near Caserta, where, under the houses close to the ancient medieval castle, a wide cave (50 m long, 17 m wide and 7 m high) has recently been discovered (Fig. 4). The thickness between the cave vault and the topographic surface is less than 3 m, and the constant water infiltration due to leaks in the underground pipelines contribute to accelerating the phenomenon of karstic dissolution of the rock, which is already in an instable condition.

Another example is represented by two collapse sinkholes in Pertosa at the base of the Alburni Mountains; when the sinking occurred is not yet known, but they are likely to have opened up in connection with one of the several karst systems developing along the basal aquifer of the massif. In this case the plan of the Salerno–Reggio Calabria motorway, which runs just a few metres from the sinkholes, has not taken into consideration either the sinkhole presence or the possibility of further sinking phenomena.

Collapse of deeply fractured limestones

Collapse phenomena can occur in highly fractured and karstic carbonatic massifs (Maffei *et al.* 2005), without necessarily the presence of a wide cave (Fig. 3).

In these cases, chemical dissolution leads to the formation of several small cavities in the bedrock, ranging in size from a few centimetres to some decimetres. When their frequency becomes particularly high, compared to the whole volume of the rock, tangential resistance can be overcome and cause the formation of a sinkhole on the surface.

Similar cases are represented by the Hill of Cancello sinkhole on Avella's mountains and by the Jala doline in the Sorrentine Peninsula (Fig. 5a). In the former case the excavation front of a stone quarry exposes the base of the sinkhole, and shows the presence of deeply fractured limestone. Similarly, the presence of deeply fractured and karstic limestone at the Jala sinkhole has been confirmed by surface geological surveys and drillings.

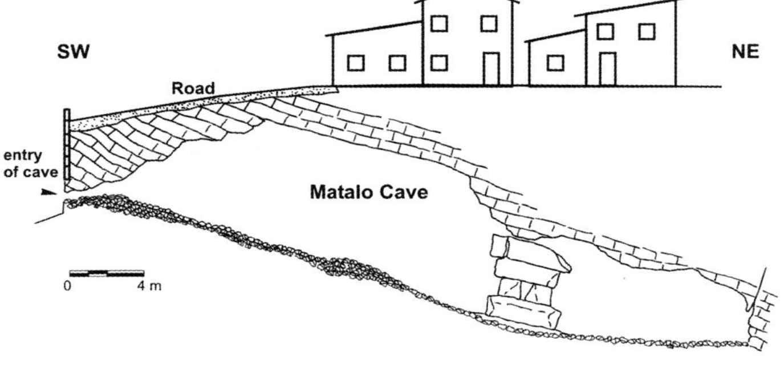

Fig. 4. An example of sinkhole risk: historical centre of the Maddaloni village.

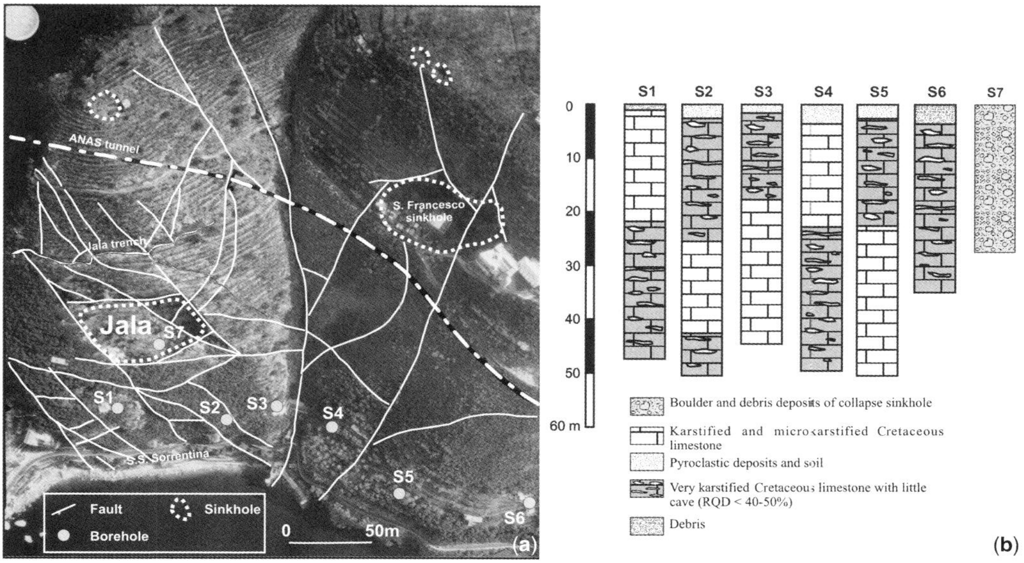

Fig. 5. (**a**) Aerial view of the 'Jala' sinkhole (Surrentine Peninsula), probably formed in the 18th century and (**b**) boreholes carried out in this area.

These have met much tectonized and karstic bedrocks with several micro-caves, filled in some cases with sand and characterized by very low Rock Quality Designation (RQD) values (<40–50%; Fig. 5b).

In this case the described morphologies have a negative impact on main road networks. In Vico Equense the presence of sinkholes caused several problems during the excavation of some tunnels because of the deeply fractured state of the karstified bedrock (Budetta *et al.* 1996; Santo & Tuccimei 1997).

Rocky pinnacle formation and slope instability

In the Campania Region, on the deeply fractured Triassic–Jurassic dolomite and limestone slopes, the numerous karstic morphologies are represented by pinnacles. They are concentrated in rock masses with low inclined bedding, having intersected highly inclined joint sets, some of which are parallel to the slope. The main erosive agent in their formation is water, which when penetrating through the joints works with a constant dissolution and

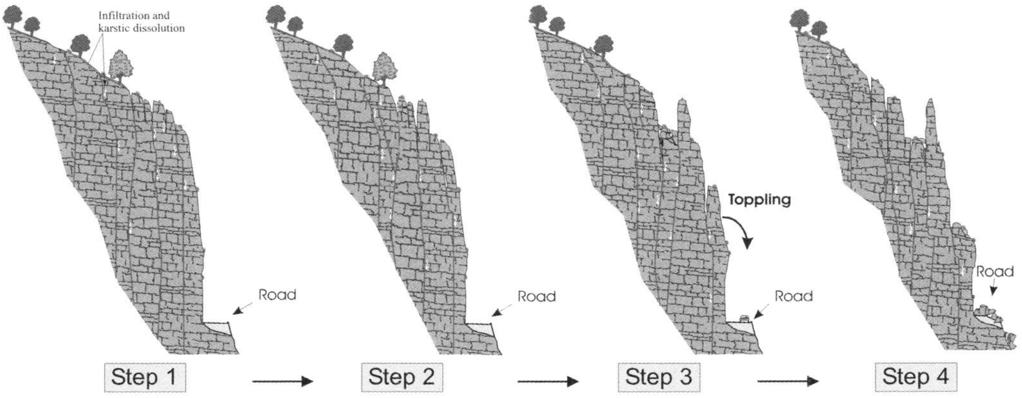

Fig. 6. Schematic representation of the various periods of the rocky pinnacle formation.

disgregation process on the masses; karst can be accompanied by effects of thermoclastic and crioclastic phenomena and of gravity. Pinnacles very often show a condition of instable balance, overlying very busy roads and creating highly risky situations.

Figures 6 and 7 show pinnacle formation schematically over time. It is evident how a slope characterized by vertical joint sets (step 1) leads through selective erosion to the formation of prismatic structures (step 2), which, with a progressive dissolution and erosion process, form unstable pinnacle structures (step 3) possibly leading to topple failures (step 4). Owing to their high, narrow form, pinnacles are inherently unstable and when adjacent to busy roads can create a hazardous situation.

In order to better understand the discontinuities pattern, failure typology and concerned volumes of the pinnacles, detailed geostructural surveys have been carried out on a sample area along Amalfi's coast, where such phenomenologies are very common (Fig. 8a).

The orientation of faults, joints and beddings has been drawn on pole plots (Schmidt equal-area stereographic projection), so that the cyclographics representative of each of the three main discontinuity sets can be found (Fig. 8b). Sets K1 and K2 transversally cut the slope, while the K3 set shows a dip direction opposite to the slope and a dip angle of 78°. The bedding appears to be generally low dipping in relation to the slope gradient, but with a very low inclination ($<10°$).

Fig. 7. Examples of rocky pinnacles along the main road between Amalfi and Positano.

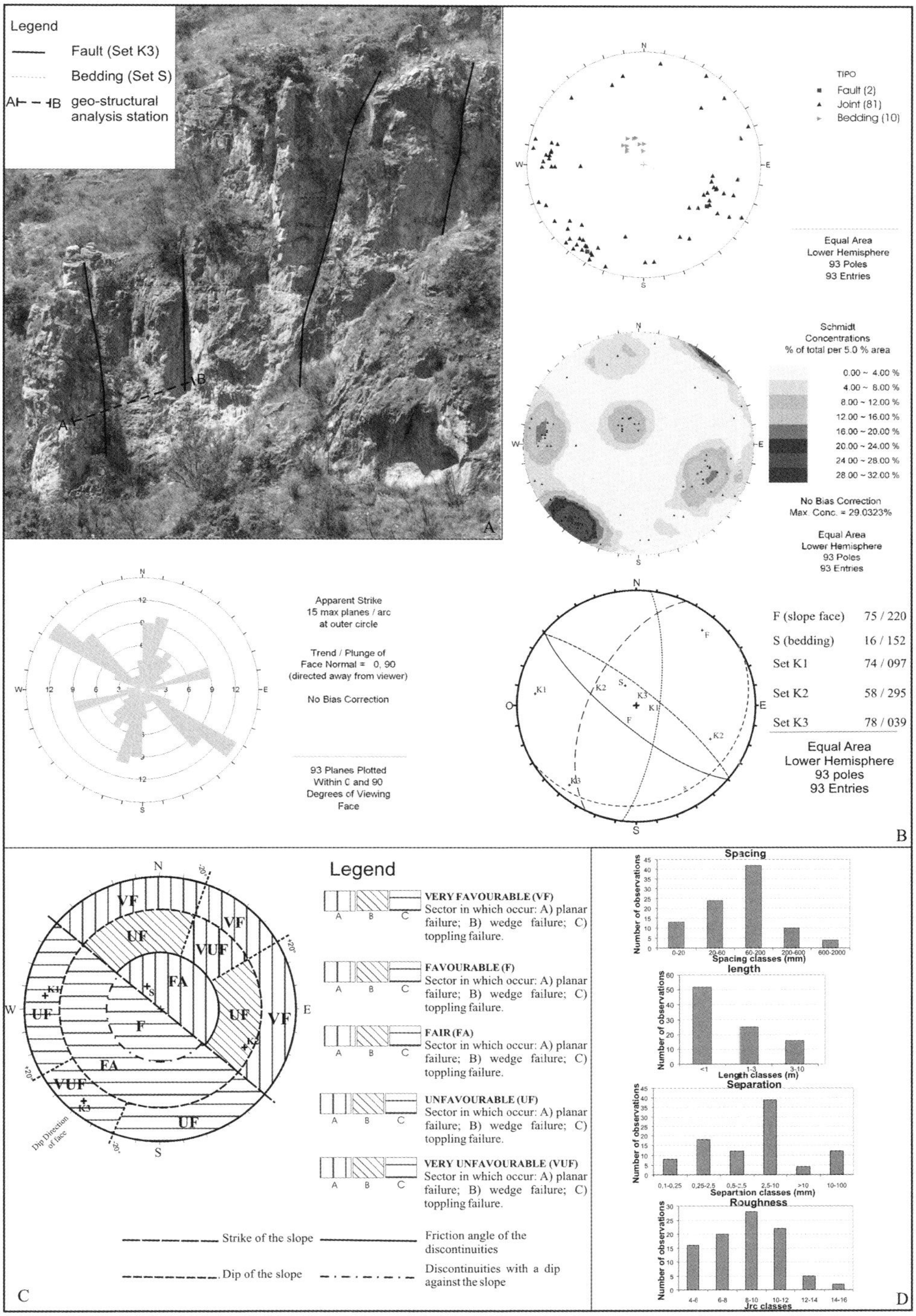

Fig. 8. (**a**) Geostructural surveys on a sample area along the Amalfi coast; (**b**) pole plot, rosette plot, Schmidt contour plot, mean poles/planes displayed for set K1, K2, K3 and bedding S; (**c**) stereonet overlay for assessment of worst-case discontinuity sets; and (**d**) other field observations of rock conditions.

Table 1. *Geomechanical values of the parameters affecting the rock masses*

Compressive strength (MPa)	Discontinuities								
	Mean spacing (mm)	RQD (%)	Length (m)	Separation (mm)	Roughness (JRC)	Infilling	Roughness profile	Groundwater	Weathering
20	60–200 Close	70	<1 Very low	2.5–10 Moderately wide	4–8; 8–12 Slightly rough to polished	Compacted clay material and calcite	Slightly undulating	None	Slightly weathered

RQD, Rock Quality Designation; JRC, Joint Roughness Coefficient.

The orientation of the discontinuity sets in relation to the slope gradient has been assessed to test the existence of degrees of looseness as a possible cause of failure (Fig. 8c) (Moon *et al.* 2001). The result attained shows a high tendency towards toppling failure.

Toppling failure is fostered by the presence of discontinuity planes at a high gradient within the rock mass, often represented by open discontinuities in the K3 set. The cyclographic of this set shows a dip direction opposite to the slope and a strike close to the slope gradient. Moreover, the masses thus isolated upwards are transversally faulted by at least another two discontinuity sets (K1 and K2), which thus isolate vertical rock prisms. These can undergo a toppling phenomenon around a rotation point connected to the stratification planes.

The field survey has also defined other parameters needed for a geomechanical rock mass assessment (Table 1, Fig. 8D) (ISRM 1978).

The geostructural analysis on the pinnacles has shown a high possibility of the occurrence of toppling failures involving volumes in the range from a few to several m^3.

Other phenomena

In addition to the cases just dealt with, in Campania there are other karstic phenomena both epigean and hypogean that are likely to favour landslide occurrence. Since historical documentation of these phenomena is lacking, we will only refer briefly to them and focus on their origin and geomorphological evolution.

Breakdown of caves along carbonate slope

In the areas of the Lattari Mountains and Cilento numerous karstic caves along slope have been surveyed. They develop on deeply fractured dolomitic limestone slopes affected by low-angle fault sets with dip direction against the slope.

Along cataclastic lines, representing the weakest points of the rock masses, there are the conditions for the formation of a 'proto-cave', that can evolve over time in response to slow karstic processes brought about by a local groundwater circulation that runs along the contact line between highly fractured zones and less affected ones. The karstic process results in the widening of joints and the occurrence of breakdown processes; the fallen blocks may be easily removed along the talus and transported outside the cave (Fig. 9). The particular joints set patterns, therefore, allow local superficial groundwater circulation and at the same time the removal of fallen material. In this way a progressive widening of a small cave can be slowly attained, sometimes with

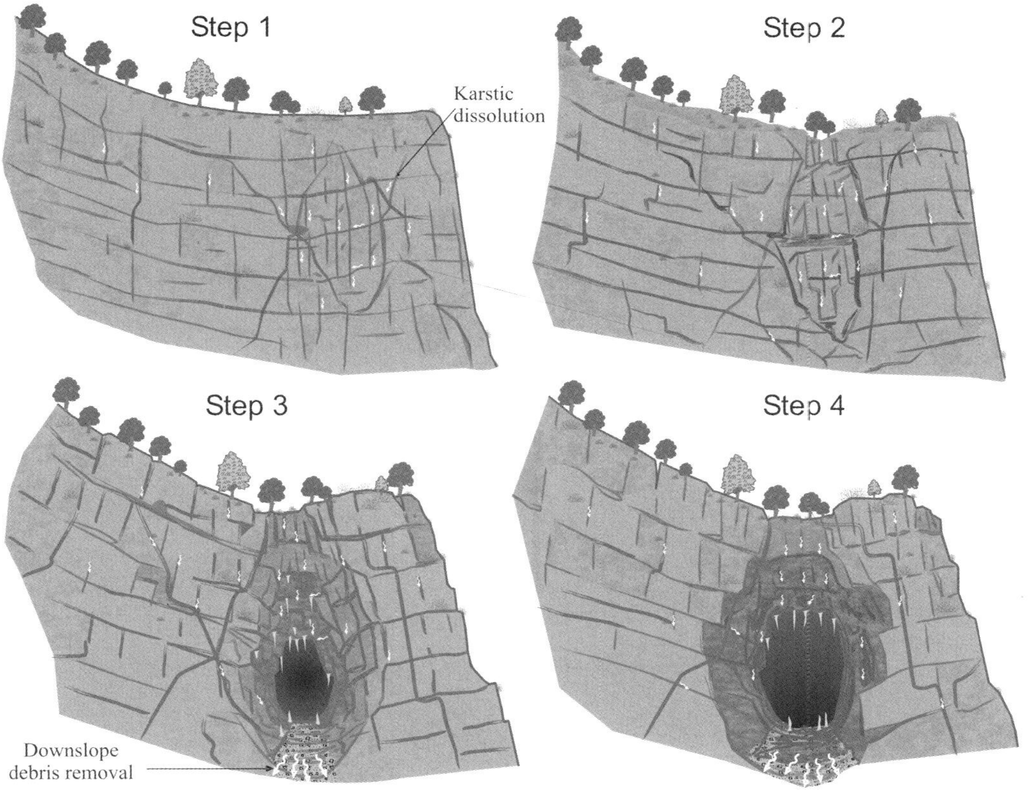

Fig. 9. Schematic representation of the various periods of a cave formation along a highly fractured carbonatic slope.

the formation of remarkably large caves that can reach vault heights of 30–40 m (Fig. 10).

Sometimes a further development of the cave upwards can lead to thinning, and thus instability, of the vault, with consequent sudden phenomena of breakdown involving the whole cave or part of it. In many cases, in fact at the base of the slopes with these cave typologies, large blocks up to several m^3 and talus can be found. In addition, on the slope surface we can find clear evidence of old caves that have fallen down because of the presence of wide hollows, still preserving relicts of stalactites and speleothems formed in a hypogean environment.

An example of a collapse involving a large cave along slope is the Amalfi landslide of 1899 (Budetta *et al.* 1994); Figure 11 shows evidence of a cave, today no longer visible, that existed on the slope before the landslide; collapse of this cave was most probably the main cause of the landslide that caused numerous casualties.

This historically documented event is not the only one, as several similar situations have been found on the slopes of Amalfi's coast and Mount Bulgheria in Cilento, some of them very close to built-up areas or overlying arterial roads. These phenomena can cause highly risky situations, so they should be studied in a more detailed way.

Breakdown of caves along a detrital slope

The foothills of carbonate massifs are characterized by the presence of highly dipping debris, with a thickness ranging from a few to several metres. When the talus is cut at the base of the slope, either for natural (faults, stream and sea erosion) or anthropogenical (road cuts, etc.) reasons, they can cause, because of karstic phenomena, some particular typologies of landslides. These have already damaged the road network at many sites.

Case histories have shown that within some talus, or at the contact between talus and limestone bedrock, a variation in the permeability can result in the local presence of groundwater. This might set

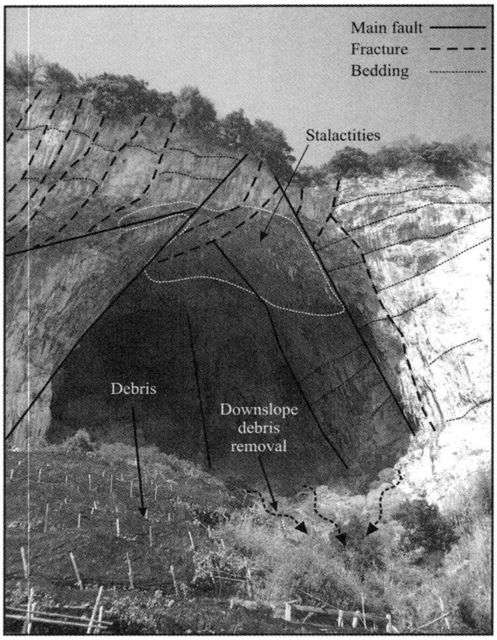

Fig. 10. The large cave of Saraceno, partially collapsed (Praiano).

off a slow karstic process affecting the debris. The formation of voids along the bedding is likely to be thus attained, which in time can collapse or initiate translational sliding of the talus material above.

The initial step (Fig. 12) is characterized by a voidless talus slope, on which we can find small crowns of rock falls. In this phase the slope is mainly susceptible to collapse of single clasts that roll down the hillslope and can cause damage only when of considerable size. The slow action of dissolution (step 2) creates a small void at an embryonic stage (proto-cave). The progressive collapse of talus, produced by the combined effect of karstic dissolution and gravity, slowly widens the cave towards the inside of the slope and upwards.

The ongoing development of caves leads to step 3, that is to the formation of bedding voids that can be as much as 30–40 m^3 wide (Fig. 13). The last phase (step 4) is that of the collapse of the cave vault, with the following rolling of blocks, or translational sliding along the contact with the limestone bedrock.

These types of landslides are strongly influenced by a slow karstic dissolution process and have caused problems on road networks. For example, the roads near Amalfi and Marina di Camerota are frequently closed to traffic because of these events.

Dissolution along large fractures

In some carbonate massifs the karstic effect along faults or tectonic fractures can slowly widen the width of surface discontinuities to create open fractures that can be some tens of metres long and more than 1 m wide. These might favour the start of rock falls.

The karstic process can occur through water rilling along fractures walls, through condensation phenomena and, most of all, through humus filling, which makes running waters more aggressive. In many cases of down-slope bedding slopes, numerous open joints occur; these show isolated large blocks of rock that foster landslides processes.

This is the case of Bikini's landslides that affect the Sorrento coastline. Limestone with fractures and karstic processes, and the presence of a marly interbed, have caused numerous translational slides that have affected the arterial road on several occasions.

Fig. 11. The 'Cappuccini' landslide occurred in 1899 in Amalfi village. In (**a**) before the event the presence of a cave is evident in the future crown zone (**b**).

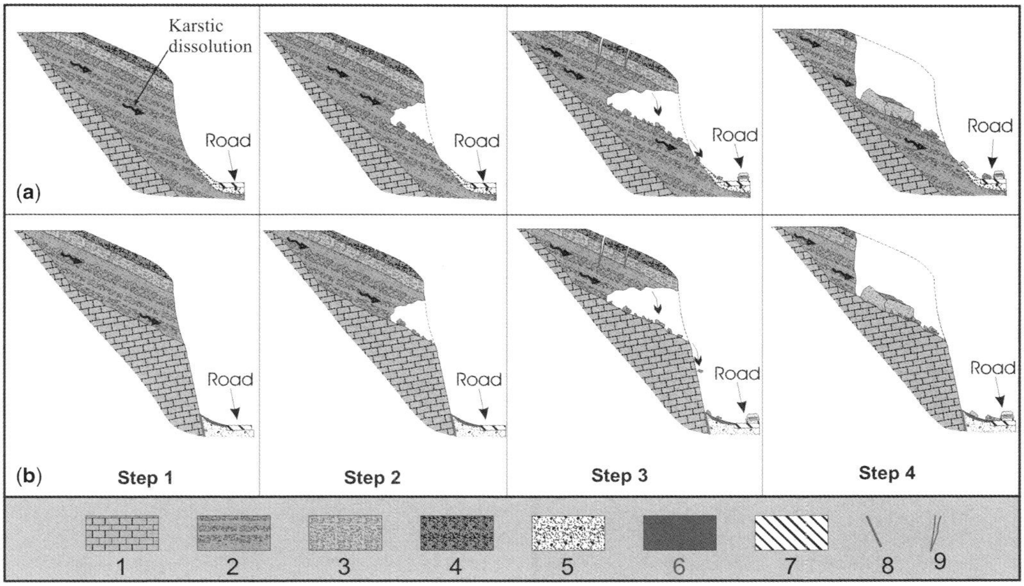

Fig. 12. Phases of cave evolution (**a**) along a talus slope and (**b**) along the contact between inclined talus and limestone: 1, carbonate bedrock; 2, incoherent or partially incoherent debris; 3, coherent debris; 4, soil; 5, talus; 6, fall deposits and soil; 7, anthropic deposits; 8, fault; and 9, fracture.

In other cases, large open joints can cause rock fall and toppling, as in the case of Mount Catiello in Positano. This rock fall occurred on 4th January 2002, and analysis of aerial photographs taken before the landslide show a wide and extended joint that isolated a rock dihedron (about 100 m high) from the surrounding rock. The failure occurred on the open joint, which is still evident (Fig. 14); it is about 2 m wide, and shows evidence of a long karstic phase (at least 2000 years BP) because of filling with reworked pyroclastic deposits and silt.

The rockfall of Mount Catiello can possibly be attributed to the effect of ice forming in the joint-filling materials. Low temperatures, much below average (−10 °C with respect to the average temperature of 10–12 °C), were recorded during the time preceding the failure (Budetta *et al.* 2002).

Collapse of caves along cliffs and natural arc formations

Along the limestone coasts of Campania we can find hundreds of karstic caves formed by the combination of karstic phenomena and erosional wave action. They are often of considerable size with planimetric developments of hundreds of metres, and are sometimes up to 20–30 m high. They originated through the dissolution and erosion of deeply fractured areas that were worked by the sea, especially when sea level was higher than present during the Middle Pleistocene and Tyrrhenian ages (Ford & Williams 1989; Esposito *et al.* 2003).

In many cases the origin of caves was favoured by the mixing of sea water and the groundwater along the coast, as occurred at Mount Bulgheria in Cilento and near Vico Equense on Sorrentine Peninsula (Fig. 15).

In some cases the progressive evolution of a cave that originates along a rock spur can extend to the opposite side, thus forming a natural arc whose origin is strongly influenced by karstic phenomena and erosional wave action. These often are very fractured and can result in rock falls when they collapse. The 'Palinuro arc' is such an example (Budetta & Santo 2000).

It has to be pointed out that many of these caves, which are heavily populated by tourists during the summer season, can be considered to be unstable.

Conclusions

The carbonate massifs of Campania show many cases of instability phenomena that can be linked

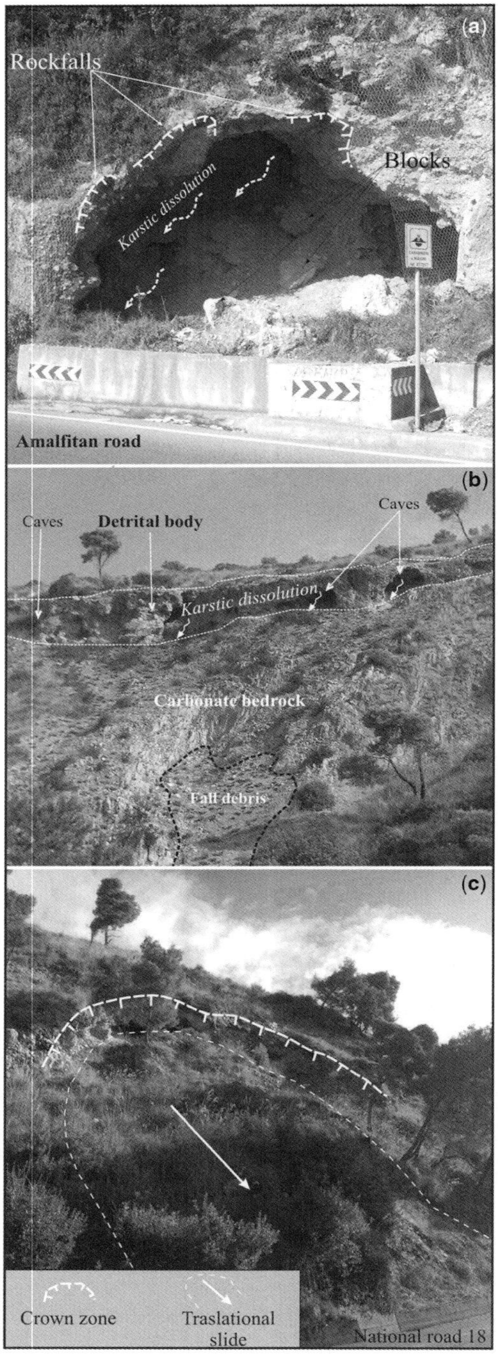

Fig. 13. Karstic dissolution and landslides along a detrital slope: (**a**) rock falls on the 'Amalfitana' road; (**b**) caves in instable conditions along the contact between inclined detritical body and limestone bedrock; and (**c**) translational slide on the National Road 18.

to karstic processes. It is evident that the determining factors are often different and concomitant, but the effect of karstic dissolution is fundamental in the processes of joint widening, clastic movement, increase of the voids and the reduction of cohesive strength. In particular, landslide types and associated fracture patterns can have very different patterns of development and can involve volumes ranging from dm^3 to hundreds of thousands of m^3.

As regards sinkholes, only recently have they been taken more into consideration for stability assessment in some regions of Italy (AA.VV. 2000), as up until a few years ago these phenomena were neglected or confused with other morphologies. In Campania, even though many cases of sinkholes are located within urban areas that are at high risk (for example, in the Sorrentine Peninsula or in Telese village), this problem is not considered seriously enough by the local management agencies. In fact, the variety of processes and geological settings of these areas, and the difficulty in the identification of subsurface karst processes, makes the identification of the possible causes of the sinkhole very difficult and, therefore, the production of detailed stability maps for the prediction of sinkholes more complex.

Moreover, the gradual subsidence that can be originated from bedrock dissolution must be considered only as a long-term management strategy. For this reason, to better define the areas at risk, it is important to gain a detailed knowledge of the hydrogeological and geomorphological history, the rate of karst dissolution of bedrock, and the likely changes in groundwater circulation that may initiate or vary the dissolution processes.

A detailed survey of rock pinnacles indicates that they occur frequently on dolomite slopes and can create hazardous situations when located close to a road network. Their fracture pattern is quite simple, so in many cases mitigation should be straightforward.

Structural detailed analyses and monitoring campaigns should be carried out on the vaults of large bedding voids, sea caves and natural arcs. In these cases sudden collapse could result in large failures. A knowledge of the joint patterns of the vaults in these caves, especially when in heavily urbanized areas as for example on Amalfi's coast, is fundamental to preventing hazardous situations. In other cases, karst has also led to limestone deposits; they are most frequently small in size (some dm^3), but when they are overlie arterial roads can still cause highly hazardous situations. The carbonate contexts of Campania karst plays an important role in local slope stability and can be one of the factors that must be considered in landslides hazard assessment.

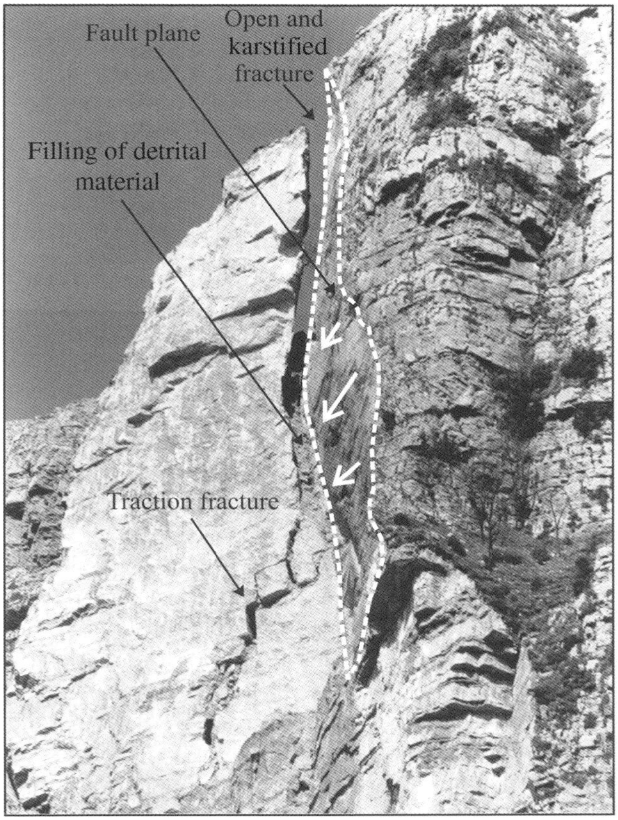

Fig. 14. Example of an open karst fracture that caused a large rock fall (Catiello Mt, near Positano village).

Fig. 15. Wave action can produce the breakdown of karstic coastal caves forming little bays and a promontory. (**a**) A partially collapsed coastal caves (carbonate coast of Marina di Camerota); (**b**) Palinuro arc: 1, bedding; 2, 3, discontinuity sets; 4, coastal caves; 5, rock fall (after Budetta & Santo 2000, modified).

Studies aimed at territory planning and proper planning of mitigation to reduce landslide risks have to take into consideration karst phenomena and the role they have in the development of some particular morphologies and the setting off of different landslide types.

The authors wish to thanks the two referees, S. Cannon and M. Parise and N. Santangelo for their suggestions and critical observations.

References

AA.VV. 2000. *Proceedings of the National Conference on 'Sinkholes – A New Problem for the Tuscany Region'. Tuscany Region, Grosseto Province and Municipality, Grosseto (Italy), 31 March 2000.*

AA.VV. 2004. *Proceedings of the Conference on 'Knowledge on Sinkhole Phenomena and the Role of Public Administration and Local Management Agencies', Italian Geological Survey, Rome (Italy), 20–21 May 2004.* APAT, Rome.

BECCARISI, L., CACCIATORE, G. *ET AL.* 2003. The Vore at Barbarano: description and speleogenesis. (In Italian.) *Thalassia Salentina*, **26**, 145–154.

BUDETTA, P. & SANTO, A. 2000. Geostructural and geomechanical survey on the Palinuro Sea arch (Campania – Southern Italy). (In Italian.) *Quaderni di Geologia Applicata*, **7**, 6–76.

BUDETTA, P., CALCATERRA, D. & SANTO, A. 1994. Engineering-geological of potentially unstable rock slopes in sorrentine peninsula (southern Italy). *In*: *Conference of the International Association of Engineering Geology*, Lisbona, 2119–2126.

BUDETTA, P., NICOTERA, P. & SANTO, A. 1996. Controlling and monitoring deforming phenomena caused by karstification in carbonatic slopes in the Southern Apennines (Campania – Italy). (In Italian.) *In*: *International Conference on 'Prevention of Hydrogeological Hazards: The Role of Scientific Research'*. C.N.R.–G.N.D.C.I. Alba (Italy), 5–7 November 1996, 383–395.

BUDETTA, P., DI CRESCENZO, G. & SANTO, A. 2002. The rockfall of Catiello Mountain (Positano): a rare phenomena in Surrentine Peninsula due to crioclastism. (In Italian.) *In*: *National Council Research Conference on 'Mountain Protection', Assisi (Italy)*, 11–12 December 2002, 401–412.

CORNIELLO, A. & DE RISO, R. 1986. Hydrogeology and hydrogeochemistry of Telese springs. (In Italian.) *Geologia Applicata e Idrogeologia*, **21**, 53–84.

CORNIELLO, A., DUCCI, D. & GUARINO, P. M. 1999. The western part of the Matese carbonatic massif and Venafro plain: hydrogeology and hydrogeochemistry. (In Italian.) *Bollettino della Società Geologica Italiana*, **118**, 523–535.

DELLE ROSE, M., FEDERICO, A. & PARISE, M. 2004. Sinkhole genesis and evolution in Apulia, and their interrelations with the anthropogenic environment. *Natural Hazards and Earth System Sciences*, **4**, 747–755.

DEL PRETE, S., DE RISO, R. & SANTO, A. 2004. First contribution on natural sinkhole in Campania Region. (In Italian.) *In*: *Conference on 'Knowledge on Sinkhole Phenomena and Role of Public Administration and Local Management Agencies', Italian Geological Survey, Rome, (Italy), 20–21 May 2004*, 361–376. APAT, Rome.

ESPOSITO, C., FILOCAMO, F., MARCIANO, R., ROMANO, P., SANTANGELO, N. & SANTO, A. 2003. Genesis and paleogeographic evolution of the Marina di Camerota sea caves (Cilento and vallo di Diano National Park, Southern Italy). (In Italian.) *Thalassia Salentina*, **26**, 165–174.

FORD, D. & WILLIAMS, P. 1989. *Karst geomorphology and hydrology*. Chapman & Hall, 601 pp.

FORTI, P. 1991. Hyperkarst processes and speleogenesis. (In Italian.) *Speleologia, Journal of the Italian Speleological Society*, **24**, 42–46, **26**, 11–15.

FORTI, P. 2002. Hyperkarst evolution in thermal aquifer and its relationships with suffusion processes. (In Italian.) *In*: *Conference on 'Sinkholes – a New Problem for the Tuscany Region', Tuscany Region, Grosseto Province and Commune, Grosseto (Italy)*, 31 March 2000, 11–26.

FORTI, P. & PERNA, G. 1986. Hyperkarst processes of Iglesias area (South-Western Sardinia, Italy). (In Italian.) *Natural Science Museum of Trento (Italy)*, **34**, 85–99.

ISRM. 1978. Suggested methods for the quantitative description of discontinuities in rock masses. *International Journal of Rock Mechanics and Mining Sciences & Geomechanical Abstracts*, **15**, 319–368.

LOLCAMA, J. L., COHEN, H. A. & TONKIN, M. J. 2002. Deep karst conduits, flooding and sinkholes: lessons for the aggregates industry. *Engineering Geology*, **65**, 151–157.

MAFFEI, A., MARTINO, S. & PRESTININZI, A. 2005. From the geological to the numerical model in the analysis of gravity-induced slope deformations: an example from the central Apennines (Italy). *Engineering Geology*, **78**, 215–236.

MOON, V., RUSSELL, G. & STEWART, M. 2001. The value of rock mass classification systems for weak rock masses: a case example from Huntly, New Zealand. *Engineering Geology*, **61**, 53–67.

NISIO, S. 2003. Sinkhole: knowledge and example in Central Italy. (In Italian.) *Italian Journal of Quaternary Sciences*, **16**, 121–132.

NOTA D'ELOGIO, E. 1979, *Mineral and thermal waters of Province of Naples*. (In Italian.) *Memorie e Note dell'Istituto di Geologia Applicata di Napoli*, **15**.

SANTO, A. & TUCCIMEI, P. 1997. Slope deformation of late Quaternary and Holocene age on the basis of geomorphological features and Th/U dating: the case of the Vico Equense area in Campania (Southern Italy). (In Italian.) *Italian Journal of Quaternary Sciences*, **10**, 477–484.

SAURO, U. 2003. Dolines and sinkholes: aspects of evolution and problems of classification. *Acta Carsologica*, **32**, 41–52.

SINCLAIR, W. C. & STEWART, J. W. 1985. *Sinkhole Type, Development and Distribution in Florida*. US Geological Survey, Florida Department of Natural Resources, Bureau of Geology Map Series, MS-110.

WALTHAM, T., BELL, F. & CULSHAW, M. 2005. *Sinkholes and Subsidence*. Springer, Berlin.

Magnetic prospection as an efficient tool for doline detection: a case study in the central Ebro Basin (northern Spain)

T. MOCHALES[1], E. L. PUEYO[2], A. M. CASAS[1] & M. A. SORIANO[1]

[1]*Dpto. de Ciencias de la Tierra, Universidad de Zaragoza, Pedro Cerbuna 12, 50009Z, Spain (e-mail: tania@unizar.es)*

[2]*Instituto Geológico y Minero de España, Fernando el Católico 59, 4C, Zaragoza 50006Z, Spain*

Abstract: The presence of alluvial dolines in the Ebro Basin causes problems to both agricultural and urban areas. At present, new urbanization of former farming areas requires new tools to detect karst zones and so diminish the hazard linked to collapses. In the surroundings of Zaragoza, dolines (developed mainly on Quaternary alluvial terraces covering a Tertiary gypsum substratum) are commonly filled with alluvial deposits, agricultural soils, urban debris, etc. Measurements of magnetic susceptibility show a remarkable contrast between host rocks and cavity fillings, demonstrating the value of magnetic surveying. A field test was made in a recently collapsed (September 2003) doline filled currently with urban debris. A magnetic survey was carried out following a 130 m^2 grid, with 1–10 m spacing between profiles. A proton magnetometer with gradiometer was utilized, and the total field intensity and gradient measurements were taken. The magnetic survey demonstrated a strong anomaly with a dipole defined by more than 650 nT and a gradient of about 100 nT m^{-1}. The 2.5-dimensional (2.5D) modelling of the magnetic anomaly fits well with the known geometrical data. Two other dolines (that are not clearly defined at the surface) were also detected during the survey. These results validate the starting hypothesis and open a new research approach to the problem. The magnetic survey output allows the construction of realistic geological models.

Geophysical prospecting is commonly used in the detection of subsoil anomalies. These surveys usually represent the most efficient way of detecting voids and cavities by covering large areas in a quick and inexpensive manner. Since pioneering work carried out by Cook (1965, 1974) seismic survey has become one of the most enduring methods of detecting caves, and its methodology has greatly developed over the last few decades (Chamon & Dobereiner 1988; Miller & Steeples 1991). Other methods comprise ground penetrating radar (GPR) as in Ballard (1983), gravimetry (Colley 1963; Neumann 1967; Butler 1984), electrical resistivity and tomography (Van Schoor 2002; Zhou *et al.* 2002). Currently multidisciplinary geophysical approaches are recognized as the best way to detect any kind of karstic cavity and collapse (Miller *et al.* 1984; Fenning *et al.* 2000; Matthews *et al.* 2000; Beres *et al.* 2001). With this in mind, and except for a few recent approaches (Armadillo *et al.* 1998; Shah *et al.* 1999; Xia & Williams 2003; Thierry *et al.* 2005), magnetic surveying is not a commonly used technique in analysing such problems. This is partly due to the need to reduce the external magnetic noise, which prevents its use in the vicinity of most urban areas (reinforced concrete, paved roads, railway lines, electric lines, etc.); in rural areas, however, these noises tend to be much reduced. In any case, the susceptibility contrast of the involved materials is the key variable for the successful application of magnetic prospecting to detect scenarios where the magnetic layering is disrupted (e.g. a karstic void or collapse) (Fig. 1). To achieve this objective an appropriate instrumental resolution and post-processing methods are necessary (Pierce *et al.* 2001).

In the Quaternary materials of the Ebro Basin the development of alluvial dolines is a common process. These dolines are generated by the dissolution of underlying gypsum, affecting both agricultural and urban areas. A great number of specific and regional studies (geomorphological, geological, geotechnical, hydrogeological, etc.), together with mapping of the distribution of dolines and hazard areas, have been developed over the last 30 years (van Zuidam 1976; Gutiérrez *et al.* 1985; Soriano & Simón 1995; Guerrero *et al.* 2004; Soriano *et al.* 2004). In this area, doline maps based on the study of aerial photographs only allows detection of a small percentage of the total number of dolines because time-spaced records of aerial photographs cannot show either their quick natural evolution or the human activity of filling the surface depressions. The high dissolution rates in gypsum karst increases the hazard associated with voids, even in semi-arid areas such as the Ebro

From: PARISE, M. & GUNN, J. (eds) *Natural and Anthropogenic Hazards in Karst Areas: Recognition, Analysis and Mitigation.* Geological Society, London, Special Publications, **279**, 73–84. DOI: 10.1144/SP279.7
0305-8719/07/$15.00

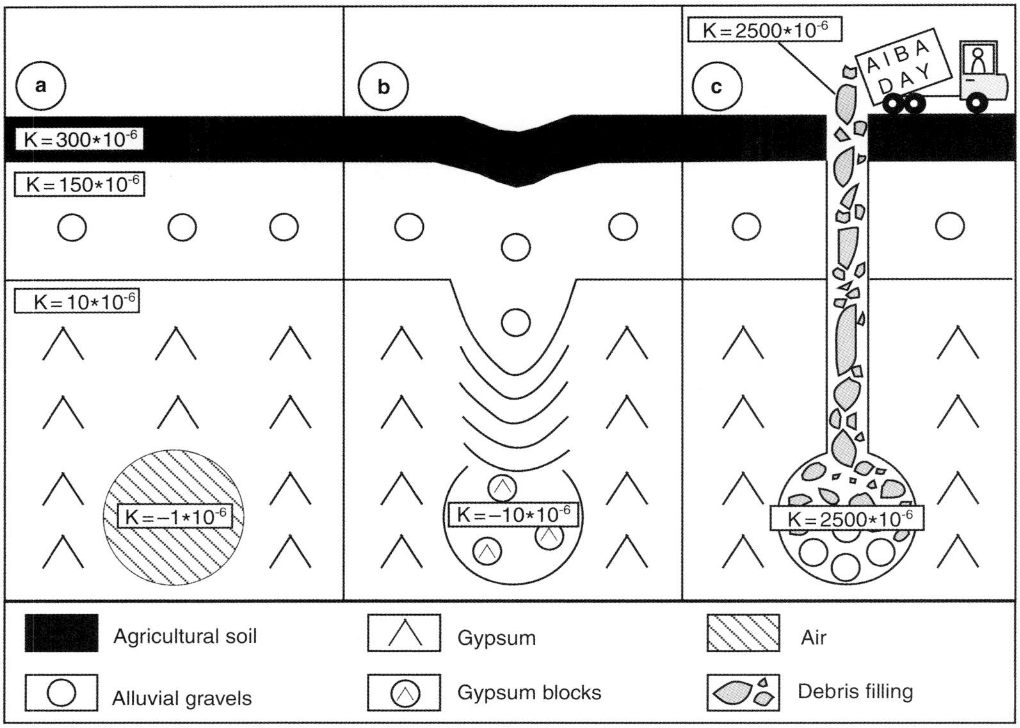

Fig. 1. (**a**) Evolution of an air-filled cavity to (**b**) subsidence or (**c**) collapse. The cavity in (b) is filled with gypsum blocks. The cavity represented in (c) is filled with urban debris. The magnetic susceptibility of each material is shown.

Basin, as the time required for their development is only tens of years (Cooper 1998; Cooper & Waltham 1999). At present, the large increase in urbanized zones occupying former rural areas makes the development of good detection methods necessary in order to diminish the hazard associated with doline development.

To check the validity of magnetic prospecting for the detection of karst, a recent doline generated by collapse (September 2003), 16 m deep and 8 m in diameter, was selected. In this paper we present the first successful magnetic survey performed in the Ebro Basin to detect and model evaporitic karstic dolines (sinkholes). The methodology used included a detailed geophysical survey, the study of magnetic properties or rocks and the modelling of anomalies, contrasting the models obtained with data from surface geology and boreholes.

Geological setting

The Ebro Basin (NE Spain) is bordered by the Pyrenees in the north, the Iberian Range in the south and the Catalonian Range in the east (Fig. 2). The basin was generated by the convergence, from the Cretaceous to the middle Miocene, of the European and the Iberian plates. The basin evolved as the southern foreland basin of the Pyrenees. In the upper Eocene the connection with the ocean was closed and from that moment the sedimentation was only continental. Fluvial and alluvial fan systems sourced in the neighbouring mountain ranges went towards the centre of the basin, where an endorheic lake environment was dominant. Consequently, most rocks exposed within the basin correspond to clastic, evaporite (mainly gypsum and halite) and carbonate facies deposited in these sedimentary environments (Muñóz *et al.* 2002). The Tertiary substratum of gypsum and marls is about 1000 m thick on average, lying on Mesozoic limestones in the surroundings of Zaragoza (Cortés-Garcia 2004). The internally drained Ebro Basin opened towards the Mediterranean Sea about 11.5 Ma ago (García-Castellanos *et al.* 2003) giving rise to the setting up of the present-day Ebro alluvial system. The Neogene rocks show near horizontal bedding. Nevertheless, gentle deformations (folds and faults) can also be recognized (Quirantes 1978; Arlegui *et al.* 1997).

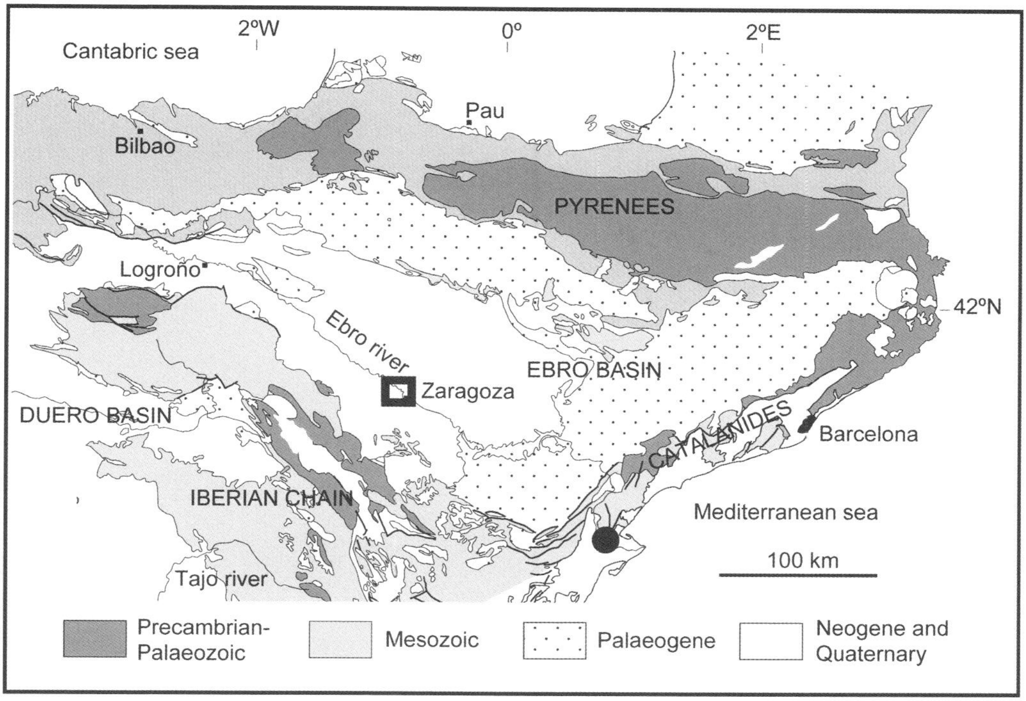

Fig. 2. Geological map. The square marks the studied area and the point is the Ebro Magnetic Observatory.

Throughout the Quaternary the activity of the main rivers developed structural landforms (on the Tertiary layers) and stepped levels of alluvial terraces. From the base of those structural landforms towards the main river plains, pediments were generated. In the sector of the Ebro Valley NW of Zaragoza, four (each one subdivided in two), six and eight stepped levels of terraces were identified by van Zuidam (1976), Mensua & Ibáñez (1977) and Soriano (1990), respectively. In addition, three and six levels of pediments were recognized by Mensua & Ibáñez (1977) and Soriano (1990), respectively. Terraces and pediments partially cover the Neogene rocks; gravels, sands and silts constitute their main deposits. Carbonate crusts are common at the top of most of terraces deposits (T_2–T_8) (Soriano 1990).

Quaternary materials are affected both by ancient and present karst processes. Karst is caused by the dissolution of underlying evaporites. In quarries and trenches the old Quaternary levels show structures that indicate the dissolution of Tertiary layers causing its deformation. The characteristics of the outcrops indicate that karst processes are both syn- and post-sedimentary with the Quaternary deposits. The morphology of this palaeokarst is varied; tubular, funnel, vault and synform (Soriano *et al.* 2004). However, present landforms are mainly represented by alluvial dolines (Fig. 3). They often develop in the central Ebro Basin, appearing preferentially on modern terraces (T_1, T_2 and T_3), but also on pediment and alluvial-fan deposits. Their development is conditioned by dissolution of evaporites by groundwater and the dragging of detrital cover. The behaviour of the cover is different depending on their non-cohesive or cohesive behaviour, slow subsidence being the usual response for the first case and sudden collapses for the second (see Fig. 1) (Yuan 1988; Benito *et al.* 1995; Buttrick & van Schalkwyk 1998; Hyatt *et al.* 2001). At the regional scale, the main factors that favour the development of dolines are: (1) *palaeovalleys* at the Miocene–Quaternary boundary; (2) low thickness of the Quaternary cover; (3) low percentage of lutites in Quaternary deposits (Soriano & Simón 1995); (4) hydrogeological factors, specifically traditional irrigation practices (field flooding, pumping) and underground flow towards the thalweg; and (5) the lithology of the underlying bedrocks.

From the study of aerial photographs taken in eight different years (between 1947 and 1993), the mapping of dolines shows a variation in size of some of them as well as their number (some of

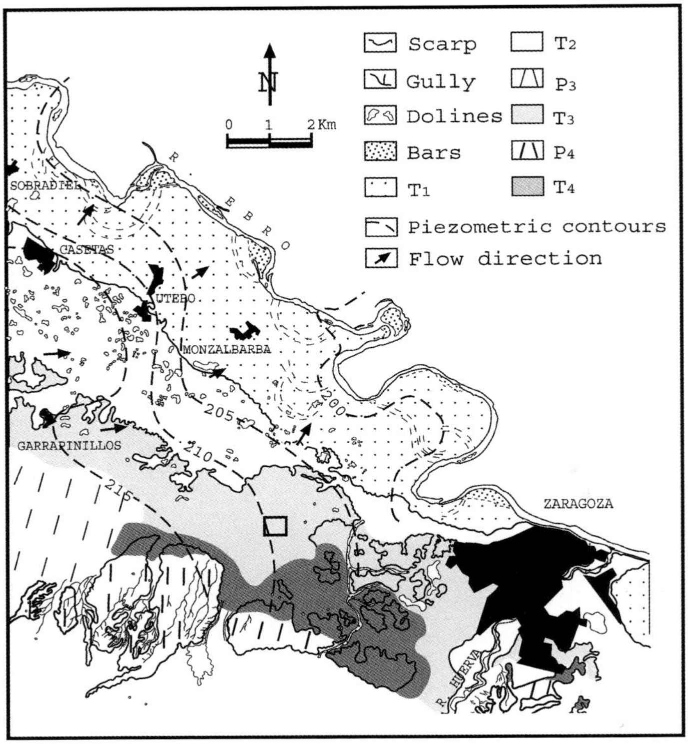

Fig. 3. Geomorphological map of the NW of Zaragoza (from 1957 aerial photographs). The square shows the location of the studied doline. Piezometric contours (according to Bielza *et al.* 1993) and flow direction (arrows) are also shown.

them have been filled and covered, and are therefore no longer visible and new ones appear because of the formation of new cavities between aerial photograph flights). These variations have become more important since 1970 when farming areas became industrial areas. Fractures and deformations have appeared in some of the buildings and roads built on the ancient dolines (Guerrero *et al.* 2004). Measured subsidence rates range between 21 and 92 mm year^{-1} (Soriano & Simón 2002).

The doline chosen for this study is located in the western part of Zaragoza on terrace level T_3 (Soriano 1990). This cavity collapsed in September 2003; its dimensions were 8 m in diameter and 16 m in depth (Fig. 4). The profile of the collapse was irregular; with a funnel shape on the surface and a vault shape at depth. Some weeks after the collapse the doline was filled with urban debris.

In the sector of the Ebro Valley where the doline was developed, only the youngest four terrace levels and two pediment levels can be identified (according to Soriano 1990).

Measurements, instruments and laboratory procedures

Magnetic survey

A magnetic gradiometer PMG-1 (GF Instruments Ltd) was used for the survey; two proton magnetometers mounted on an aluminium pole at 150 and 200 cm from the surface gave readings of 0.1 nT for resolution and 1 nT for accuracy. Measurements were taken at discrete intervals every 4 s.

An east–west pilot transect (80 m in length) in the northern part of the doline displayed a total intensity anomaly of greater than 200 nT in intensity and 40 m in length. Therefore, a wider squared frame of 130 m was defined to accurately describe the anomaly and to avoid any border effects during the modelling (Fig. 5). Fixed references (timber sticks) were installed during the sampling days (between 21 September and 13 December 2004). Total intensity was measured in 64 sections with different lengths (between 60 and

Fig. 4. Caidero doline pictures taken (**a**) after the collapse and (**b**) after the anthropogenic filling.

128 m). Measurements were taken every 2 m. The central zone (60 × 60 m) was characterized by a denser net (30 sections with 30 measurement per section). The initial net was enlarged laterally owing to the unexpected discovery of two other cavities. More than 2500 measurements of the magnetic field intensity were taken, an equal number of magnetic gradient data was simultaneously acquired (Fig. 5).

Several drift loops passing by a base station in the working area were performed to control the daily variation of the Earth's magnetic field (EMF). Furthermore, the measuring was carried out during the solar afternoon to avoid the large field depression. The intensity of the EMF is available from the 'Ebro Magnetic Observatory' located 150 km eastwards of the survey area. This station records (and supplies) the EMF total intensity per min with a 0.1 nT accuracy. Intensity differences during the measurement hours and days never exceeded 10 nT and no significant solar storms occurred during the prospecting.

Magnetic susceptibility

Magnetic susceptibility of the involved materials (background rocks, soils, urban debris filling) is a first-order variable to model the local magnetic anomaly. Low-field susceptibility (KLF) was measured at 20 °C, applying a direct weak magnetic field (0.4 mT in a a.c. field of 920 Hz). This variable is a standard measurement and represents the contribution of all minerals (dia-, para- and ferromagnetic). Measurements were taken with two different instruments; a KLY-3 (Agico Ltd) was used in the AMS laboratory (Universidad de Zaragoza) and a portable susceptometer in the field (SM-20; GF Instruments). The susceptibility measurements were carried out in two different ways.

- surface measurements to know the value of the agricultural soil around the doline and the debris from the doline filling. These data were used together with other surface data around Zaragoza.
- Depth soundings from the surroundings of the doline area. Samples of about 20 g in weight were taken every 10 cm from three continuous borehole cores (66 m in total length) located in the vicinity of the doline.

Magnetic properties of the involved materials

Susceptibility was monitored in the doline surroundings along a few sections (sampling sites every 4–5 m), as well as in several discrete points; five or six measurements were taken directly from the surface in every location with a portable susceptometer (SM20). With respect is the agricultural modified soils (intensive growing of *Medicago sativa*; Fam. *Leguminosae*), no significant variations were found; a stable mean of 268.4×10^{-6} S.I. ($\pm$ 6.2×10^{-6} S.I.) out of more than 300 measurements (Fig. 6a, Table 1). This value falls within the expected value for this kind of antropogenic soil in which fertilizers are commonly used. Conversely, susceptibility measurements in the debris filling of the doline range over five orders of magnitude (with an average of $26\,566 \times 10^{-6}$ $\pm$ 7869×10^{-6} S.I.), this high variability is owing

Fig. 5. The magnetic survey grid over the Caidero area.

to the diverse nature of materials (bricks, pottery, concrete, plastics, iron pieces, etc.).

Susceptibility was also monitored in three continuous borehole cores (with a total length of more than 100 m), all three were drilled in the vicinity of Zaragoza by geotechnical companies. One of them was located near the survey area (Fig. 6b). More than 1000 measurements were taken directly by means of the portable susceptometer. Discrete samples (about 10–20 g) were also sampled from every 10 cm of the cores, which were measured in the laboratory using the KLY-3 susceptometer. The boreholes cut through all the rock types present in the area and evince a significant magnetic contrast; Miocene marls display a very stable and moderate signal ($137 \times 10^{-6} \pm 8 \times 10^{-6}$ S.I.), the gypsiferous marls and pure gypsum (together with air) present a diamagnetic signal ($-1.5 \times 10^{-6} \pm 3 \times 10^{-6}$ S.I.), Quaternary alluvial and terrace deposits oscillate between $96.2 \times 10^{-6} \pm 25 \times 10^{-6}$ (marls with pebbles) and $227.5 \times 10^{-6} \pm 59 \times 10^{-6}$ S.I. (alluvial gravels). The uppermost part cuts a complete soil section (all horizons) and displays higher values ($1550 \times 10^{-6} \pm 380 \times 10^{-6}$ S.I.).

This remarkable magnetic contrast between the different involved materials allows the application of magnetic prospection for doline detection due to the expected layering disruption in the case of collapse or subsidence morphologies (working hypothesis). This susceptibility divergence is especially strong when considering urban debris (not cut in the borehole).

Magnetic anomalies

As previously stated, daily variations were filtered in a few cases and never produced perturbations larger than 3 nT. Besides, these small variations are not comparable to the large anomaly caused by urban debris filling dolines. Therefore, we consider that there is no significant feature or anisotropy in the map caused by solar activity.

Several sections display a strong anomaly, both in total intensity and gradiometer, especially spectacular were those crossing the doline. The total

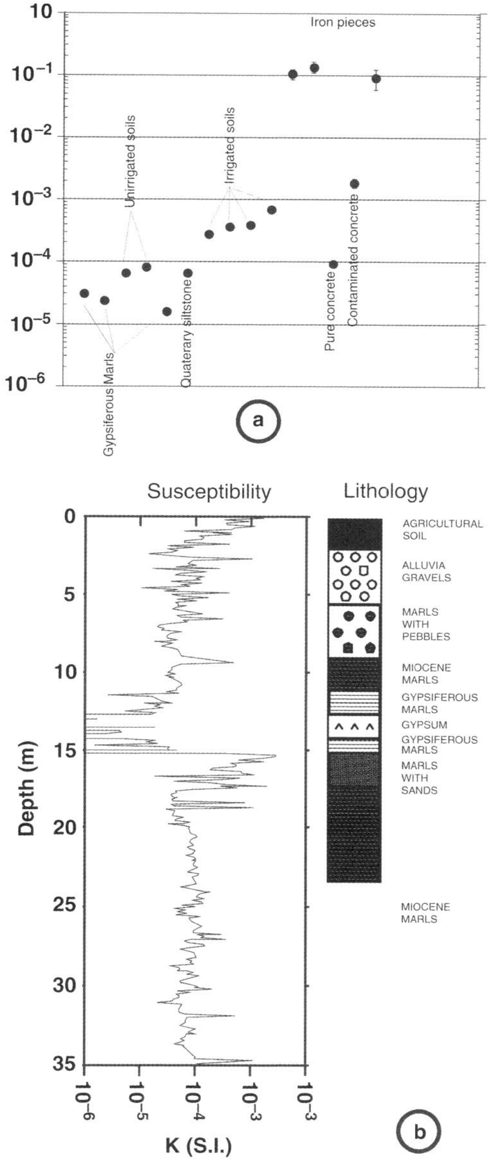

Fig. 6. (**a**) Magnetic susceptibility (S.I.) in the soils around the doline (surface data). Error bars fall within the points in most cases. (**b**) Susceptibility log in one core from the vicinity.

intensity shows up to 650 nT and a strong asymmetric dipolar anomaly that expands 30 m to each side around the centre of the doline (Fig. 7a). The gradient is obtained by subtracting the total intensity measured in both magnetometers and notes the high intensity of shallow perturbations. The gradiometer evinces a 150 nT m^{-1} maximum anomaly (500 nT smaller than the total field) and extends 20 m to each side around the central point (Fig. 7b). Inflection points of anomalies due to total intensity and the gradient of the Earth's magnetic field meet at the doline middle point.

Two other unexpected dolines were also detected during the survey; they do not show any clear surface evidence of their existence. A strong anomaly 80 m SW of the centre of the main doline was characterized by dipolar anomalies of up to 450 nT (total intensity) and 90 nT m^{-1} (gradient). Interviews with local farmers confirmed the occurrence and activity of this doline, with an original diameter of about 25 m. In contrast with the main doline data, inflection points of both anomalies are not coincident. The age of this doline (which is more than 20 years old) together with agricultural practices may cast some light to explain this observation. During that time the filling of the subsiding depression has been scattered over the surface, giving rise to the present-day differences between total intensity and gradient.

Another small anomaly located 20 m from the main doline centre to the NE was also detected. It is defined by a maximum dipolar anomaly in the total intensity of 110 nT and 40 nT m^{-1} in the gradiometer. In this case local farmers also confirmed the activity of this small doline, formed more recently with dimensions of 1 m in diameter and 13 m deep. Both the total intensity and gradient fit with the doline central point. These results validate the starting hypothesis and open a new research approach to the problem. A three-dimensional (3D) reconstruction of magnetic anomalies indicates that all three (even minor anomalies) show an alignment along a NE–SW direction (Fig. 7). This trend is parallel to the groundwater gradient, which is trending NE–SW (N060°E: Bielza *et al.* 1993) and perpendicular to the piezometric contours in this area, oriented in a NW–SE direction according to their regional pattern (see Fig. 3).

Modelling

Modelling of the anomaly obtained was carried out by means of the British Geological Survey Gravmag program. This program allows modelling of magnetic and gravimetric anomalies in 2.5 dimensions. Therefore, the third dimension may be introduced into the cross-section modelling by giving the magnitude of the half-strike length of bodies in the direction perpendicular to the cross-section. As only the half strike is introduced as a parameter and not the total length of the body, the modelling can not be consider as 3D, but only 2.5D. The parameters needed to model a magnetic

Table 1. *Magnetic susceptibility robust means from different materials involved in the karst development from the Zaragoza surroundings*

Material	Mean (10^{-6} S. I.)	Standard error	Number of measurements
Gypsum	−4	1	20
Gypsiferous marls-1	29.6	3	81
Gypsiferous marls-2	23.5	2	55
Gypsiferous marls-3	15.6	1	47
Soil (not irrigated)-1	65	6	27
Soil (not irrigated)-2	81.3	5	40
Soil (moderately irrigated)	272.1	10	140
Pediments	65.4	2	53
Soil (intensively irrigated)-1	362.2	19	54
Soil (intensively irrigated)-2	377.1	29	16
Soil (intensively irrigated)-3	689	40	66
Concrete	91.7	11	15
Contamined concrete	1855	328	4
Bricks (and pottery)	2936.6	947.9	14
Iron pieces-1	90 388	31 800	6
Iron pieces-2 (fence)	1.019×10^5	17 922	3
Iron pieces-3 (poles)	1.355×10^5	29 500	2

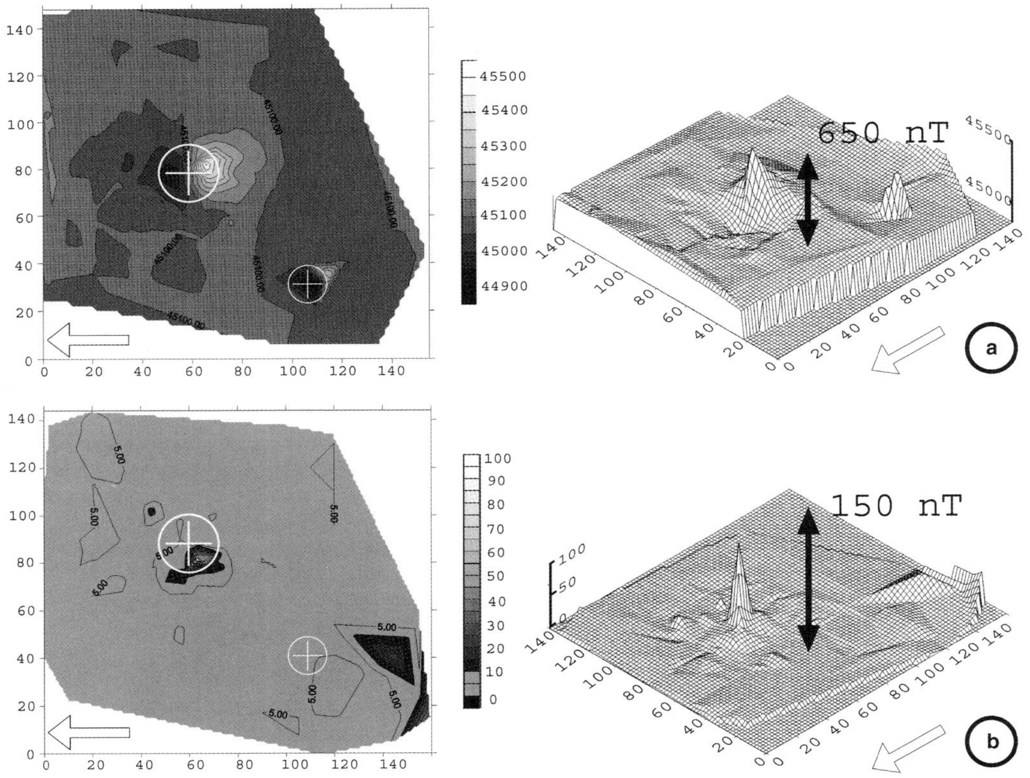

Fig. 7. Magnetic anomalies. (**a**) Total intensity. (**b**) Gradiometer. The crosses indicate the location of the test and the main secondary dolines.

anomaly are the susceptibilities of the different bodies (entered as polygons into the modelled cross-section) and their geometry. Other inputs such as declination, inclination and intensity of the present-day magnetic field can be considered as constants. Remanent magnetization was not considered because in Quaternary terraces remanence is expected to have very low values according to the low susceptibility (typically paramagnetic), and the chaotic nature of the filling of cavities precludes the existence of a consistent remanence pattern in these fillings.

The modelled profile shows a north–south orientation and crosses through the centre of the main doline filled with urban debris (Fig. 8). The main positive anomaly reaches values of 650 nT over the background, and its associated negative anomaly north of it reaches negative values of −100 nT with respect to the background. The arrangement of bodies chosen to fit the anomaly is very simple, with a very thin cover of agricultural soil, up to 1 m thick, and an infinite length in the direction perpendicular to the cross-section with susceptibility values of 270×10^{-6} S.I. The

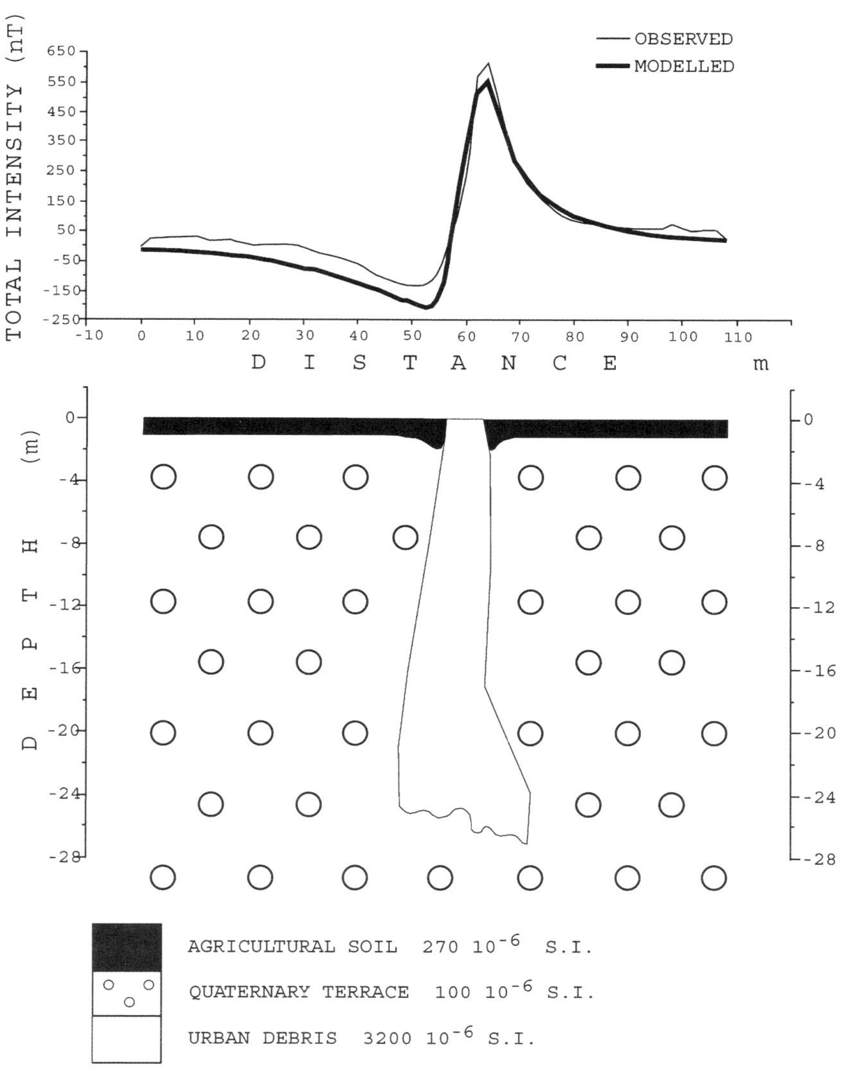

Fig. 8. Magnetic modelling based on geometrical and susceptibility data, and its good fit with the observed magnetic anomaly.

background is considered to be the Quaternary terrace, consisting mainly of gravels and sands, with averaged susceptibility values of about 100×10^{-6} S.I.

The body responsible for the main anomaly must show an average susceptibility of about 3200×10^{-6} S.I., an abnormally high value for natural materials and probably caused by pottery, bricks, concrete and metallic debris filling the void (Fig. 6a). This high value is constrained by the maximum intensity of the anomaly that depends strongly on the properties of this material and not so much on its geometry. Conversely, the detailed shape of the anomaly curve is strongly dependent on the morphology of the causative body. The filling can reach depths of about 28 m according to the model fitting the anomaly, although the proven depth is 16 m. Nevertheless, several tests performed during forward modelling showed that the influence of the lowermost 10 m of the filling of the cavity only slightly modifies the shape of the anomaly detected at surface. The width of the modelled anomaly strongly constrains the width of the filling, that must not widen at depth within the shallowest first 15 m, and therefore must have a vertical pipe geometry with a diameter of between 5 and 10 m. The deepest part of the filling of the cavity was reconstructed according to field observations, indicating the vault shape of the cavity at depth. However, as no borehole data coinciding exactly with the cavity exist, this part of the model must be considered as speculative.

The northernmost part of the section does not fit exactly with the modelled anomaly, probably due to the existence of other bodies with different susceptibility values below the surface. However, because its interpretation may be speculative owing to the lack of precise subsurface data in this sector, we have not attempted to exactly fit this part of the curve. The occurrence of voids at depth below the filling cannot be determined in this profile because the intensity of the main anomaly precludes the determination of smaller anomalies linked to voids or facies changes at depth.

In summary, the profile obtained allows both the properties and the geometry of the main causative body to be reasonably attained.

Conclusions

A magnetic survey has been carried out with a proton gradiometer in a doline with known location and depth. The survey evinces a strong anomaly centred in the doline and with a dipole defined by more than 650 nT and a gradient of about $150\,\mathrm{nT\,m^{-1}}$. Two more dolines, which show no field evidence and were completely unknown at the beginning of the survey, were detected by the same method very close to the studied doline.

The sampling procedure used, with 2 m grid spacing, has proved to be optimum with respect to the accurate and reliable characterization of the magnetic anomaly. To avoid border effects and to entirely define the anomaly the sampling area must be, at least, 10 times larger than the surface expression of the collapse.

The modelling of the magnetic anomaly has allowed the geometry of the cavity to be reproduced, with vertical walls (funnel-shaped) near the surface and a vault shape at depth, consistent with field observations made immediately after the collapse. Magnetic modelling is consistent with the magnetic properties (susceptibility) of its filling determined from field and laboratory analysis.

The contrasts in magnetic properties observed in the rocks and soils involved in the evaporite karst covered with alluvial deposits and anthropogenic materials point towards promising results in detecting other type of dolines (especially naturally filled collapses) by using the magnetic survey methodology because of the proven results shown in this paper. Further studies applying other geophysical methods, now in progress, will shed more light in the detection and modelling procedures of this phenomenon.

The authors are very grateful to J. J. Curto from the 'Observatorio magnético del Ebro' for providing the daily variation of the magnetic field. This work was supported by projects from the Spanish Ministry of Sciences and Technology (PROFIT: FIT-310200-2004-181), as well as the Aragonian Government (PM 045) and Geotransfer. T. Mochales benefited from a PROFIT fellowship and E. L. Pueyo from a 'Ramón y Cajal' research contract.

References

ARLEGUI, L. E., SIMÓN, J. L. & SORIANO, M. A. 1997. Estructuración neógena del sector central de la Cuenca del Ebro. *In*: CALVO, J. P. & MORALES, J. (eds) *Avances en el conocimiento del Terciario Ibérico*. Museo Nacional de Ciencias Naturales, Madrid, 33–36.

ARMADILLO, E., MASSA, F., CANEVA, G., GAMBETTA, M. & BOZZO, E. 1998. Modelling of karst structures by geophysical methods; an example: the doline of S. Pietro dei Monti (western Liguria). *In*: BOZZO, E., MELONI, A. & PATELLA, D. (eds) *VI Workshop, Geo-electromagnetism. Annali di Geofisica*, **41**, 389–397.

BALLARD, R. F. 1983. *Cavity Detection and Delineation Research. Report 5, Electromagnetic (Radar) Techniques Applied to Cavity Detection.* Technical Report, **GL, 83–1**, 90.

BENITO, G., PÉREZ, P., GUTIÉRREZ, M. & SANCHO, C. 1995. Natural and human-induced sinkholes in gypsum terrain and associated environmental problems in NE Spain. *Environmental Geology*, **25**, 156–164.

BIELZA, V., MARTÍNEZ-GIL, F. J., CARCELLER, T., GARRIDO, E., NERÍN, C., SÁNCHEZ, Y. & DEL VALLE, J. 1993. *Contaminación del Manto Freático del Corredor del Ebro*. Fundación Nueva Empresa, Zaragoza.

BERES, M., LUETSCHER, M. & OLIVIER, R. 2001. Integration of ground-penetrating radar and microgravimetric methods to map shallow caves. *Journal of Applied Geophysics*, **46**, 249–262.

BUTLER, D. K. 1984. Microgravimetric and gravity gradient techniques for detection of subsurface cavities. *Geophysics*, **49**, 1084–1096.

BUTTRICK, D. & VAN SCHALKWYK, A. 1998. Hazard and risk assessment for sinkhole formation on dolomite land in South Africa. *Environmental Geology*, **36**, 170–178.

CHAMON, N. & DOBEREINER, L. 1988. An example of the use of geophysical methods for the investigation of a cavern in sandstones. *Bulletin of the International Association of Engineering Geology*, **38**, 37–43.

COLLEY, G. C. 1963. The detection of caves by gravity measurements. *Geophysical Prospecting*, **XI**, 1–9.

COOK, J. C. 1965. Seismic mapping of underground cavities using reflection amplitudes. *Geophysics*, **30**, 527–538.

COOK, J. C. 1974. Yes, we can locate solution cavity boundaries. *In*: COOGAN, A. H. (ed.) *Rock Mechanics and Geophysics, Fourth Symposium on Salt*, Volume 2, 33–40.

COOPER, A. H. 1998. Subsidence hazards caused by the dissolution of Permian gypsum in England: geology, investigation and remediation. *In*: MAUND, & EDDLESTON, (eds) *Geohazards in Engineering Geology*. Geological Society, London, Engineering Geology, Special Publications, **15**, 265–275.

COOPER, A. H. & WALTHAM, A. C. 1999. Subsidence caused by gypsum dissolution at Ripon, North Yorkshire. *Quarterly Journal of Engineering Geology*, **32**, 305–310.

CORTÉS-GRACIA, A. L. 2004. Geometría del subsuelo de la Cuenca del Ebro en el campo del Belchite (Zaragoza). *Geotemas*, **6**, 225–228.

FENNING, P. J., BROWN, A. J. & NIND, D. 2000. Geophysical surveys across a ground subsidence feature. *In*: POWERS, M. H., IBRAHIM, A. B. & CRAMER, L. (eds) *Proceedings of the Symposium on the Application of Geophysics to Environmental and Engineering Problems (SAGEEP)*, 2000, 857–866.

GARCÍA-CASTELLANOS, D., VERGÉS, J., GASPAR-ESCRIBANO, J. & CLOETINGH, S. 2003. Interplay between tectonics, climate, and fluvial transport during the Cenozoic evolution of the Ebro Basin (NE Iberia). *Journal of Geophysical Research*, **108**, 2347.

GUERRERO, J., GUTIÉRREZ, F. & LUCHA, P. 2004. Paleosubsidence and active subsidence due to evaporite dissolution in the Zaragoza area (Huerva River valley, NE Spain): processes, spatial distribution and protection measures for transport routes. *Engineering Geology*, **72**, 309–329.

GUTIÉRREZ, M., IBÁÑEZ, M. J., PEÑA, J. L., RODRIGUEZ, J. & SORIANO, M. A. 1985. Quelques exemples de karst sur gypse dans la Depression de l'Ebre. *Karstologia*, **6**, 29–36.

HYATT, J. A., WILSON, R., GIVENS, J. S. & JACOBS, P. M. 2001. Topographic, geologic, and hydrogeologic controls on dimensions and locations of sinkholes in thick covered karst, Lowndes County, Georgia. *In*: BECK, B. F. & HERRING, J. G. (eds) *Geotechnical and Environmental Applications of Karst Geology and Hydrology*, Balkema, Rotterdam, 37–45.

MATTHEWS, M. C., CLAYTON, C. R. I. & RIGBY-JONES, J. 2000. Locating dissolution features in the Chalk. *Quarterly Journal of Engineering Geology*, **33**, 125–140.

MENSUA, S. & IBÁÑEZ, M. J. 1977. *Sector central de la Depresión del Ebro. Mapa de terrazas fluviales y glacis*. III Reunión Nacional del grupo de trabajo del Cuaternario. Dipartimento de Geografía, Universidad de Zaragoza.

MILLER, C. H., EGE, J. R., ODUM, J. K. & GOLOB, J. J. 1984. Preliminary Seismic-Velocity and Magnetic Studies of a Carbonate Rock-Sinkhole Area in Shelby County, Alabama. US Geological Survey, Report, OF 84-0409.

MILLER, R. D. & STEEPLES, D. W. 1991. Detecting voids in 0.6 m coal seam, 7 m deep, using seismic reflection. *Geoexploration*, **28**, 109–119.

MUÑOZ, A., ARENAS, C., GONZÁLEZ, A., LUZÓN, A., PARDO, G., PÉREZ, A. & VILLENA, J. 2002. Ebro basin (northeastern Spain). *In*: GIBBONS, W. & MORENO, T. (eds) *The Geology of Spain*. Geological Society, London, 370–385.

NEUMANN, R. 1967. La gravimétrie de haute precision application aux recherches de cavités. *Geophysical Exploration*, **15**, 116–134.

PIERCE, C. J., EVERETT, M. E., CARTER, J. L., MOORE, M. & BRIGHT, R. 2001. Magnetics as a tool for urban site characterization, Shreveport Convention Center, Louisiana. *Abstracts with Programs, Geological Society of America*, **33**, 366.

QUIRANTES, J. 1978. *Estudio sedimentológico y estratigráfico del Terciario continental de los Monegros*. Ph.D. thesis, Institución Fernando el Católico (CSIC), Zaragoza.

SHAH, A., CORMIER, M. H., RYAN, W. B. F., JIN, W., BRADLEY, A. M. & YOERGER, D. R. 1999. High resolution 3-D map of the earth's magnetic field at the EPR reveals shallow dike outcrops and large-scale void space in extrusive layers. *Eos, Transactions of the American Geophysical Union*, **80**, 1074, 16.

SORIANO, M. A. 1990. *Geomorfología del sector centromeridional de la Depresión del Ebro*. Diputación Provincial de Zaragoza.

SORIANO, M. A. & SIMÓN, J. L. 1995. Alluvial dolines in the central Ebro Basin, Spain: a spatial and developmental hazard analysis. *Geomorphology*, **11**, 295–309.

SORIANO, M. A. & SIMÓN, J. L. 2002. Subsidence rates and urban damages in alluvial dolines of the central Ebro Basin (NE Spain). *Environmental Geology*, **42**, 476–484.

SORIANO, M. A., SIMÓN, J. L., ARLEGUI, L. E., LIESA, C. L. & POCOVÍ, A. 2004. *Problemas causados por el karst aluvial en el centro de la cuenca del Ebro (España)*. Proceedings of the Conference on 'Knowledge on Sinkhole Phenomena and the Role of Public Administration and Local Management Agencies' Rome, 20–21 May 2004. APAT, Rome, 521–532.

THIERRY, P., DEBEBLIA, N. & BITRI, A. 2005. Geophysical and geological characterization of karst hazard in urban environments: application to Orléans (France). *Bulletin of Engineering Geology and Environment*, **64**, 139–150.

VAN SCHOOR, M. 2002. Detection of sinkholes using 2D electrical receptivity imaging. *Journal of Applied Geophysics*, **50**, 393–399.

VAN ZUIDAM, R. A. 1976. *Geomorphological Development of the Zaragoza Region, Spain. Processes and Landforms Related to Climatic Changes in a Large Mediterranean River Basin*. International Institute for Aerial Survey and Earth Sciences (ITC), Enschede.

XIA, J. & WILLIAMS, S. L. 2003. High-resolution magnetic survey used in searching for buried brine wells in Hutchinson, Kansas. *In*: JOHNSON, K. & NEAL, J. T. (eds) *Evaporite Karst and Engineering/Environmental Problems in the United States*, Oklahoma Geological Survey, Circular, Report, **109**, 169–175.

YUAN, D. 1988. Environmental and engineering problems of karst geology in China. *Environmental Geology, Water Sciences*, **12**, 79–87.

ZHOU, W., BECK, B. F. & ADAMS, A. L. 2002. Effective electrode array in mapping karst hazards in electrical resistivity tomography. *Environmental Geology*, **42**, 922–928.

Coastal karst geomorphosites at risk? A case study: the floods of 6–11 December 2004 in central-east Sardinia

A. COSSU[1], J. DE WAELE[2] & F. DI GREGORIO[3]

[1]*Servizio Agrometeorologico Regionale per la Sardegna, Viale Porto Torres 119-07100 Sassari, Italy*

[2]*Dipartimento di Scienze della Terra e Geologico-Ambientali, Università di Bologna, Via Zamboni 67, 40126, Bologna, Italy (e-mail: dewaele@geomin.unibo.it)*

[3]*Dipartimento di Scienze della Terra, Università di Cagliari, Via Trentino 51, 09127 Cagliari, Italy*

Abstract: Extreme rainfall causing floods and great damage occurred in many areas of central-east Sardinia in the period 6–11 December 2004. A total of approximately 700 mm of rain was measured during this extreme event, with a maximum reaching 510 mm of rainfall in 1 day at the rain gauge of Villagrande (Ogliastra). During and immediately after the event all fluviokarstic canyons were activated for at least 1 week, reaching the highest water levels in at least 50 years and reversing great quantities of sediment-loaded water onto the coast and with important geomorphical modifications. There was public fear that serious damage to the natural resources would occur, such as the famous Cala Luna beach that was almost completely destroyed by the flooding of the Codula Ilune River and by the coinciding sea storm. The river, in fact, eroded the longshore bar (beach) and destroyed the small backshore lagoon. A monitoring study has been initiated in order to analyse the natural evolution of this littoral system and to define the resilience of this interesting geomorphosite. The observations have shown that the flood, albeit impacting negatively in the moments immediately after the disaster, almost completely restored the natural equilibrium of this coastal karst geo-ecosystem within a season.

The entire central and SE coastal area of Sardinia is characterized by a mean annual rainfall of 700–1000 mm (standard deviation 350 mm), with the maximum occurring in December (mean monthly rainfall 130–170 mm). It is not unusual to have hot and moist currents coming from Africa during the autumn causing convective motions that give rise to rain storms following each other at short intervals and with the same epicentre. One of the last extreme events of this type dates back to 1951 and caused severe damage in several parts of Sardinia and, in particular, Ogliastra, not far south of the Gulf of Orosei. Extreme rainfall causing floods and great damage also occurred in many areas of central-east Sardinia in the period 6–11 December 2004. Major hazards occurred east of the Gennargentu mountainous region, in the villages of Ogliastra (e.g. Villanova Strisaili) and Barbagia–Baronia (Oliena, Dorgali, etc.) where major floods caused two human casualties, and in the coastal alluvial plains of the east coast along the main rivers (e.g. the Cedrino River). Hydrogeological risk was less important in the less populated littoral area of the Gulf of Orosei, a very well known coastal karst area with high landscape and natural value; nevertheless, flooding occurred in localized coastal areas. The impact of this flood on the natural and geomorphological heritage in this coastal area has been assessed, with special emphasis on the famous Cala Luna beach, one of the most outstanding geomorphosites of the entire Gulf of Orosei.

Geological context

The Gulf of Orosei coastal karst area is characterized by a thick sequence of dolomites and limestones that reach a total thickness of about 800 m above a Palaeozoic granite or biotitic schist basement (Carmignani *et al.* 2001).

The Mesozoic transgression occurred after a Permian–Lower Jurassic continental period and invaded an undulating Variscan peneplain (Bajocian–Bathonian). This sedimentary sequence is composed of quartz conglomerates and sandstones, followed by marls, dolostones containing limestone lenses, oolitic limestones, and massive fossil-rich outer-shelf, reef, inter-reef and backreef limestones ranging in age from Bathonian to Berriasian (Dieni & Massari 1985).

The first sediments locally recognizable on the Mesozoic rocks are alluvial conglomerates and quartz sands, related to an intense erosion–deposition cycle caused by an uplifting phase of Middle Pliocene age. These sediments precede

From: PARISE, M. & GUNN, J. (eds) *Natural and Anthropogenic Hazards in Karst Areas: Recognition, Analysis and Mitigation.* Geological Society, London, Special Publications, **279**, 85–95. DOI: 10.1144/SP279.8
0305-8719/07/$15.00

and are contemporary to the Pliocene basalts that have given K–Ar ages of 3.56 ± 0.23–1.99 ± 0.08 Ma.

On these basalts alluvial sands and conglomerates are found (Cala Gonone, San Pietro valley, Biddunie, Cala Luna) probably deposited during the Early Pleistocene. Periglacial stratified slope-waste deposits (éboulis ordonnés), composed of limestone fragments in a reddish clay matrix forming coastal cliffs of more than 40 m height and developing up to 600 m a.s.l (above sea level), and aeolian sands visible in karst pockets along the limestone cliffs up to several metres a.s.l. and also on the near-shore continental shelf up to 40 m b.s.l. (below sea level) near Cala Gonone are also of Pleistocene age.

From a structural point of view the Jurassic carbonate cover is tilted towards the east (centre of the Gulf) with an almost constant dip of 20–30°. This monocline structure is disturbed by two main transcurrent arcuate fault systems directed NNW–SSE and NNE–SSW (Pasci 1997). These faults, often of reverse character, are related to the continental collision between the Apulian and the South European plates (Carmignani *et al.* 2001). The eastwards lowering of the Jurassic carbonates is thus compensated by this tectonic style, and the sediments crop out continuously up to a bathymetric depth of almost 100 m on the continental shelf (Orrù & Ulzega 1987).

Geomorphology

The entire coastal karst area covers a surface of more than 150 km^2 and forms spectacular high limestone cliffs cut by some major streams. From north to south these are the Codula Fuili, Codula Ilune and Codula Sisine, each forming canyons opening directly into the sea. This drainage network is most probably a relic of wetter and warmer conditions in the Late Quaternary and can be traced offshore for several kilometres to a depth of at least 120 m (Orrù & Ulzega 1987). This corresponds to the maximum lowstand of the Mediterranean Sea during the Last Glacial Maximum, around 22 ka ago (Lambeck & Bard 2000; Siddall *et al.* 2003).

Present surface drainage is active only after heavy rain, typically twice a year on average. It almost exclusively involves the major rivers, namely Codula Ilune and Sisine, and only exceptionally (seldom more than once every 30 years) other tributaries such as Codula Fuili, Codula Goloritzé, Bacu Maore and Bacu Mudaloro. This reflects the presence of granite beneath more than 60% of the upper drainage basin of Codula Ilune, with the Sisine draining part of the basaltic San Pietro Plain. Almost all other rivers are underlain by limestone, characterized by well-developed underground drainage.

The fluviokarstic landscape is enriched by typical karst landforms such as dolines, well developed on the dolostones close to the contact with the Palaeozoic granites (De Waele 2004). Other karst landforms such as the rock arches, which may attain impressive dimensions as in the Bacu Addas rock arch which is more than 50 m high and wide, are well-known landmarks (e.g. Goloritzé coastal arch) typical of the Gulf of Orosei (Barca *et al.* 1995).

Karst micromorphologies are locally well developed on the massive microcrystalline limestones, with solution flutes (*rillenkarren*), solution pans (*kamenitze*) and *lapiez* fields.

Morphology in the coastal areas is the product of a combination of structural, karst and littoral processes. Erosion is locally enhanced by corrosion that becomes prevalent where mixing between fresh and salt water occurs leading to hyperkarst phenomena (De Waele *et al.* 2001). Many erosion sea caves, formed by wave action on joints and structurally weaker areas, are located along the coast. Where fresh water outlets occur instead, mixing phenomena have led to the development of important coastal karst caves developing to a size several kilometres (De Waele & Forti 2003).

The important Bue Marino cave is worth mentioning here; located on the coast between Cala Fuili and Cala Luna, it has more than 15 km of passages, some of which are open to the public. Prehistoric anthropomorphic graffiti dated to 6000 years BP have been discovered in the wide entrance hall (Lo Schiavo 1978), in which the Tyrrhenian relict tidal notch at 9.5 m a.s.l. is easily recognizable (Carobene 1972; Antonioli *et al.* 1999).

This huge cave system contains important fluvial sedimentary deposits composed predominantly of granites, dolostones and limestones. Passages cut older phreatic conduits that are completely filled with basalt (Mahler 1979) related to the Pliocene volcanic period and contemporary with the basalts cropping out in the upper parts of Cala Luna, Cala Fuili and on the San Pietro Plain, which have been dated to 2–3.5 Ma.

The nearby Codula Ilune canyon contains the largest cave system in Sardinia, first explored in the early 1980s and comprising more than 42 km of passages, with interesting cave morphologies and sediments (Forti & Rossi 1991; De Waele *et al.* 1997). The present submarine spring of this cave system, located immediately south of the Cala Luna beach, comprises of a narrow drowned passage. This has been explored for 630 m to a maximum water depth of 37 m, but mostly lies at depths of about 10 m (Hovorka 1993). Inside this submarine karst spring remnants of aeolian-cemented sediments related to an arid and cold period have been found (Antonioli & Ferranti 1992).

North of the beach six big cave entrances are located practically at present sea level. Formed by hyperkarst phenomena due to the mixing of sea water with meteoric water infiltrating along joints, they show evidence of the high Tyrrhenian sea stand (tidal notch at 9.2 m a.s.l.) and an important continental phase with the deposition of a thick sequence of aeolian sediments. A phreatic conduit occupied by Pliocene basalt boulders opens in the roof of the passage in the northernmost of these caves (Cave No. 6 of Cala Luna) (Assorgia *et al.* 1968).

Cave development clearly started before the emplacement of the Pliocene basalts, as demonstrated by the infilling of well-developed pressure conduits by these volcanic rocks in the Bue Marino cave (Mahler 1979). The initial stages of speleogenesis might date back to the Eocene, when the Mesozoic sediments first emerged and surface denudation started. Karst processes have enabled the development of pressure tubes under phreatic conditions, especially during the Mio-Pliocene, and relics of this karst can be found at various altitudes or have sometimes been filled with Pliocene basalts (De Waele 2004). The sea level dropped drastically during the Messinian salinity crisis around 6.3–5.2 Ma BP, caused by the closure of the Gibraltar Strait and the consequent drying up of the Mediterranean (Cita & Corselli 1993; Krijgsman *et al.* 1999). Although the climate was semi-arid it is generally believed that minor karst processes did occur enabling the development of phreatic tubes at altitudes far below present sea level (Perna 1994; Audra *et al.* 2004).

Another important aspect of coastal morphology is the tidal notch, well developed along the entire carbonate coast of the Gulf, and indicating present and past mean sea levels. Such tidal notches are described from depths of 10 m b.s.l. to heights of 10.5 m a.s.l. (Carobene 1972, 1978; Carobene *et al.* 1973; Carobene & Pasini 1982; Antonioli & Ferranti 1992). The relict tidal notch, that has a decreasing height from north to south of between 10.5 and 7.7 m a.s.l., is attributed to the isotopic stage (IS) 5e (125 ka BP). Its continuity along the 37 km of limestone coast testifies to relative tectonic stability since that time, with the exception of the slight N–S tilting (Forti & Rossi 1991; Antonioli *et al.* 1999). The −10 m tidal notch, corresponding also to a well-recognized base level with coastal caves, also developed at this water depth, formed during a period of sea-level stability between isotopic stages 5e (125 ka BP) and 2 (around 20 ka BP). The aeolian sand deposits cropping out on the continental shelf up to a depth of 120 m b.s.l. are attributed to this same period (IS 5e–2), probably characterized by climatic pulses with frequent arid intervals (Antonioli & Ferranti 1992). A sea-level still-stand is also documented at a depth of 45–50 m b.s.l., correlated by most authors (Orrù & Ulzega 1987) to the pre-Versilian transgression (approximately 10 ka BP). Relative tectonic stability has prevailed in the Gulf since IS 5, as demonstrated by this Tyrrhenian tidal notch well preserved along more than 50% of the carbonate coastline. Although it decreases in height from 10.5 m a.s.l. in the north to 7.7 m a.s.l. in the south, it does not show abrupt changes in height (Antonioli *et al.* 1999).

The geomorphological setting of the Cala Luna beach and its surroundings is presented in Figure 1.

The December 2004 extreme meteorological event

The central-eastern coast of Sardinia experiences a mean rainfall of 700–1000 mm year^{-1}, with a high standard deviation of 350 mm year^{-1}. Maximum rainfall is in December, with mean monthly values of between 130 and 170 mm (Chessa & Delitala 1997). The topography of the region, with substantial altitude changes over short distances from the coast, often causes intense rainfall especially during autumn. In fact, the warm and moist currents coming in from the SE (from Africa) create convective motions leading to a continuous series of heavy rainfalls with more or less the same epicentre. One of these extreme events, known as the Gairo Flooding, caused several deaths and severe damage to two villages in the Riu Pardu river basin, only few kilometres east of Villagrande Strisaili in 1951 (Ulzega & Marini 1977; Di Gregorio 1985; Ciccu *et al.* 1994).

Maximum rainfall of 700 mm, with a mean for the whole area of 400 mm, was recorded in central-east Sardinia during the period 6–11 December 2004. There were particularly extreme values on 6 December, with the epicentre at Villanova Strisaili (Table 1).

The meteorological radar instrument of the SAR (Servizio Agrometeorologico Regionale –Regional Agro meteorological Service) positioned on Monte Rasu (646 m a.s.l.), approximately 55 km WNW of Cala Luna, allowed an estimate to be made of the amount of water that fell on the region on the 6 December (Fig. 2) and providing information additional to that from rain gauges. Some 60 million m^3 of water is calculated to have fallen in the drainage basin of Codula Ilune, based on geomorphological and karst hydrogeological knowledge (De Waele 2004), but this is believed to be an underestimate of the total volume of water introduced into the system.

To analyse this extreme event of 6 December 2004 on a statistical basis, the maximum daily

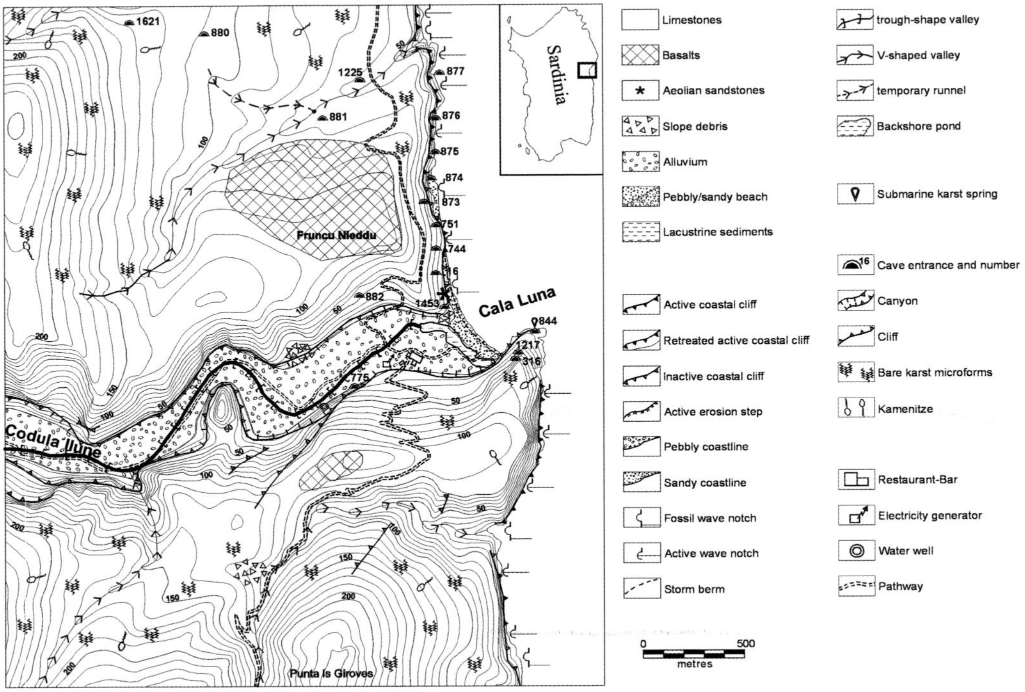

Fig. 1. Geomorphological map of the Cala Luna area.

rainfall data for each year measured by the rain gauge at Villanova Strisaili for the interval 1950–2000, and the distributions and the recurrence intervals according to Gumbel (extreme value) and to Fréchet (general extreme value), have been used (Coles 2001). Figure 3 reports the results of these statistical analyses. Both methods show that the probability not to exceed these extreme daily rainfalls is very high (0.999 and 0.992 for Gumbel and Fréchet, respectively), leading to very diversified recurrence intervals of 715 and 127 years. These great differences are related to the different behaviour of the statistical distributions, especially for the extreme rainfalls. Recurrence intervals for extreme events should, therefore, be calculated using the Fréchet method that gives a more pessimistic view, leading to more frequent recurrence intervals and thus higher probability that extreme events could occur in the future. It is, however, obvious that these purely statistical analyses take into account only the heaviest daily rainfalls of the past 50 years and does not consider major unknown variables such as global climate change leading to more frequent and severe climatic events.

Table 1. *Total rainfall registered at six different rain gauges of the SAR in a 6-day period (6–11 December 2004) and on 6 of December 2004, and climatic annual values (1961–1990)*

Rain gauge	Elevation (m a.s.l.)	Climatic annual rainfall (mm)	Rainfall (mm) 6–11 December 2004	Rainfall (mm) 6 December 2004
Nuoro	490	743	181.8	40.2
Dorgali	86	750	433.8	149.2
Oliena	124	799	384.6	254
Orgosolo	290	702	202.4	40.8
Villanova Strisaili	813	935	699.6	517.4
Jerzu	46	903	124.2	16.4

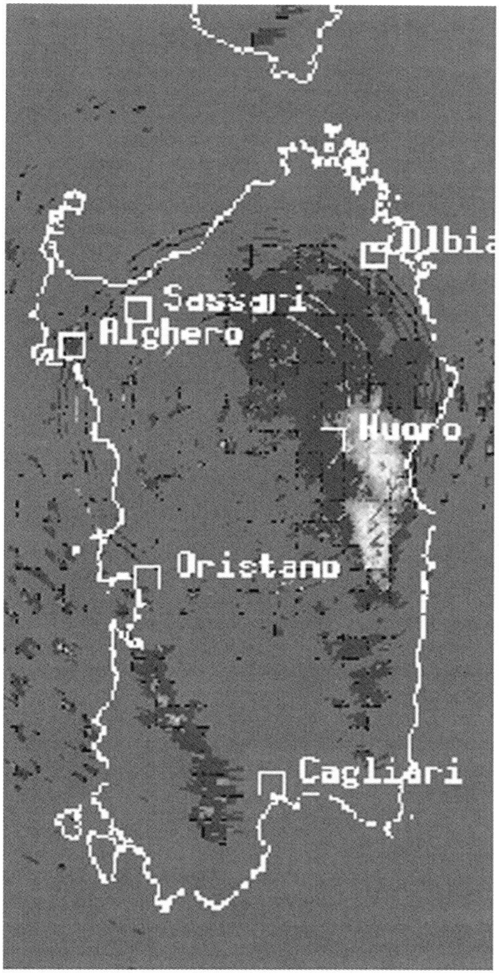

Fig. 2. Meteorological radar image (by courtesy of SAR) of the 6 December 2004 extreme rainfall. The major precipitation occurred in the central-eastern part of Sardinia (clear parts in the image).

Cala Luna case study

Cala Luna is located at the mouth of the Codula Ilune canyon, and forms an arch-like beach with a maximum width of 50 m and a shore length of approximately 500 m (Fig. 4). Cala Luna and the entire Gulf of Orosei have been repeatedly proposed as a Natural Regional Park since 1975 and Law 979/82 has listed the coastal area of the Gulf of Orosei–Capo Monte Santo as a possible Marine Protected Area. However, this marine park status has never been instituted. Using Law 394/91, the National Park of Gennargentu–Gulf of Orosei was designated, but again this park never really became established due to numerous protests and a lack of agreement with local communities. The coastal area of the Gulf of Orosei has nevertheless been designated a Special Area of Conservation (SAC ITB 000014) by the Project BioItaly (the Habitat's Directive 92/43/CEE).

At present the area is only protected by Regional Law No. 17 of 19 May 1981, valid for the entire Sardinian coastline, that prohibits construction of residences (houses, restaurants, hotels, etc.) within 300 m range of the shoreline. Recently, with Regional Law No. 8 of 25 November 2004, this range has been extended to 2 km.

At Cala Luna the only important and stable construction is a restaurant-bar, with annexed services, built on an old construction of stone probably erected in the 19th century by foresters. The restaurant is completely isolated and the few small paths that connect this restaurant to the inland are the old tracks used by foresters. Because of its remoteness, located in the middle of a wilderness area, and its multiple valences (geomorphology, coastal caves, canyon, flora and fauna, beach and backshore pond, archaeology, etc.) it is one of the most desired natural places of interest of this part of the island, visited by thousands of tourist every year, mainly during the period April–October.

The beach sediment is a pale–light yellow (5Y7/2–5Y8/2) medium sand ($0 < Mz < +2\phi$) composed of a mixture of granite and limestone and minor quantities of basalt (Cristini *et al.* 1992). The coarse fraction is mostly related to the limestone component even though, especially after the flood, the largest boulders are mostly granites. The absence of the important fine fraction can be related to the regular occurrence of storm waves and relevant coastal currents. The mean energy of the waves comes from the NE, while overall littoral drift is from north to south exposing thus the northern portion of the beach to a greater risk of erosion, while the southern part has a tendency to withhold sediments (Atzeni *et al.* 2002).

Every year during seasonal floods the Codula Ilune River breaches the longshore bar (i.e. beach), but normally floods last only a couple of days and the coastal dynamics restore the beach conformation in less than a couple of weeks.

During the flood of December 2004 the volume of water discharged by the Codula Ilune River was enormous. Large volumes of water continued to flow for at least 2 weeks, as shown in the picture taken on 16 December 2004 (Fig. 5), and the river continued to flow into the Tyrrhenian Sea for almost 2 months. Furthermore, a second flood occurred towards the end of January hindering still further the natural reconstruction of the beach. Local stakeholders and politicians were alarmed by this situation, fearing economic

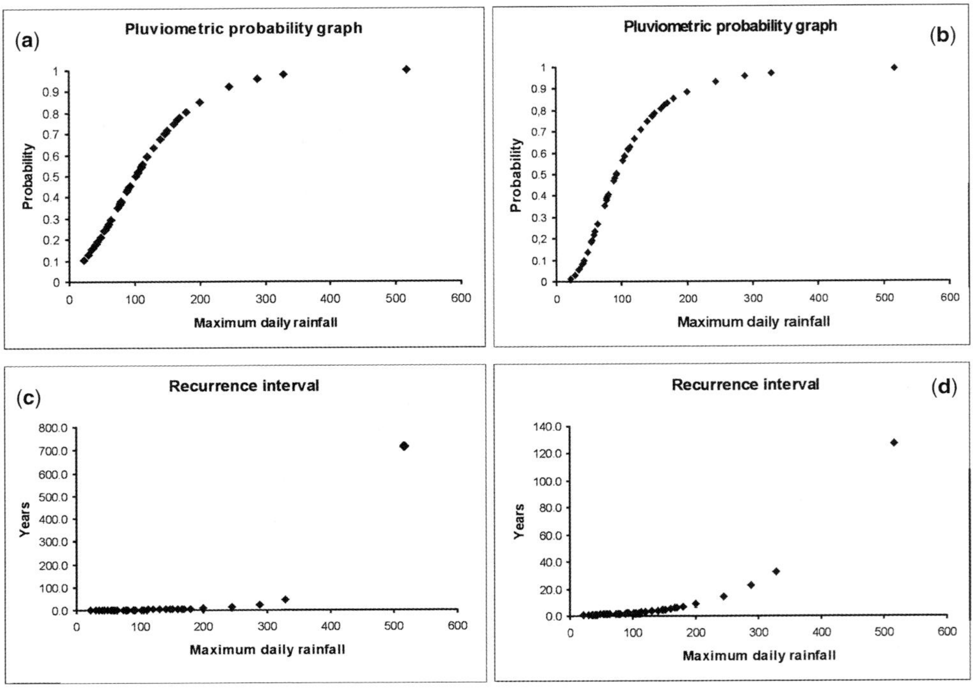

Fig. 3. Probability graphs of the extreme rainfall of 6 December 2004 according to the methods of (**a**) Gumbel and (**b**) Fréchet, and their respective recurrence intervals: (**c**) Gumbel and (**d**) Fréchet.

Fig. 4. Aerial photograph of 29 June 2004 of Cala Luna, showing the geo-environmental situation prior to the extreme flood in December. (Courtesy of Autonomous Region of Sardinia.)

Fig. 5. The complete destruction of Cala Luna during the flood: the picture was taken from the southern tip of Cala Luna on 16 December 2004. (Photograph: Aldo Nieddu.)

repercussions related to the erosion or the poor quality of the beach.

A geomorphological survey of the beach prior, during and after the flood instead has highlighted the medium-term positive effects of this extreme event on the whole geo-ecosystem. Figure 6 shows the natural evolution of the beach geo-ecosystem in the period April 2004–July 2005.

A discharge of several tens of $m^3 s^{-1}$ has been estimated during the peak flow of the river. This probably occurred around 10 December 2004. It was during this maximum flood, in particular, that large volumes of rock and vegetation were transported along the river bed. The geomorphological survey along the river bed close to the beach has enabled the large dimensions of the boulders (up to m^3 in volume) and tree trunks (of over 1 m in diameter) that were transported in the turbulent flow (Fig. 7) to be seen. A 2 m-deep river escarpment has been cut into the old soil-covered river terraces and the river has widened its bed from 2 to over 10 m. The sediment load was mainly composed of granite boulders, pebbles and sands, soils and minor quantities of limestone and basalt, and this was deposited in front of the river mouth on the near-shore continental shelf (Fig. 8). The fine fraction was transported offshore, while marine and coastal dynamics brought large amounts of the coarser sediments back to the shoreline resulting in significant accretion of the beach (Fig. 9). The large boulders that were present on the shoreline immediately after the flood have been covered with medium–coarse sands by wave action, although grain size is still relatively coarser than the original sediments.

Conclusions

Extremely heavy rainfall caused severe damage to infrastructures and great discomfort to the population in central-east Sardinia in the beginning of December 2004. In addition, many natural sites were subjected to important and, sometimes, permanent changes that have modified the landscape and the morphology, especially along the coast of the Gulf of Orosei. An emblematic case is given by the Cala Luna beach, which was completely destroyed during the flooding of Codula Ilune, alarming local stakeholders and tourists who were afraid that this geomorphosite would have been partially lost.

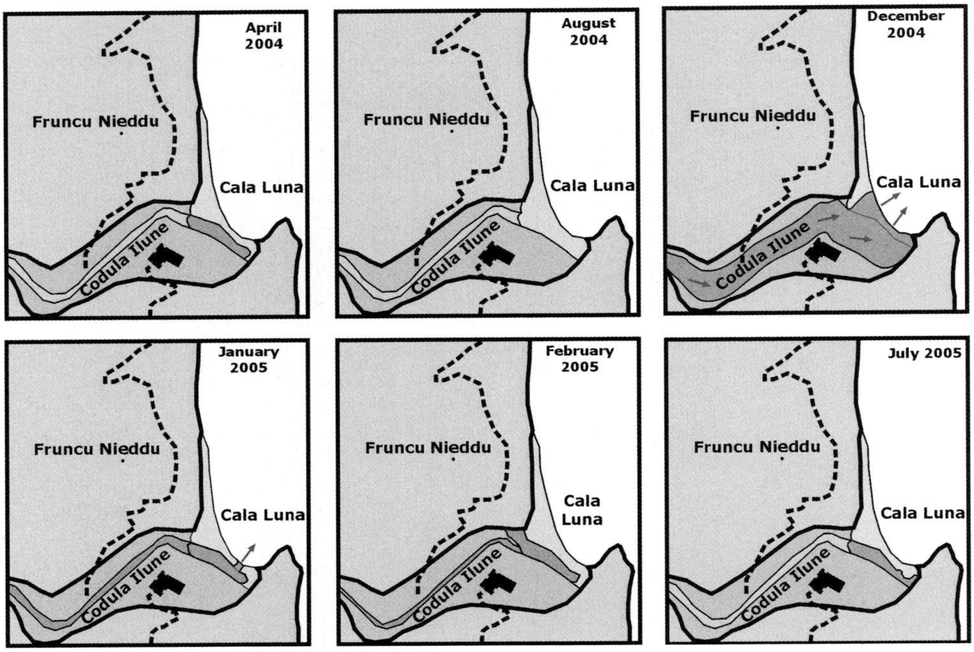

Fig. 6. Schematic sketch maps of Cala Luna before, during and after the December flood. The dark areas correspond to the water flow; arrows indicate discharge into the Tyrrhenian Sea.

Fig. 7. Large tree trunks were transported for kilometres downstream by the Codula Ilune in its peak flow during the December flood. (Photograph: Aldo Nieddu).

Fig. 8. Cala Luna beach seen from above the southern tip on 14 July 2005, 7 months after the flood. A large dead tree is still stuck in the sediments about 10 m from the shoreline. The granite boulders deposited offshore in front of the river mouth are also clearly visible (dark colour).

Fig. 9. Two photographs of Cala Luna taken in July before and after the extreme December flood, showing the increase in beach sediment in the northern section, confirmed by periodical monitoring measurements of the beach.

A total of 700 mm of rainfall was measured by the rain gauge at Villanova Strisaili in the period 6–11 December 2004, with 517 mm on 6 December alone. Statistical analysis of these data with the Fréchet method and analysis of the Meteorological Radar Images of SAR have shown that recurrence time of such an extreme event is of the order of 130 years or more, and that the total volume of water that fell in the drainage basin of Cala Luna probably exceeded 60 million m^3 on 6 December 2004.

Monitoring of the Cala Luna beach has been carried out within the framework of a National Research Project PRIN 2004–2006 on geological heritage (national responsible Prof. M. Panizza, local responsible Prof. F. Di Gregorio) in order to follow the natural evolution of this karst geo-ecosystem and to define the resilience of this most interesting geomorphosite after this centennial-scale flood. These preliminary studies have allowed the natural evolution of this site to be followed in the medium–long term. It is evident that, 6 months after the extreme meteorological event, the beach has restored its shape and seems to have benefited from the natural disaster, with an improvement both in quality and quantity of beach sediment. Further investigations are still in progress to follow the evolution of this system in time.

This Research has been carried out in the framework of the National Research Project PRIN 2004–2006 on geological heritage (national resp. Prof. M. Panizza, local resp. Prof. F. Di Gregorio).

The authors are greatly indebted to the many excursionists who have made their pictures of various flooding phases at Cala Luna available S. Cabras, I. Cabras, C. Conca, V. Crobu, B. Ibba, A. Nieddu, M. Pappacoda, S. Perra, S. Pillai and D. Porcu. Special thanks to the Forestry Department of Sardinia for their help during our campaign surveys, and to the regional office of RAI television for the videos on the Cala Luna flood.

Finally the many corrections and suggestions from M. Simms and some comments from an anonymous referee were much appreciated and greatly improved this paper.

References

ANTONIOLI, F. & FERRANTI, L. 1992. Geomorfologia costiera e subacquea e considerazioni paleoclimatiche sul settore compreso tra S. Maria Navarrese e Punta Goloritzé (Golfo di Orosei, Sardegna). *Giornale di Geologia*, **54**, 66–89.

ANTONIOLI, F., SILENZI, S., VITTORI, E. & VILLANI, C. 1999. Sea level changes and tectonic mobility: precise measurements in three coastlines of Italy considered stable during the last 125 ky. *Physics and Chemistry of the Earth* (*A*), **24**, 337–342.

ASSORGIA, A., BENTINI, L. & DERNINI, C. 1968. Nuove conoscenze sulle grotte costiere del settore di Cala di Luna (Dorgali–Sardegna Orientale). *In*: *Atti Congresso Internazionale Studi Sardi 10, Cagliari*, 4–31. Fossataro, Cagliari.

ATZENI, A., DE MURO, S., DI GREGORIO, F. & PIRAS, G. 2002. *Map of the Geoenvironmental Hazard on the Coast of Sardinia (Italy)*. SELCA, Florence.

AUDRA, P., LUDOVIC, M., CAMUS, H., GILLI, E., CLAUZON, G. & BIGOT, J.-Y. 2004. The effect of the Messinian Deep Stage on karst development around the Mediterranean Sea. Examples from Southern France. *Geodinamica Acta*, **17**, 389–400.

BARCA, S., DI GREGORIO, F. & MULAS, G. 1995. Natural rock arches of the Orosei Gulf, Sardinia, Italy. *In*: *Medcoast 95, Volume 1, Ankara*, 217–229. Middle East Technical University, Ankara.

CARMIGNANI, L. & OGGIANO, G. *ET AL*. 2001. Geologia della Sardegna. Note illustrative della Carta Geologica della Sardegna a scala 1:200 000. *In*: *Memorie Descrittive della Carta Geologica d'Italia*, **60**. Istituto Poligrafico e Zecca dello Stato, Roma.

CAROBENE, L. 1972. Osservazioni sui solchi di battente attuali ed antichi nel Golfo di Orosei in Sardegna. *Bollettino della Società Geologica Italiana*, **91**, 583–601.

CAROBENE, L. 1978. Valutazione di movimenti recenti mediante ricerche morfologiche su falesie e grotte marine del Golfo di Orosei. *Memorie della Società Geologica Italiana*, **19**, 641–649.

CAROBENE, L. & PASINI, G. C. 1982. Contributo alla conoscenza del Pleistocene superiore e dell'Olocene del Golfo di Orosei (Sardegna orientale). *Bollettino della Società Adriatica di Scienze*, **64**, 5–35.

CAROBENE, L., PASINI, G. & SELLI, R. 1973. Ancient coastlines in the Gulf of Orosei (western Sardinia). *In*: *Symposium sur la Géodynamique de la région Méditerranéenne*. Commission Internationale pour l'Exploration Scientifique de la Mer Méditerranée, Paris, France.

CHESSA, P. A. & DELITALA, A. 1997. Il clima della Sardegna. *In*: *Collana Agrometeorologia per la Sardegna, Nota Tecnica 2*. Edizioni Chiarella, Sassari.

CICCU, R., MANCA, P. & DI GREGORIO, F. 1994. Propensione al dissesto idrogeologico in Sardegna: analisi storica degli eventi ed aspetti metodologici. *In*: *Convegno Internazionale di Geoingegneria, Difesa e Valorizzazione del Suolo e degli Acquiferi*, Volume 1, Torino, 73–85. Politechnico di Torino, Torino.

CITA, M. B. & CORSELLI, C. 1993. Messiniano: vent'anni dopo. *Memorie della Società Geologica Italiana*, **49**, 145–164.

COLES, S. 2001. *An Introduction to Statistical Modelling of Extreme Values*. Springer, London.

CRISTINI, A., DI GREGORIO, F. & FERRARA, C. 1992. Sedimentological and geochemical characteristics of the carbonatic beaches of the Gulf of Orosei (East-Central Sardinia). *In*: LANG, P. (ed.) *International Coastal Congress*. Kiel, 671–682. Peter Lang, Frankfurt.

DE WAELE, J. 2004. Geomorphologic evolution of a coastal karst: the Gulf of Orosei (Central-East Sardinia, Italy). *Acta Carsologica*, **33**, 37–54.

DE WAELE, J. & FORTI, P. 2003. Estuari sotterranei. *In*: CICOGNA, F., NIKE BIANCHI, C., FERRARI, G. & FORTI, P. (eds) *Grotte Marine: cinquant'anni di ricerca in Italia*. Ministero per la Difesa dell' Ambiente, Rapallo, 91–104.

DE WAELE, J., DI GREGORIO, F., FOLLESA, R. & PIRAS, G. 1997. Complesso sotterraneo di Codula Ilune, Sardinia, Italy. *International Caver*, **20**, 3–10.

DE WAELE, J., FORTI, P. & PERNA, G. 2001. Hyperkarstic phenomena in the Iglesiente mining district (SW-Sardinia). *In*: CIDU, A. (ed.) *Water–Rock Interaction 2001*. Balkema, Rotterdam, 619–622.

DI GREGORIO, F. 1985. L'uomo, le acque e il dissesto idrogeologico in Sardegna. *In*: *III Convegno*

Internazionale di Studi Geografico-Storici 'La Sardegna nel Mondo Mediterraneo', Sassari. Patron Editore, Bologna.

DIENI, I. & MASSARI, F. 1985. Mesozoic of Eastern Sardinia. *In*: *19th European Micropaleontological Colloquium – Guide Book*. AGIP Sardinia, Cagliari, 66–78.

FORTI, P. & ROSSI, G. 1991. Idrogeologia ed evoluzione carsica della Codula di Luna (Sardegna). *Atti e Memorie della Commissione 'E. Boegan'*, **30**, 53–79.

HOVORKA, J. 1993. Cave diving exploration on Sardegna. *Czech Speleological Society*, **30**, 30–36.

KRIJGSMAN, W., HILGEN, F. J., RAFFI, I., SIERRO, F. J. & WILSON, D. S. 1999. Chronology, causes and progression of the Messinian salinity crisis. *Nature*, **400**, 652–655.

LAMBECK, K. & BARD, E. 2000. Sea-level change along the French Mediterranean coast for the past 30 000 years. *Earth and Planetary Science Letters*, **175**, 203–222.

LO SCHIAVO, F. 1978. Figurazioni antropomorfe nella grotta del Bue Marino-Cala Gonone (Dorgali). *In*: *Sardegna centro-orientale. Dal Neolitico alla fine del Mondo Antico, Sassari.* 53–55. Dessì Editore, Sassari.

MAHLER, A. 1979. Verkarstung der Karbonatgebiete am Golfo di Orosei (Sardinien). *Geologischer Palaeontologischer Mitteilungen Innsbruck*, **7**, 1–49.

ORRÙ, P. & ULZEGA, A. 1987. Rilevamento geomorfologico costiero e sottomarino applicato alla definizione delle risorse ambientali (Golfo di Orosei, Sardegna orientale). *Memorie della Società Geologica Italiana*, **37**, 471–479.

PASCI, S. 1997. Tertiary transcurrent tectonics of North-Central Sardinia. *Bulletin de la Société Géologique de France*, **168**, 301–312.

PERNA, G. 1994. Il carsismo profondo nel Sulcis-Iglesiente (Sardegna sud occidentale) e nel Trentino-Veneto (Alpi orientali italiane). 'Carsismo Messiniano': esempi di carsismo profondo correlati con il livello del Mediterraneo nel Messiniano. *Annali dei Musei Civici di Rovereto*, **10**, 327–378.

SIDDALL, M., ROHLING, E. J., ALMOGI-LABIN, A., HEMLEBEN, C., MEISCHNER, D., SCHMELZER, I. & SMEED, D. A. 2003. Sea-level fluctuations during the last glacial cycle. *Nature*, **423**, 853–858.

ULZEGA, A. & MARINI, A. 1977. L'évolution des versants dans la vallée du Rio Pardu (Sardaigne centre-orientale). *Zeitschrift fur Geomorphologie*, **21**, 466–474.

Contributory area definition for groundwater source protection and hazard mitigation in carbonate aquifers

J. GUNN

Limestone Research Group, The University of Huddersfield, Queensgate, Huddersfield HD1 3DH, UK (e-mail: j.gunn@hud.ac.uk)

Abstract: Carbonate aquifers provide important sources of potable water but are known to be particularly prone to pollution owing to rapid transfer of pollutants from the surface to springs or boreholes. Source protection zones and groundwater vulnerability maps are commonly used to mitigate against the pollution hazard but cannot be applied simplistically to carbonate aquifers, which are usually highly heterogeneous with overlapping groundwater divides that may vary with water levels. Divergent flow and disjunct contributory areas provide further complexity. Under these conditions, water-tracing experiments, repeated under different flow conditions, are the only tool capable of identifying those areas that contribute recharge to a particular source. Examples of water pollution affecting disjunct and overlapping source contributory areas are presented from the Waitomo area (New Zealand), Cuilcagh Mountain (Ireland) and the Peak District (UK). Source protection zones (SPZ), that have been defined by the Environment Agency in the Buxton area of the Peak District using equivalent porous medium models, are shown to be deficient. Further water-tracing experiments are essential if carbonate aquifers are to be adequately protected from pollution.

Groundwater provides the base flow of rivers and is an increasingly important source of drinking water in many countries. For example, over 30% of drinking water in England and Wales comes from groundwater (Environment Agency 2005). Carbonate aquifers are amongst the most important and it has been estimated that they occupy about 35% of the European land surface (Goldscheider 2005) and that around a quarter of the world's population derives its drinking water from carbonate aquifers (Ford & Williams 1989). Hence, protection of groundwater quality is a major priority for hydrogeologists. There are two main approaches to groundwater protection: resource protection which has the general aim of protecting the whole groundwater body; and source protection which has the specific aim of protecting an individual spring, well or borehole source.

The primary tool for resource protection is aquifer vulnerability mapping, and the particular problems of vulnerability mapping in karst terrains have been addressed in Europe by two European Commission COST (Co-Operation in Science and Technology) Actions. Action 65 addressed hydrogeological aspects of groundwater protection in karst areas (COST 65 1995) and Action 620 was entitled 'Vulnerability Mapping for the Protection of Carbonate (Karst) Aquifers'. Action 620 proposed definitions for intrinsic and specific vulnerability, and put forward a European approach to intrinsic vulnerability mapping in karst (Daly *et al.* 2002) that is partly based on the PI method (Goldscheider 2005). In the United Kingdom a series of 53 groundwater vulnerability maps have been published covering the whole of England and Wales, and similar maps are also available for Scotland and Northern Ireland (Environment Agency 1999). However, these maps do not consider the particular vulnerability problems of carbonate aquifers.

The problems of delineating source protection zones in carbonate rock aquifers were addressed in the United States in a report by Eckenfelder Inc. (1996) but have not been the subject of specific research in Europe where protection zones are largely delineated on the basis of theoretical travel times to the source. An essential first step for protection zone delineation, and for all other hydrogeological analyses in karst, is identification of the area of the land surface that contributes recharge to the source. The term 'catchment' is sometimes applied to this area (e.g. Bonacci 1995, 2004; Environment Agency 2005) but this is not entirely appropriate as the catchment of a surface stream is usually a well-defined area of land that drains in only one direction, downslope, whereas in karst the sources (springs, wells and boreholes) may receive recharge from overlapping and/or disjunct areas on the surface. In these circumstances it may be more appropriate to use the term 'contributory areas' rather than 'catchment'. This paper addresses the problem of defining contributory areas in carbonate aquifers in the wider context of groundwater source protection zone delineation.

From: PARISE, M. & GUNN, J. (eds) *Natural and Anthropogenic Hazards in Karst Areas: Recognition, Analysis and Mitigation.* Geological Society, London, Special Publications, **279**, 97–109. DOI: 10.1144/SP279.9
0305-8719/07/$15.00

Groundwater source protection zones (SPZ)

Groundwater SPZ are usually defined on the basis of the time it takes for a pollutant to travel to the source (a well or spring used for abstraction). In Europe various zones and travel times are used ranging from 10 days in Switzerland to 100 days in Ireland. In England and Wales the Environment Agency (EA) commonly define three protection zones, with a fourth zone being defined in special cases (Environment Agency 1999, 2005). Zone 1, the inner protection zone, has a minimum 50 m protection radius around the source and a 50 day time of travel to the source. The time and distance are based principally on biological decay criteria, and are designed to protect against the transmission of toxic chemicals and water-borne disease. Zone 2, the outer protection zone, is defined as the 400 day time of travel zone, or 25% of the total 'source catchment area', whichever is the larger. The 400 day travel time is thought to be the minimum amount that slowly degrading pollutants need to be diluted, reduced in strength or delayed by the time they reach the source. Zone 3 is the total 'source catchment area' for a spring, or the area needed for effective rainfall to provide long-term recharge equivalent to the protected yield from a borehole. Where the aquifer is confined beneath lower permeability strata the source catchment area may be some distance from the spring or borehole (Environment Agency 1999). A fourth zone may be defined where local conditions mean that industrial sites and other polluters could affect the groundwater source even though they are outside the normal 'catchment'. These SPZ 'do not have statutory status, but enable the Agency to respond to various statutory and non-statutory consultations in a consistent and uniform manner' (Environment Agency 2005). In practice this means that they are designed to safeguard specific sources of water from pollution.

Defining source protection zones

Where aquifers are isotropic and homogeneous, topographic boundaries may provide a useful guide for defining groundwater basins feeding individual springs, and porous media or equivalent porous media (EPM) models may be used to establish the SPZ. However, many aquifers have a secondary, fracture, permeability and, although these are often modelled using EPM assumptions, this becomes increasingly less sustainable as the degree of heterogeneity increases. Carbonate aquifers pose particular problems because in addition to their primary (matrix) permeability, which is usually low, and higher secondary (fracture) permeability they usually possess a tertiary (channel) permeability resulting from dissolution processes. The tertiary permeability increases over time and when apertures exceed 10 mm turbulent flow is likely and the enlarged channels are termed conduits (Worthington & Smart 2004*a*). With continued dissolution, conduits may achieve sufficient size to be entered by human explorers and are then called caves. Seen in this way caves have no special hydrogeological significance and there is no basis for equating cavernous permeability with karst. Indeed, it can be argued that all carbonate aquifers that possess well-developed tertiary porosity and permeability should be regarded as karstic. These aquifers usually receive point recharge from dolines or stream-sinks and have rapid flow-through times, with the potential to transmit pollutant pulses that have little or no dilution. Microbial pollutants provide a particularly serious hazard because there is insufficient time for decay, and very high concentrations have been measured at karst springs (Gunn *et al.* 1998). Microbial contamination of a water supply borehole in carbonate rocks at Walkerton, Ontario, resulted in several deaths and more than 2000 illnesses (Worthington *et al.* 2003). Under these conditions definition of the contributory area must be as precise as is reasonably possible while recognizing that most boundaries are fuzzy.

Conceptual models for contributory area definition

In order to assess the area that contributes recharge to a spring or borehole it is necessary to have a clear conceptual model of the aquifer (Environment Agency 1999). The models used in defining SPZ in England and Wales all assume that the aquifer is homogenous and isotropic. In contrast, most models of permeability development in carbonate aquifers are based on groundwater flowing through a dendritic network of conduits and targeting a single, integrative, outlet spring. It is also recognized that a single conduit may discharge through a number of distributary springs and that these may exhibit underflow and overflow characteristics (Worthington & Smart 2004*b*). In addition to underflow/overflow behaviour within a single outlet area, the long-term evolution of carbonate aquifers often results in stacked conduits, with relict conduits at higher-levels than active conduits. During high recharge events the capacity of the lower conduit may be exceeded and the relict conduit(s) reactivated. If the higher-level relict conduits originally drained to a different outlet area from that targeted by the modern conduits then the contributory area may change (Bonacci 1995, 2004). It is also possible for point recharge to flow in two or more completely divergent directions such that springs several kilometres apart may have part of their contributory areas in common (Smart 1976 and examples later in this paper). Hence, it is clear that the natural consequences of permeability development in carbonate aquifers mean that they differ markedly from EPMs. Human activities in karst areas may add to the natural inhomogeneity of carbonate aquifers. Drainage levels and tunnels that are constructed to cause groundwater to flow in a particular direction may also have inadvertent effects, and the construction of dams/reservoirs may also have the inadvertent effect of causing divergent drainage (Bonacci 2004).

Methodologies for contributory area delineation

Three broad methods may be used to define the areas draining to springs and boreholes in carbonate rocks: (i) hydrogeological mapping; (ii) hydrological, hydrogeological and hydrochemical observations; and (iii) tracer testing (Eckenfelder Inc. 1996). Each requires knowledge and a significant input of time and resources. Groundwater models are designed to make the process simpler but it will be argued below that they must be used with great care in carbonate aquifers, and in some cases their use cannot be justified.

Hydrogeological mapping

Although the boundaries of recharge areas in carbonate rocks rarely coincide precisely with surface watersheds, an initial assessment of the likely area draining to a particular point can sometimes be obtained from consideration of topographic maps, as groundwater always flows from high points to low points. However, topographic maps become progressively less useful as the processes of karstification disrupt surface drainage. In fluviokarst areas, conduits may follow the direction of dry valleys but this is by no means always the case. Any initial assessment must also consider the overall geological structure (anticlines and synclines, faults and joints), the local structure (dip and strike), the nature of any non-carbonate interbeds and the possible location of conduit inception horizons (Lowe 2004). If there is a sufficiently dense network of boreholes in which water levels are recorded on a reasonably regular basis it may be possible to construct a piezometric map, to evaluate variations over time and to estimate the position of groundwater divides. Probably the most outstanding example is from Kentucky (USA) where Quinlan & Ray (1981) identified 28 different groundwater basins and seven sub-basins on the basis of piezometric measurements and tracer testing. However, it has been estimated that there is only a 1–2% probability of a borehole intercepting a major conduit (Smart & Worthington 2004*a*) and total reliance on maps of the piezometric surface may result in serious errors. For example, in the same area of Kentucky, Quinlan & Ewers (1989) proved by tracer testing that effluent from a sewage treatment plant was not picked up in any of the 23 wells that it passed as it flowed through a conduit en route to a major spring.

Hydrological, hydrogeological and hydrochemical observations

If the discharge of a spring, or the abstraction from a borehole, has been continuously measured for a period of time (ideally at least 1 water year) and the rainfall and evapotranspiration are also known, then it is possible to use a water balance approach to estimate the contributory area. Greater accuracy can be achieved if there are springs in the vicinity of the point of interest whose contributory areas are already well known as in this case the normalized discharge (discharge per unit area) can be used instead of the difference between estimated rainfall and evapotranspiration. The normalized discharge of a nearby surface river could also be used but care is needed as the normalized discharge of limestone springs tends to be higher than that of surface rivers because of reduced evapotranspiration. Although this method gives an approximation of the size of the area supplying recharge to the source it does not give any indication of where the boundaries actually are. The results may also be misleading if a recharge area contributes to several different groundwater sources such that the area is much larger than would be predicted from a water balance assessment on an individual source. However, where the recharge boundaries are constrained by geology, water balance studies may provide useful additional information.

Where discharge data are available for at least a groundwater year and cover a number of storm events of various sizes, then storm hydrograph and recession curve analysis may provide additional information on the flow within the aquifer (Petric 2002). This can be supplemented by analysis of water conductivity and turbidity. Conventional hydrogeological techniques such as pumping or slug tests provide useful information on matrix and, in many cases, fracture permeability but this is unlikely to be pertinent to contributory area definition in those carbonate aquifers where there is channel and conduit flow (Smart & Worthington 2004*a*).

Tracer testing

The only reliable way to establish a connection between two or more points in an aquifer is to undertake a tracer experiment, and there is a substantial literature on water tracing in carbonate aquifers (Smart & Worthington 2004*b*). Tracing is the only reliable method for establishing drainage basin boundaries for karst springs (e.g. Paviĉić & Iviĉić 1997) and for assessing protection zones around boreholes (e.g. Eckenfelder Inc. 1996; Ward *et al.* 1997). However, a single tracing experiment is of limited value because velocities vary and flow paths can change with water level in the aquifer. Hence, tracer experiments should ideally be repeated under a range of flow conditions (Stanton & Smart 1981). When designing water tracing experiments it is essential to monitor all possible outlet points and where possible to undertake quantitative tracing (Smart 1976), although this may be difficult if multiple outlets need to be monitored. Considerable care is also needed in assessing the amount of tracer to use. Equations exist to predict the amount of tracer required to produce a given concentration at a spring based on input discharge, spring discharge, distance and velocity (Worthington & Smart 2003). However, most workers aim to use the minimum amount of tracer to avoid the possibility of visual coloration at the spring and this can result in a failure to detect divergent flow. For example, if 90% of water entering sink 'X' flows to spring 'A' and the remainder to springs 'B' and 'C', then if the amount of tracer injected at 'X' is just sufficient for detection at 'A' it is likely to be below the limit of detection at 'B' and 'C'. The situation is made worse if the discharge from 'B' or 'C' is larger than from 'A'. In the example from Kentucky cited earlier, the groundwater divides were tested and verified by over 500 water-tracing experiments. These showed that most springs were part of distributary systems such that pollutants from point sources could be dispersed to up to 53 springs over a 19 km-stretch of river (Quinlan & Ewers 1989).

Case studies

Four case studies from contrasting areas illustrate the difficulties of delineating contributory areas, and hence SPZ, in carbonate aquifers.

Contributory area definition in a polygonal karst area: Waitomo, New Zealand

In the Waitomo district of central North Island, New Zealand, non-carbonate rocks overlie a gently dipping, carbonate sequence of Oligocene age that comprises an upper limestone (Otorohanga Limestone, 35–50 m), a thin sandstone (Waitomo Sandstone, *c.* 3 m) and a lower limestone (Orahiri Limestone, 35–50 m) that overlies a thick sequence of non-carbonate rocks. A six-stage sequence of landform evolution from surface drainage on the overlying non-carbonates through karst to surface drainage on the underlying non-carbonates is described in Gunn (1986). The fourth stage of this sequence is an outstanding example of polygonal karst in which the entire surface is pitted by dolines with no trace of the original surface drainage remaining. In this type of environment the only method of defining recharge boundaries is to undertake a water tracing programme, the tracer being injected at the base of dolines where there is point recharge of the aquifer. Figure 1 shows the results of such a programme in the upper Managapohue catchment (Gunn 1981). Three important conclusions can be drawn. First, the recharge boundaries do not show any relationship to the major (explorable) conduits that underlie the area. Secondly, the explorable conduits clearly have a tributary network of smaller conduits draining from the dolines. Smart (1999) refers to these as subsidiary conduits. Thirdly, the contributory area of the Cymru Stream comprises two disjunct areas between which the dolines drain to different streams. Owing to time constraints it was not possible to trace every individual doline so the results of the tracing experiments were supplemented by topographic survey to define recharge boundaries. For example, doline 34 is shown as being outside the Cymru Stream contributory area because it is a shallow doline that is topographically within the larger doline 35 which drains to Max's Cave. Similarly, dolines 54, 56 and 59 are shallow dolines that probably drain to doline 17 and JP Cave (Fig. 1). Hydrometeorological observations over a water year were used to obtain a very good water balance (imbalance <2% of rainfall) suggesting that the derived contributory areas are reliable (Gunn 1981). As this work was undertaken in a wilderness area the results are of purely academic interest but the importance of defining contributory areas in polygonal karst was well illustrated less than 1 km away at Awatiro where tracing demonstrated that piggery waste discharged into a doline was contaminating a domestic water supply spring. Fortuitously, the adjacent doline drained in a completely different direction and discharged into a large river allowing 'dilute and disperse' (Gunn 1976).

Divergent drainage from allogenic recharge: Cuilcagh Mountain, Ireland

Cuilcagh Mountain, which lies some 20 km SW of Enniskillen in County Fermanagh, Northern Ireland, forms a distinctive ridge profile. The summit (667 m) is the highest point in the county, and the summit ridge forms the international border between the United Kingdom and County Cavan in

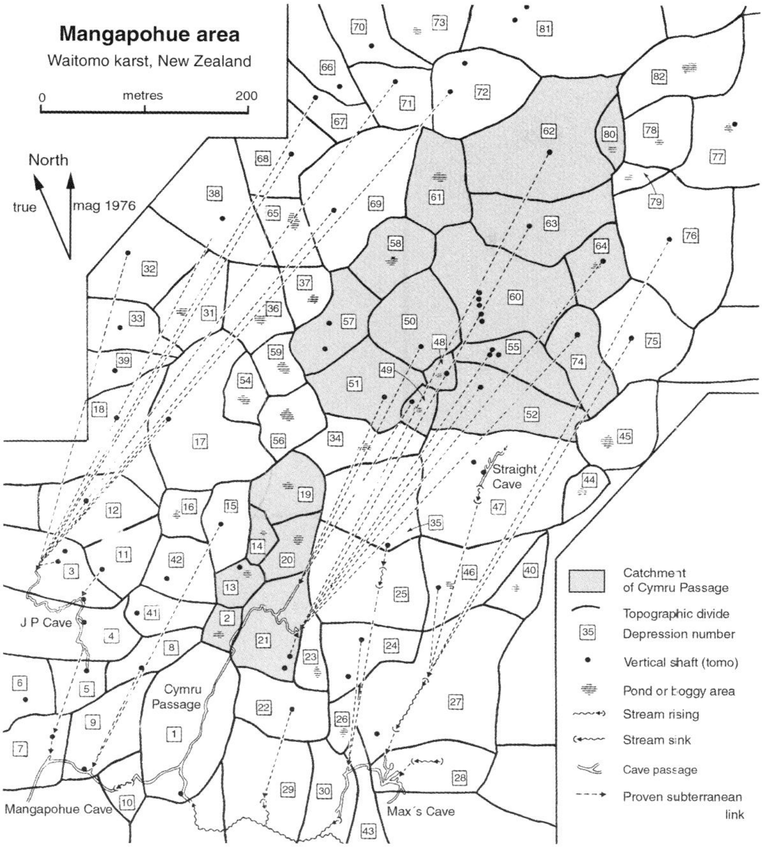

Fig. 1. Dolines and contributory area boundaries in part of the Waitomo karst, New Zealand (from Gunn 1981).

the Irish Republic. The rocks are all of Carboniferous age, with Leitrim Group sandstones and shales underlain by the Dartry Limestone Formation that has a maximum thickness 320 m and crops out over an area of approximately 58 km^2, about 50% of the northern flank of Cuilcagh Mountain. The Dartry Limestone is underlain by the Glencar Limestone Formation, a shaley limestone that has a markedly lower permeability. Karst landforms are almost entirely located within the Dartry Limestone Formation with only a few in the Glencar Limestone Formation. Although the limestone outcrop is largely continuous, Trien Mountain splits the escarpment into two distinct plateaus, East Cuilcagh and the Marlbank. The morphology of these limestone plateaus reflects the diverse lithologies that compose the Dartry Limestone Formation. On East Cuilcagh, well-bedded limestones form the top of the Dartry Limestone Formation and, as a result, the top of the plateau forms a gently dipping planar surface that is parallel to the bedding. However, on the Marlbank, the topography is dominated by knoll-type hills formed by mud-mound accumulations that reflect a Dinantian sea-floor topography with mounds that grew to over 60 m in height.

The hydrology of the Marlbank area appears deceptively simple. Numerous small streams have their origins on the Leitrim Group sandstones and shales that have a low permeability and most drain in a northerly direction to form three main rivers, the Owenbrean, Aghinrawn and Sruh Croppa, which flow for about 5–6 km before crossing on to the Dartry Limestone (Fig. 2). Surface flow is maintained for a variable distance before the rivers sink at the end of major blind valleys and the water from these sinks drains to the Marble Arch Rising, close to the base of the Dartry limestones. There are many other springs, the largest being Tullyhona, Cascades, Cladagh Glen, Hanging Rock, Marlbank, Ture and Shannon Pot, the legendary source of Ireland's longest river (Gunn 1996). These are fed by smaller sinking streams and by both concentrated and dispersed recharge to the limestone. An initial attempt to define contributory areas for the larger springs was made by Gunn (1982) using water-tracing experiments. However, further tracing experiments in the 1990s showed that the drainage is much more complicated (Gunn 1997). In particular, drainage from a stream that sinks at Pigeon Pot on East Cuilcagh (Site 16, Fig. 2) and from a nearby smaller sink (Badger Pot, Site 18, Fig. 2) has been shown to flow in completely divergent directions targeting the Sumera (SE) and Gortalughany (ESE) risings in the East Cuilcagh karst, as well as draining west to Shannon Pot and NW to Cascades Rising in the Marlbank karst (Fig. 2). The direction of drainage, distances, hydraulic gradients and straight-line flow velocities are given in Table 1, and it can be seen that conduits from Pigeon Pot fan out over a 225° arc and extend over distances in excess of 10 km. On the basis of exploration of the Pigeon Pots it is known that the vadose zone is more than 55 m thick under normal conditions and at this depth the conduits are too small for human access. The conduits are also unable to transport all the recharge under flood conditions and water rises up the pots reducing the thickness of the vadose zone to approximately 20 m. These changes in water level do not appear to affect the direction of water movement and in this case the divergent drainage appears to be a consequence of the elevation of the piezometric surface, local geology (particularly fracturing and faulting) and glacial down-cutting, which has lowered local base levels (Brown 2005).

The conduits from East Cuilcagh to Shannon Pot and to Cascades Rising are of particular interest as they pass beneath extensive areas of non-carbonate rock supporting surface drainage that sinks lower on the mountain and eventually targets the Marble Arch Rising. Thus, both Shannon Pot and Cascades Rising have two disjunct contributory areas, a proximal area that extends out from each rising and a distal area on East Cuilcagh. The distal contributory areas overlap in the region of Pigeon Pots, but the East Cuilcagh contributory area for Cascades is larger and extends further to the north than that for Shannon Pot.

The contributory area for the Cascades Rising is further complicated because it has been shown that under low-flow conditions stream-sinks close to the sandstone–limestone boundary capture the entire flow of the upper Owenbrean River and divert it to Cascades Rising (Fig. 2). As discharge increases the capacity of the conduit(s) that drain to the Cascades Rising is exceeded and the excess flow is discharged from a series of springs about 200 m downstream of the Upper Sinks. These springs are adjacent to the channel of the Owenbrean River,

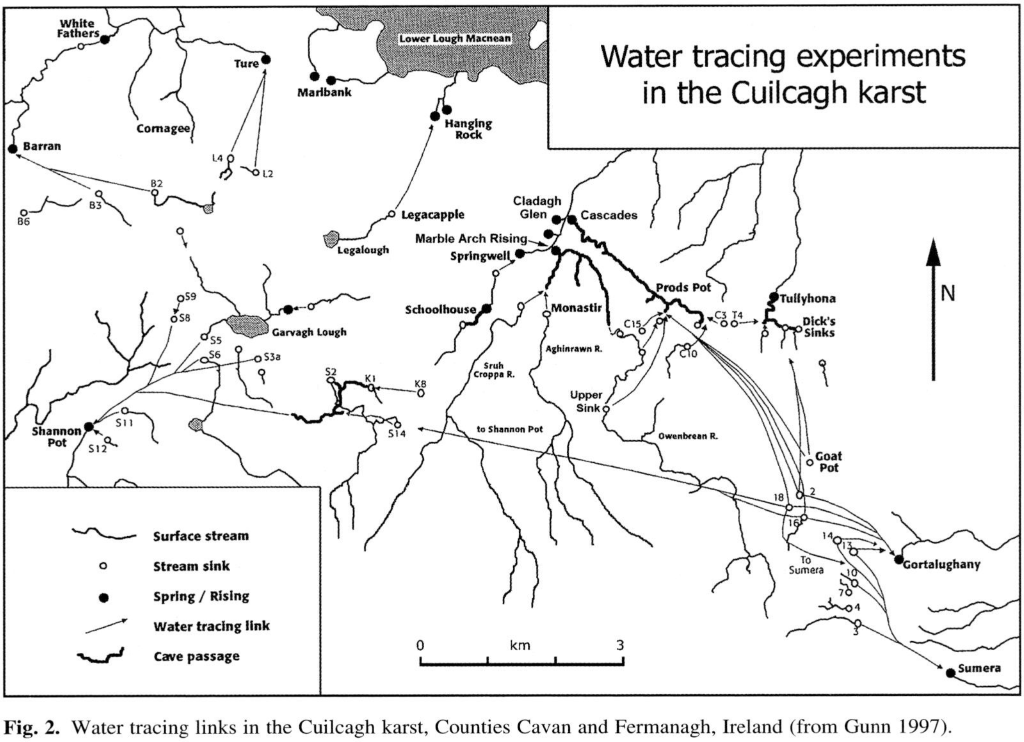

Fig. 2. Water tracing links in the Cuilcagh karst, Counties Cavan and Fermanagh, Ireland (from Gunn 1997).

Table 1. *Springs in the Cuilcagh Karst, Ireland that are targeted by flow sinking at Pigeon Pot (direction, distance and flow velocity are measured on a straight line from sink to spring)*

Spring	Direction (degrees from N)	Distance (m)	Hydraulic gradient (%)	Flow velocity (m h^{-1})
Gortalughany Risings	120	1500	3.9	41
Sumera Risings	140	3115	5.1	51
Shannon Pot	275	10386	1.8	32
Cascades Rising	325	5623	3.2	20

which is dry over the reach between the springs and the Upper Sinks. If the discharge of the upper Owenbrean River continues to increase, the capacity of the conduit(s) taking flow to the springs is also exceeded, the Upper Sinks are overtopped and submerged, and continuous surface flow drains to the Pollasumera sink and, thence, to the Marble Arch Rising. Thus, the contributory area for the Cascades Rising is constant but that of the Marble Arch Rising varies temporally with flow in the upper Owenbrean River. Until 2000 the Cascades Rising was an important public water source (PWS) but concern over catchment protection led to it being withdrawn from use. Subsequently, the water was used by a fish hatchery and in January 2001 there was a major pollution incident caused by slurry entering a doline in the proximal catchment. This resulted in a fish-kill at the hatchery but the consequences would have been many times more serious had the spring still been a PWS.

Divergent drainage from autogenic recharge: Castleton, Derbyshire, UK

Castleton lies close to the NE boundary of the English 'White Peak' District, an area underlain by limestones of Dinantian (Carboniferous) age. To the north the limestones dip steeply beneath Namurian sandstones and shales that are overlain by Quaternary solifluction deposits. A series of streams with a combined catchment of approximately 5 km^2 (Hardwick 1995) flow over these deposits and sink close to the limestone boundary. Repeated tracing experiments have shown that water from the sinks flows through the Speedwell Cavern system and emerges from two springs on opposite sides of the Peakshole Water, Slop Moll and Russet Well (Gunn 1991). The Peakshole Water is a tributary of the River Noe and has its source upstream of Slop Moll and Russet Well, at Peak Cavern Rising. Under low–moderate flows this rising discharges water of solely autogenic origin from an approximately 8.4 km^2 contributory area in which the limestones are overlain by soils of loessic origin (Hardwick 1995). Peak Cavern and Speedwell Cavern are part of the same, multi-level, cave system, the present-day (active) Speedwell conduit being developed on a lower level to that on which the present-day Peak conduit has formed. Under conditions of high recharge the lower part of the Speedwell conduit is unable to transport all the flow and water backs up and enters the Peak conduit at a number of discrete points, eventually emerging from Peak Cavern Rising. Hence, this rising may be regarded as being an overflow spring for the Speedwell water. Thus, under high-flow conditions the three risings have a common allogenic catchment. However, the autogenic (limestone) contributory area is more difficult to define.

The first tracing experiments in the limestone area were undertaken at Eldon Hole, an open shaft, and at a doline in an area of Eldon Hill used as a motorcycle scrambling track. In both cases it was necessary to inject tracer using a bowser and flow was towards Speedwell Cavern proving that this conduit receives autogenic recharge in addition to allogenic recharge from the stream-sinks. No tracer was recovered from Peak Cavern. The third trace was undertaken to the east of Eldon Hill with the aim of determining the route taken from Dirtlow Pit, an open-cast fluorspar mine (Fig. 3). In addition to the Castleton springs fluocapteurs were placed in springs further to the east in the Bradwell area, and it was shown that drainage targeted both Peak Cavern and three sites near Bradwell: Kronstadt Sough (a sough is a lead mine drainage level), Springfield Rising and Brookside Sough. However, no tracer was recovered from Bagshaw Resurgence, the main spring in the Bradwell area (Fig. 3).

In 1999, cavers reported an orange–red staining in the water emerging from Ink Sump into Lake Passage, a major tributary in Peak Cavern (Fig. 3). The water flows through the Peak Cavern system, and the main streamway was stained orange–red for approximately 1000 m downstream of Ink Sump. The area around the emerging water in Lake Passage was covered in a gelatinous biofilm that sloughed off when disturbed. Cave divers confirmed that the material was present in two sumps downstream, Buxton Water Sump and the Peak Cavern Rising Sump. Although the latter is the source of the Peakshole Water no visible evidence of the pollution was detectable in the watercourse outside of the cave. Detailed examination of the autogenic contributory area suggested that the most likely source of the pollution was 'agricultural material' being stored on unimproved pasture prior to spreading on the land. This material was a mixture of pulp from the production of paper and organic-rich peat sludge from a water treatment works. Runoff from the material was similar in colour to that in Peak Cavern and was observed to sink in an area of closed depressions (Wood *et al.* 2002). The Environment Agency commissioned a tracing experiment to determine the pathway of drainage from the depressions, and to ensure an unequivocal result two tracer dyes, sodium fluorescein (CI 45350 Acid Yellow 73) and Rhodamine-WT (CI Acid Red 388), were injected simultaneously into a doline that received effluent runoff (Fig. 3). As in the Dirtlow Pit experiment, water entering the closed depressions moved in two main directions, eastwards towards Bradwell and northwards towards Peak Cavern (Fig. 3). Very small amounts of tracer were also recovered from two other sites. Flow velocities were of the order of 240–500 m day^{-1}. Two of the Bradwell recovery points are lead mine soughs and it is possible that the third is also fed by a sough. Hence, it is possible that water which originally targeted Peak Cavern via Dirtlow Rake and Ink Sump has been partially captured by soughs driven west from the Bradwell area. Alternatively, there may have been natural drainage to the Bradwell area that has been captured by the soughs. Irrespective of the cause, it is clear that the Castleton and Bradwell springs, which are about 3 km apart, have part of their autogenic contributory areas in common. However, on the basis of the four experiments it would appear that the autogenic contributory area for Slop Moll and Russet Well does not overlap with the autogenic contributory area for the Peak Cavern Rising.

Source protection zones in the Buxton area, Derbyshire, UK

Buxton lies to the west of Castleton at the NW boundary of the White Peak. The hydrogeology is broadly similar, with allogenic recharge from streams that rise on Namurian sandstones and shales and sink close to the limestone boundary, and autogenic recharge through a soil cover (Fig. 4). Consultants acting for the Environment Agency (EA) have drawn SPZ for four potable water sources in this area: the Portobello Borehole, St Anne's Well, the Stanley Moor Borehole

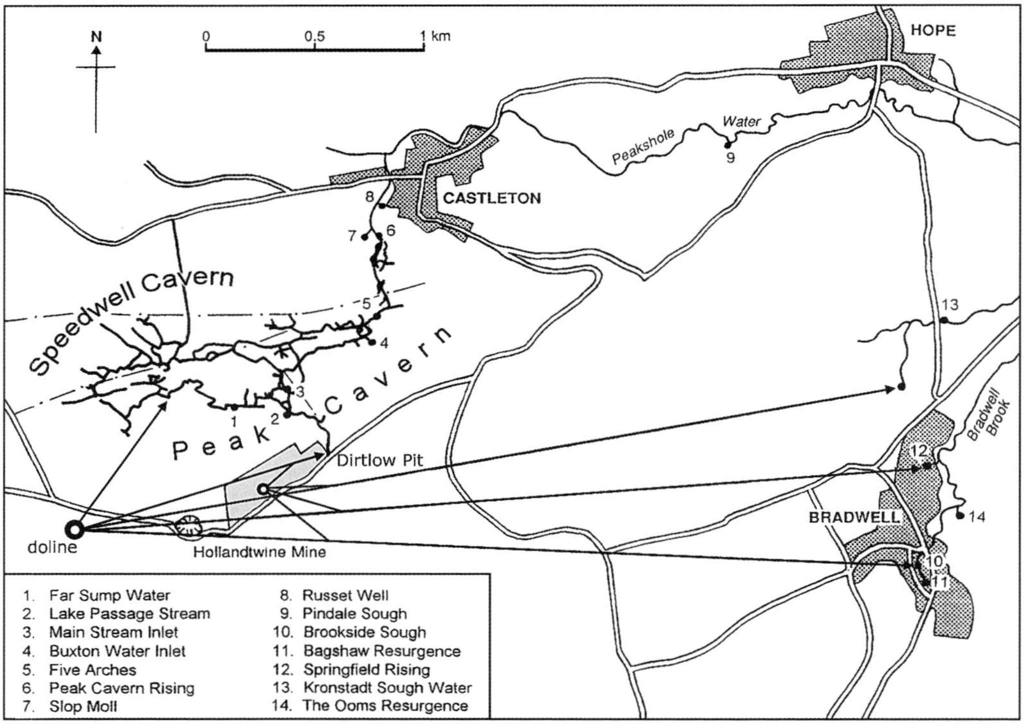

Fig. 3. Water tracing links in the Castleton–Bradwell area, Derbyshire, UK.

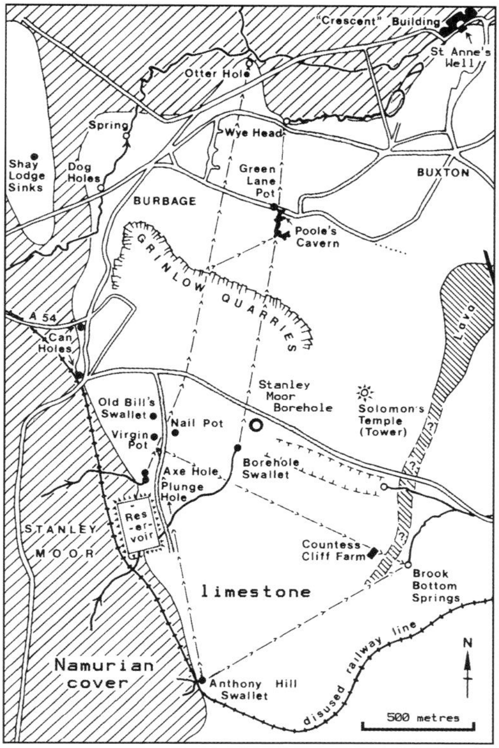

Fig. 4. Water tracing links in the Buxton karst, Derbyshire, UK (from Ford & Gunn 1992).

and the Staden Borehole (referred to as the H&H Borehole in the EA files). The reports are unpublished but available for consultation at the EA Sutton Coldfield Office. The zones are shown on the EA website (Environment Agency 2005) and Figure 5 provides a simplified representation.

The Portobello Borehole is 205 m deep and is entirely in Namurian sandstones and shales. On the EA website the 'total catchment' is shown as overlapping with the 'total catchment' for the Stanley Moor Borehole (Fig. 5). However, as there is a major fault between the two boreholes, and the Namurian strata to the west have been downthrown by approximately 1 km relative to the limestones to the east, it seems very unlikely that the catchments do overlap. Indeed, there is no conceivable mechanism by which the Portobello Borehole could draw water from the limestones and, hence, this source is not considered further in the present paper. However, it is notable that the same assumptions have been used to model flow in the sandstones and in the limestones.

St Anne's Well is a thermal complex in the centre of Buxton (Figs 4 & 5) that is thought to comprise a single deep source with multiple outlets, including at least two major fissures in the underlying limestone (Barker *et al.* 2000). The source has been famous since Roman times and part of the discharge is bottled as 'Buxton Mineral Water', one of the leading natural mineral water brands in the UK. The spring has had a constant temperature of 27 °C since it was first measured and isotope measurements show that it is several thousand years old. Calculations in Barker *et al.* (2000) indicate that the water circulates to a depth of more than 1 km. They suggest that recharge is from the limestone outcrop of the surrounding Peak District and, although the area is not specified, an unpublished report by Hydrotechnica shows a catchment to the SE of Buxton. Gunn *et al.* (2006) obtained a broadly similar depth of circulation (1–1.5 km) but suggest, on the basis of isotope and geochemical data, that the Buxton thermal water has migrated from deep sandstone aquifers below the high ground to the west of Buxton before finally discharging through fissured limestone at a topographic low. On the basis of other, unpublished, information the EA consultants arrived at a similar conclusion and suggested that the majority of the discharge from St Anne's Well is likely to be groundwater from a deep, inter-regional flow system. They also argued: (a) that it is not possible to define protection zones for this distant catchment; (b) that 'the travel times involved (>5000 years) indicate that the application of standard SPZ delineation procedures will not be wholly appropriate for this source'; and (c) that there may be mixing of this distant water with water from a localized flow system and hence protection zones should be derived for the local catchment. To do this they followed a water balance approach, defining the catchment on the assumption that local flow contributes 5% of an assumed yield. This figure was divided by the estimated recharge to obtain a catchment area of 0.116 km^2 that was assumed to be circular. A circle with an area of 0.116 km^2 would have a radius of 193 m. However, as this radius would have included areas underlain by Namurian sediments it was progressively increased to 330 m, giving a circle of area 0.342 km^2 that encloses 0.116 km^2 of limestone. It is interesting that although an attempt was made to exclude areas underlain by non-limestones from the catchment, similar consideration was not given to the extent of the urban area in which the source is located. Most of the derived protection zone (Fig. 5) is overlain by tarmac and other impermeable surfaces that direct drainage to the local storm sewer system so the amount of direct recharge to the limestone will be very small and a much larger area would be required to provide the necessary recharge. The SPZ were calculated using a saturated thickness of

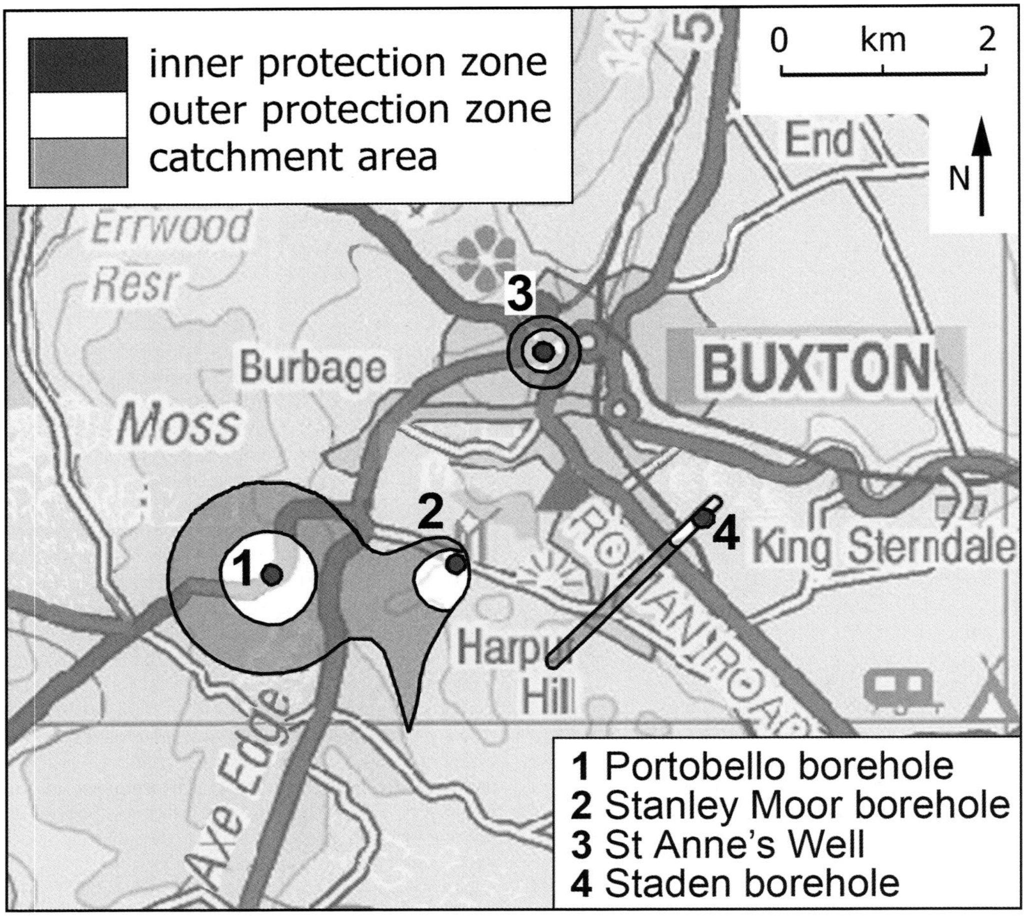

Fig. 5. Source protection zones in the Buxton area (from Environment Agency 2005).

75 m, a porosity of 1%, and 50 and 400 day time of travel for the Inner and Outer zones, respectively. The results are shown as circles with radii of 68 and 193 m, respectively, implying hydraulic conductivities of 1.36 and 0.48 m day^{-1}. These are typical of values for fractures but the hydraulic conductivity in channels is several hundred times higher (Worthington & Smart 2004*a*), and water tracing experiments in the Buxton area (Gunn & Edmans 1989) have shown that the hydraulic conductivity in conduits may be over 1000 times higher. However, the size of the zones is probably irrelevant as there is no evidence of any mixing of the deeper water with local recharge.

Staden Borehole is 2.2 km to the SE of St Anne's Well and approximately 2 km from the nearest sandstone–shale outcrop (Fig. 5). There are no sinking streams in the area and the only surface evidence of karst processes is provided by a few shallow dolines, so it is reasonable to assume that the area is dominated by dispersed autogenic recharge through the soil and loessic cover deposits that range from 1 to 3 m in thickness. The borehole is 150 m deep and water levels range from 60 to 82 m below ground level. The EA consultants estimated the total catchment zone to be 0.252 km^2 using a water balance approach (annual abstraction/annual effective recharge). The geometry of the protection zones was determined using two formulae:

$$YL = Q/kbi \quad (1)$$

$$XL = Q/2\pi kbi \quad (2)$$

where YL is the width of the catchment (calculated as 128 m); XL is the distance to stagnation point (calculated as 20.3 m); Q is abstraction rate (480 m^3 day^{-1}); k is hydraulic conductivity

(2.5 m day^{-1}); b is aquifer thickness (75 m); and i is hydraulic gradient (2%). The EPA Well Head Protection Area (WHPA) model was used to confirm the dimensions and to provide guidance on the form of the zone. This model gives an elongate flow tube aligned parallel with the groundwater flow direction and bounded at the upgradient end by an approximate semi-circle. The inner zone was calculated to have a length of 16 m that was manually adjusted to give a 50 m radius circle around the borehole. The outer zone was calculated to have an area of 63 203 m^2, and hence a length of 494 m, and the length of the total catchment was estimated as 1970 m (Fig. 5).

As part of recent developments in the inner SPZ the author undertook two tracer tests from soakaways situated to the SW of the borehole and within the inner protection zone. The tracer was not recovered from the borehole but took less than 7 days to reach a spring approximately 800 m away on the banks of the River Wye, giving a velocity of at least 110 m day^{-1}. According to oral accounts (Mr P. Hockenhull, owner, pers. comm.) the borehole was dry when it was originally drilled and a small amount of explosive was detonated at the base in an effort to improve the fracturing. Immediately following the detonation, water entered the borehole and rose up it to approximately 75 m below ground level suggesting that a conduit or conduits had been intercepted. At the time of the tracing experiment the water level in the borehole was over 70 m below ground level and it is suggested that the tracer followed a lateral route through the epikarst/vadose zone past the borehole rather than descending essentially vertically to the 'water table'. Thus, the SPZ is not serving to protect the borehole. Further unpublished research suggests that the borehole has intersected a large body of deep groundwater that drains broadly SE from Buxton. The concept of a shared catchment is important here as the deeper groundwater provides base flow to the River Wye over a reach of several kilometres via springs in its banks and bed. It is probable that most of the catchment is outside the SPZ that has been defined by the EA, although part of catchment may be within the outer SPZ.

The Stanley Moor Borehole is 177 m deep and is located approximately 2 km to the SSW of St Anne's Well (Figs 4 & 5). The licensed maximum daily abstraction (q) is 2182 m^3 day^{-1} and the SPZ was derived using a FLOWPATH 5.12 model with an assumed saturated thickness (b) of 300 m and porosity (n) of 0.01. The inner zone (50 day travel time) was assumed to be a circle with a theoretical area (A_i) derived by:

$$A_i = (q \times 50)/(b \times n). \quad (3)$$

This yields an area of 36 366 m^2, giving a circle of radius 107 m. The EA 'Source Evaluation Report' gives the theoretical area of the outer zone (A_o, 400 day travel time) as 149 000 m^2 but it is not clear how this has been derived as application of equation (3) yields an area of approximately 291 000 m^2. The FLOWPATH model shows an ellipsoid shape with its axis orientated NE–SW. The Source Evaluation Report states that the FLOWPATH 'total catchment zone' was adjusted so that its boundary coincides with the major faulted boundary between the limestone and Namurian sandstones and shales. However, the EA website map (Fig. 5) shows the boundary as extending across the fault and overlapping with that of the Portobello Borehole.

The area within the 'total catchment zone' of the Stanley Moor borehole is the most obviously karstic of the three SPZ discussed in this paper as it contains many dolines and several sinking streams. Tracing experiments summarized by Gunn & Edmans (1989) demonstrate divergent flow to three main springs (Otter Hole, Wye Head and Brook Bottom) at distances of up to 2.8 km (Fig. 4) and with velocities of up to 1750 m day^{-1}. It was not possible to monitor the borehole during these experiments as no pumping was being undertaken at the time. However, one of the sinks, Borehole Swallet, is located a few tens of metres from the borehole and the flowpath from the sink to the Wye Head Resurgences is via Poole's Cavern. The manager of Poole's Cavern reported that when pumping was taking place from the Stanley Moor borehole the stream flowing through the cave reduced markedly in volume (Mr D Alsop, former cave manager, pers. comm.). This indicates a degree of connectivity between the sinking streams and the borehole that renders the SPZ almost meaningless unless the entire allogenic catchment is included in the Inner Protection Zone.

Discussion and conclusions

This paper has highlighted the problems of defining contributory areas for springs, and hence SPZ, in carbonate aquifers. Examples from the Peak District and Ireland show that concentrated recharge, both allogenic and autogenic, may target multiple springs several kilometres apart. Hence, SPZ need to take into account the potential for both shared and disjunct source contributory areas. The idea of disjunct contributory areas to boreholes and springs may seem novel but it is implicit in the work of Tóth (1963) who identified local, intermediate and regional flow paths. Recharge onto the contiguous contributory area contributes to local flow and recharge onto the disjunct areas contributes to intermediate flow (as is the case on East Cuilcagh) or regional flow (as is the case at St Anne's Well). One important consequence is that in carbonate aquifers the simple water balance equation 'area = yield/recharge' must be regarded as giving the minimum contributory area for a particular source and in many cases the actual area will be much larger because recharge is targeting several discharge/abstraction points. In these aquifers tracer testing represents the only reliable method for defining those areas on the land surface that contribute recharge to a particular source, and for determining the direction and velocity of aqueous pollutant movement. It is particularly important to undertake tracing experiments under different flow conditions as disjunct contributory areas are likely to be dynamic, only contributing flow to a particular source when a threshold is exceeded. The contiguous contributory area for a source may also be dynamic. For example, in the Cuilcagh karst the Upper Owenbrean River is always part of the contributory area for the Cascades Rising but is only part of the contributory area for the Marble Arch Risings when the discharge exceeds the capacity of the conduits that take flow to the Cascades Rising.

It is sometimes assumed that carbonate aquifers can be divided into those that have been karstified (as evidenced by surface landforms) and those that have not. In published accounts attention has been focused primarily on the more obviously karstified aquifers that receive concentrated recharge from sinking streams and dolines, and most authors would accept that there can be no justification for applying EPM models to these aquifers. Most of the examples presented in this paper are in areas that are widely accepted as being karstic but it is notable that the immediate area around the Staden Borehole has dispersed autogenic recharge and very few surface karst landforms. However, tracer testing has demonstrated very rapid flow and it seems likely that the epikarst and shallow vadose zone comprise a perched aquifer that drains laterally in a different direction to channels/conduits in the deeper phreatic zone. Those carbonate aquifers in which there is only dispersed recharge and few, or no, surface karst landforms are sometimes assumed to be 'non-karstic' but are still likely to possess triple porosity/permeability and overlapping contributory areas, as shown by Smart (1976) in the Jurassic limestone of the Cotswold Hills in western England. Hence, application of EPM models to define SPZ in these aquifers may lead to serious error.

In conclusion, it is suggested that the majority of carbonate aquifers have been subjected to some degree of karstification, even if this is not manifest by surface landforms. Hence, unthinking use of EPM models to define SPZ may lead to serious error. An alternative approach of using repeated tracing experiments to define contiguous and disjunct contributory areas is advocated.

Thanks to T. Waltham for preparing the figures for publication and to L. Brown, D. Lowe and S. Worthington for critical and constructive comments on an earlier draft.

References

BARKER, J. A., DOWNING, R. A., GRAY, D. A., FINDLAY, J., KELLAWAY, G. A., PARKER, R. H. & ROLLIN, K. E. 2000. Hydrogeothermal studies in the United Kingdom. *Quarterly Journal of Engineering Geology and Hydrogeology*, **33**, 41–58.

BONACCI, O. 1995. Ground water behaviour in karst: example of the Ombla Spring (Croatia). *Journal of Hydrology*, **165**, 113–134.

BONACCI, O. 2004. Hazards caused by natural and anthropogenic changes of catchment area in karst. *Natural Hazards and Earth System Sciences*, **4**, 655–661.

BROWN, L. 2005. *Inception and subsequent development of conduit in the Cuilcagh karst, Ireland.* Ph.D. thesis, University of Huddersfield.

COST 65. 1995. *Hydrogeological Aspects of Groundwater Protection in Karstic Areas.* Final Report (COST action 65). European Commission Directorate-General XII Science, Research and Development, Report EUR 16547 EN, Brussels.

DALY, D. & DASSARGUES, A. *ET AL.* 2002. Main concepts of the 'European approach' to karst-groundwater-vulnerability assessment and mapping. *Hydrogeology Journal*, **10**, 340–345.

ECKENFELDER INC. 1996. *Guidelines for Wellhead and Springhead Protection Area Delineation in Carbonate Rocks.* Prepared for US EPA Region 4.

ENVIRONMENT AGENCY. 1999. Groundwater Source Protection Zones. Only available as a free download from: http://publications.environment-agency.gov.uk/epages/eapublications.storefront/

ENVIRONMENT AGENCY. 2005. Groundwater source protection zones. http://www.environment-agency.gov.uk/maps/info/groundwater/ (SPZ for particular areas can be obtained from this site).

FORD, T. D. & GUNN, J. 1992. *Caves and karst of the Peak District* (2nd edn). British Cave Research Association. Cave Studies Series, No 2.

FORD, D. C. & WILLIAMS, P. W. 1989. *Karst Geomorphology and Hydrology*. Unwin Hyman, London.

GOLDSCHEIDER, N. 2005. Karst groundwater vulnerability mapping: application of a new method in the Swabian Alb, Germany. *Hydrogeology Journal*, **13**, 555–564.

GUNN, J. 1976. Water pollution in caves. *Bulletin of the New Zealand Speleological Society*, **99**, 557–562.

GUNN, J. 1981. Limestone solution rates and processes in the Waitomo District, New Zealand. *Earth Surface Processes*, **6**, 427–445.

GUNN, J. 1982. Water tracing in Ireland: A review with special reference to the Cuilcagh Karst. *Irish Geography*, **15**, 94–106.

GUNN, J. 1986. Solute processes and karst landforms. *In*: TRUDGILL, S. T. (ed.) *Solute Processes*. Wiley, Chichester, 363–437.

GUNN, J. 1991. Water tracing experiments in the Castleton karst, 1950–1990. *Cave Science*, **18**, 43–46.

GUNN, J. 1996. Source of the River Shannon, Ireland. *Environmental Geology*, **27**, 110–112.

GUNN, J. 1997. A brief summary of the karst hydrology of Cuilcagh Mountain. *In*: JONES, G. L., BURNS, G., FOGG, T. & KELLY, J. (eds) *The Caves of Fermanagh and Cavan*. Lough Nilly Press, Florencecourt, Northern Ireland, 60–61.

GUNN, J. & EDMANS, A. M. 1989. The Wye Head systems: some hydrological observations. *Caves & Caving*, **45**, 35.

GUNN, J., TRANTER, J., PERKINS, J. & HUNTER, C. 1998. Sanitary bacterial dynamics in a mixed karst aquifer. *In*: LEIBUNDGUT, C., GUNN, J. & DASSARGUES, A. (eds) *Karst Hydrology*. IAHS Publication, **247**, 61–70.

GUNN, J., BOTTRELL, S. H., LOWE, D. J. & WORTHINGTON, S. R. H. 2006. Deep groundwater flow and geochemical processes in limestone aquifers: evidence from thermal waters in Derbyshire, England. *Hydrogeology Journal*, **14**, 868–881.

HARDWICK, P. 1995. *The impact of agriculture on limestone caves with special reference to the Castleton Catchment, Derbyshire*. Ph.D. thesis, Manchester Metropolitan University.

LOWE, D. J. 2004. Inception of caves. *In*: GUNN, J. (ed.) *Encyclopedia of Caves and Karst Science*. Fitzroy Dearborn, London, 437–441.

PAVIĈIĆ, A. & IVIĈIĆ, D. 1997. Drainage basin boundaries of major karst springs in Croatia determined by means of groundwater tracing in their hinterland. *In*: KRANJC, A. (ed.) *Tracer Hydrology 97*. Balkema, Rotterdam, 273–278.

PETRIC, M. 2002. *Characteristics of Recharge–Discharge Relations in Karst Aquifers*. ZRC SAZU, Postojna-Ljubljana.

QUINLAN, J. F. & EWERS, R. O. 1989. Subsurface drainage in the Mammoth Cave area. *In*: WHITE, W. B. & WHITE, E. L. (eds) *Karst Hydrology: Concepts from the Mammoth Cave Area*. Van Nostrand Reinhold, New York, 65–103.

QUINLAN, J. F. & RAY, J. A. 1981. *Groundwater Basins in the Mammoth Cave Region, Kentucky*. Friends of the Karst, Occasional Publication, **1** (map).

SMART, C. C. 1999. Subsidiary conduit systems: A critical hiatus in aquifer monitoring and modelling. Modelling in karst aquifers. *In*: PALMER, A. N., PALMER, M. V. & SASOWSKY, I. D. (eds) *Karst Modelling*. Karst Waters Institute, Charles Town, W.V., 146–157.

SMART, C. C. & WORTHINGTON, S. R. H. 2004*a*. Groundwater in Karst: Borehole hydrology. *In*: GUNN, J. (ed.) *Encyclopedia of Caves and Karst Science*. Fitzroy Dearborn, London, 397–398.

SMART, C. C. & WORTHINGTON, S. R. H. 2004*b*. Water tracing. *In*: GUNN, J. (ed.) *Encyclopedia of Caves and Karst Science*. Fitzroy Dearborn, London, 769–771.

SMART, P. L. 1976. Catchment delimitation in karst areas by the use of quantitative tracer methods. *In*: *Proceedings of the Third International Symposium of Underground Water Tracing*, Bled, Yugoslavia, 291–298.

STANTON, W. I. & SMART, P. L. 1981. Repeated dye traces of underground streams in the Mendip Hills, Somerset. *Proceedings of the University of Bristol Speleological Society*, **16**, 47–58.

TÓTH, J. 1963. A theoretical analysis of groundwater flow in small drainage basins. *Journal of Geophysical Research*, **68**, 4795–4812.

WARD, R. S., WILLIAMS, A. T. & CHADHA, D. S. 1997. The use of groundwater tracers for assessment of protection zones around water supply boreholes – A case study. *In*: KRANJC, A. (ed.) *Tracer Hydrology 97*. Balkema, Rotterdam, 369–376.

WOOD, P. J., GUNN, J. & PERKINS, J. 2002. The impact of pollution on aquatic invertebrates within a subterranean ecosystem – out of sight out of mind. *Archiv für Hydrobiologie*, **155**, 223–237.

WORTHINGTON, S. R. H. & SMART, C. C. 2003. Empirical determination of tracer mass for sink to springs tests in karst. *In*: BECK, B. F. (ed.) *Sinkholes and the Engineering and Environmental Impacts of Karst. Proceedings of the Ninth Multidisciplinary Conference, Huntsville, Alabama*. American Society of Civil Engineers, Geotechnical Special Publications, **122**, 287–295.

WORTHINGTON, S. R. H. & SMART, C. C. 2004*a*. Groundwater in Karst: Conceptual models. *In*: GUNN, J. (ed.) *Encyclopedia of Caves and Karst Science*. Fitzroy Dearborn, London, 399–401.

WORTHINGTON, S. R. H. & SMART, C. C. 2004*b*. Springs. *In*: GUNN, J. (ed) *Encyclopedia of Caves and Karst Science*. Fitzroy Dearborn, London, 699–703.

WORTHINGTON, S. R. H., SMART, C. C. & RULAND, W. W. 2003. Assessment of groundwater velocities to the municipal wells at Walkerton. *In*: STOLLE, J., PIGGOTT, A. R. & CROWDER, J. J. (eds) *Proceedings of the 55th Canadian Geotechnical and 3rd Joint IAH-CNC and CGS Groundwater Specialty Conferences*, Niagara Falls, Ontario, October 20–23, 2002. Southern Ontario Section, Canadian Geotechnical Society, 1081–1086.

Rapid karstic bypass flow in the unsaturated zone of the Yorkshire chalk aquifer and implications for contaminant transport

S. J. L. ALLSHORN, S. H. BOTTRELL, L. J. WEST & N. E. ODLING

Institute of Geological Sciences, School of Earth and Environment, University of Leeds, Leeds LS2 9JT, UK (e-mail: s.allshorn@see.leeds.ac.uk)

Abstract: Tracer tests have been performed on the unsaturated zone of the East Yorkshire chalk aquifer, UK. Rapid tracer travel times through significant thicknesses of unsaturated chalk (15–38 m) indicate that bypass flow must occur through fractures. Transport processes in the unsaturated zone of the chalk aquifer thus have similarities to those in the vadose zone of more typically karstic limestone aquifers. Modelling of tracer breakthrough curves indicates that bypass flow is sufficiently rapid to significantly inhibit diffusional loss of tracer into the porous matrix of the chalk. The presence of rapid karstic bypass flow in the unsaturated zone of the chalk will limit the potential for attenuation of groundwater contaminants in this zone.

Groundwater quality is increasingly under threat from rising levels of pollution, especially from nitrates and pesticides. The Cretaceous chalk aquifer (Fig. 1) is the most important in the UK, supplying some 55×10^6 m^3 year^{-1} through public and private abstractions. Nitrate and pesticide concentrations in this aquifer have been steadily rising over past decades, probably as a result of the increased use of these agrichemicals that took place in the middle part of the last century. Prediction of future trends in groundwater nitrate concentration requires understanding of the mechanisms by which this contaminant is delivered to the saturated zone of the aquifer from which groundwater is abstracted.

Much of the chalk aquifer is characterized by a thick unsaturated zone and flow processes that transport contaminants within that zone are the subject of this study. In particular, the chalk aquifer matrix is often highly fractured and these fractures provide potential pathways for unsaturated groundwater flow. Solutional enlargement of fractures by flowing groundwater may enhance bypass fracture flow. The unsaturated zone of the chalk thus has many similarities with the vadose zone of more typical limestone karst aquifers.

We have employed tracer tests, similar to those that have often been utilized with great success in karst systems, to investigate the role of flow in the fracture network of the unsaturated chalk. However, because the carbonate matrix of the chalk is highly porous, the unsaturated zone of the chalk is more hydrogeologically complex than typical limestone karst aquifers, particularly in respect of solute transport. The small pore sizes of the chalk matrix (typically <1 μm) ensure that even in the unsaturated zone the matrix is close to fully water saturated. There is thus potentially highly mobile groundwater in fractures and far less mobile groundwater in the matrix blocks between them. Solutes can be exchanged between these two types of groundwater by diffusion across the fracture surfaces.

Nitrate problem

Figure 2 shows the temporal trend in nitrate concentration in saturated zone groundwater abstracted at a typical chalk borehole in East Yorkshire. There is a clear seasonality with elevated concentrations occurring during the spring months, although this signal is suppressed in some years relative to others. There is also a general rise in annual minimum nitrate concentrations after 1997. The strong seasonal signal can potentially be explained by two very different contaminant transport mechanisms for nitrate (and other agrichemical pollutants) from application at surface to the watertable.

- Annual fluctuations in the watertable can lead to a flushing of contaminants from matrix pore water into fracture water during periods when the watertable is elevated (see Fig. 3a–c) (Chilton *et al.* 1997). At times of high watertable, nitrate is eluted from the matrix blocks into the 'clean' fracture waters that then migrate towards groundwater abstractions. This requires that the matrix porewater close to the watertable is recharged with nitrate-contaminated water by gradual 'piston flow' from the surface. Recharge to the chalk aquifer is highly seasonal, occurring mostly between January and March, thus high watertable and flux of nitrate from the matrix occur in the spring months, producing the trend seen in Figure 2.

From: Parise, M. & Gunn, J. (eds) *Natural and Anthropogenic Hazards in Karst Areas: Recognition, Analysis and Mitigation.* Geological Society, London, Special Publications, **279**, 111–122. DOI: 10.1144/SP279.10
0305-8719/07/$15.00

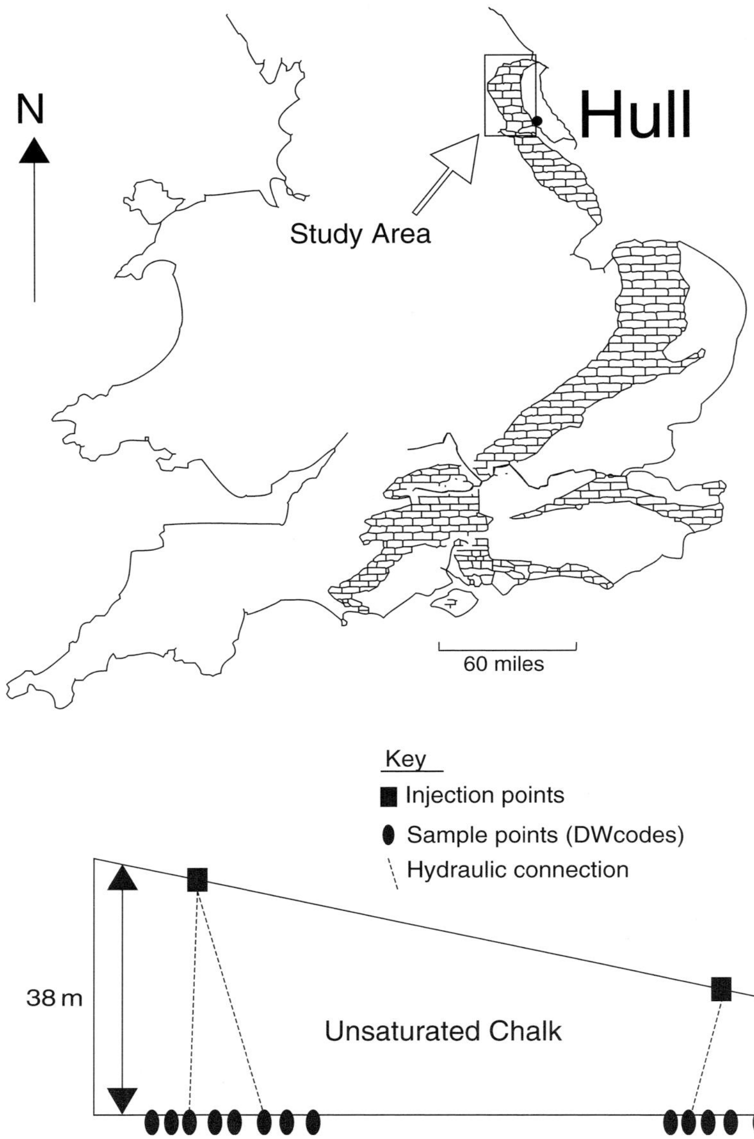

Fig. 1. Location map of the southern British Isles with the chalk outcrop marked. The East Yorkshire chalk is the area in the box. The schematic diagram shows the relation of surface injection points to groups of monitored subsurface inlets located around a point vertically below injection.

- Bypass flow within fractures could transport nitrates direct from the surface/near surface to depth (Fig. 3d). In this scenario, the seasonal nitrate maxima result from direct recharge of nitrate-contaminated water from the soil zone during the winterspring recharge season. This mechanism of contaminant transport by fracture flow through the unsaturated zone is directly analogous to flow mechanisms in the vadose zone of karstic limestone aquifers.

These two mechanisms are not mutually exclusive, but bypass flow will always reduce any buffering or attenuation of contaminant that the unsaturated zone matrix may provide. However, the existence and

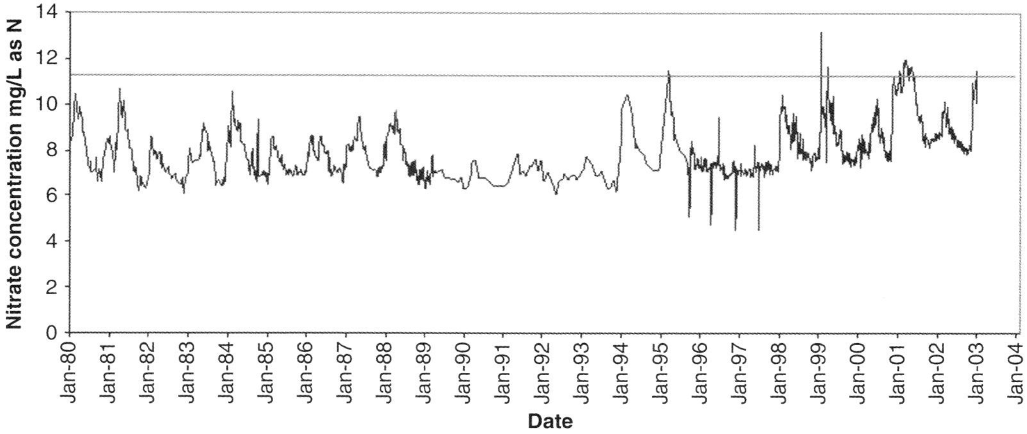

Fig. 2. Nitrate time series from an East Yorkshire chalk aquifer borehole from 1980 to 2003, with strong seasonality and the increased trend since 1997 (from Knapp 2005). The horizontal line indicates the WHO nitrate drinking water limit at 11.3 mg l^{-1} N.

significance of bypass flow in the chalk unsaturated zone remains controversial. For example, different chemical and isotopic tracer investigations of the chalk unsaturated zone by Barraclough *et al.* (1994) provided strongly ambivalent evidence regarding bypass flow. Their tracer recovery data indicate that significant flow did not occur, but blurring of seasonal deuterium signals indicates that it must. Other studies (for example of nitrate and tritium migration: Foster & Smith-Carington 1980) provide profiles that can be interpreted as consistent with bypass flow, but fail to produce conclusive evidence for its existence. However, other studies conclude that alternative mechanism to bypass flow can explain the observed hydrogeological characteristics of the chalk unsaturated zone (Price *et al.* 2000). The aim of this project is to determine whether bypass flow occurs in the chalk unsaturated zone and, if it does, whether it is a significant pathway for the transport of nitrate and other contaminants.

Rationale of this investigation

This study uses tracer test methodologies to investigate bypass flow through fractures in the chalk unsaturated zone. Such flow would produce rapid transport of solute tracer through the unsaturated zone, compared to piston flow through the matrix that would have downward velocities of approximately 500 mm $year^{-1}$ (Downing *et al.* 1993). Thus, unsaturated zone tracer velocities significantly higher than this would provide good evidence for bypass flow. The impact of bypass fracture flow on contaminant transport would be significantly reduced if travel times were long enough to permit diffusive loss of contaminant from waters flowing in fractures to pore water in the matrix blocks between fractures. Measuring tracer breakthrough concentration profiles and modelling them to estimate the proportion of tracer solute lost from fracture waters to matrix pore waters will give an indication of the extent of the attenuation in the unsaturated chalk matrix. The objectives of the study are thus:

- to measure tracer travel times and thus estimate flow velocities in the chalk unsaturated zone through the use of tracers in both qualitative and quantitative tests;
- to measure tracer breakthrough concentration profiles in quantitative tracer tests;
- to model tracer breakthrough concentration profiles in order to determine the extent of diffusional loss of tracer by exchange with chalk matrix pore water.

Materials and methods

The field site in East Yorkshire (Fig. 1) has subsurface access along an essentially linear tunnel network totalling 2.5 km where sampling is permitted, but the location cannot be disclosed under the terms of the access agreement. Sampling takes place at variable distances horizontally and through vertical ranges of 12–38 m of unsaturated chalk (Fig. 1). The installations at the site are above the water table and there are no nearby groundwater abstractions.

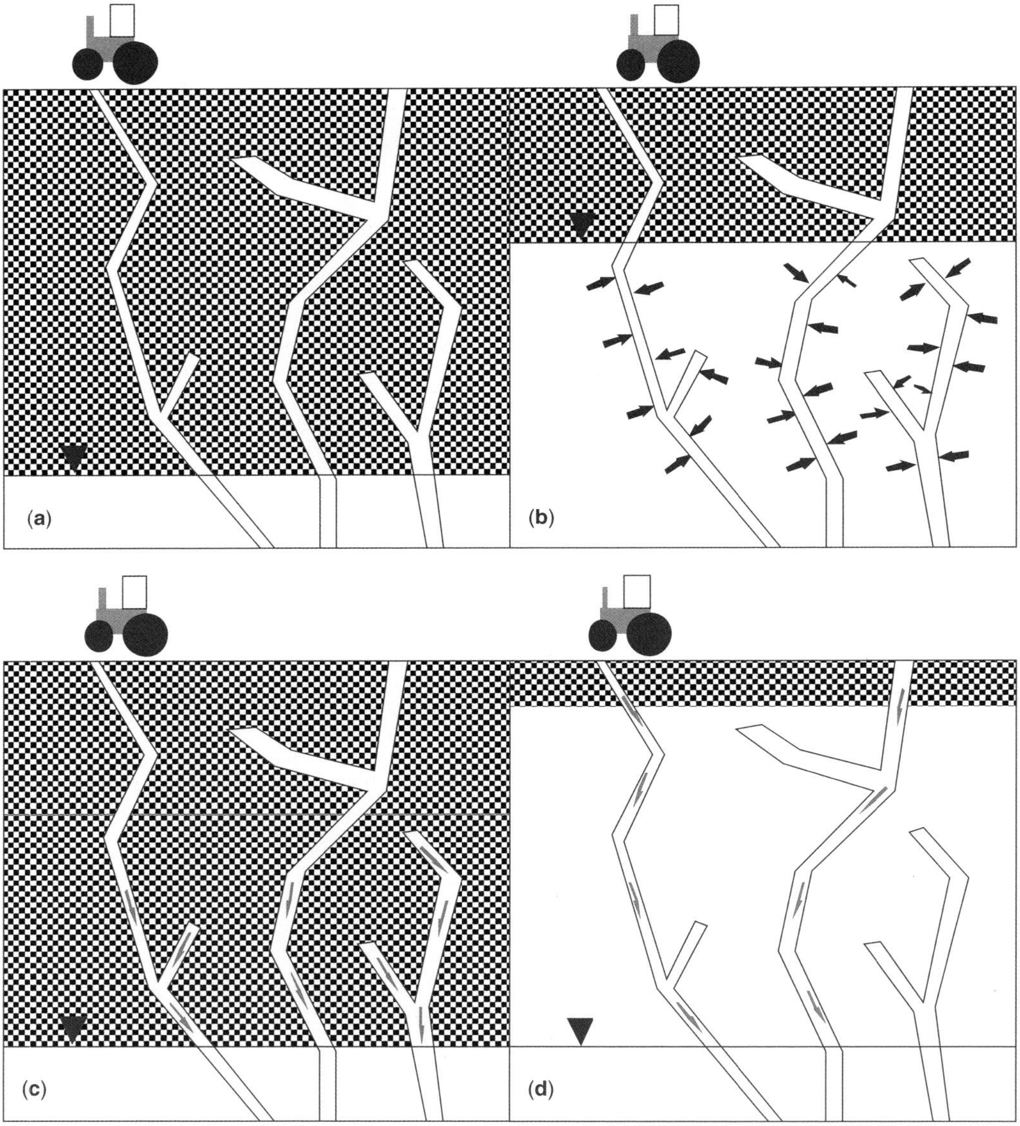

Fig. 3. Hatched areas represent chalk pore waters contaminated with agrichemicals. (**a**)–(**c**) illustrate piston flow through the matrix as caused by gravity and diffusion. The water table rises during the recharge period, contaminant diffuses into the 'clean' fracture water (b) which then drains into the saturated zone as the water table falls again (c). (**d**) 'Karstic' bypass flow as films/plug flow in fractures causes direct and rapid contamination. (a–c, adapted from Chilton *et al.* 1997.)

Methods used in this study were similar to those used by Bottrell & Atkinson (1992). Two types of tracer test were performed:

- qualitative tests, used to establish surface to subsurface hydraulic connections and approximate travel times;
- quantitative tests, performed along previously established connections, where travel times and dye concentration at breakthrough were measured.

Within the installations, 37 seepage inlets were selected for monitoring and permanently numbered

Table 1. *Summary of the injected tracers used and the solution they were made into (CI name/number were obtain from the Full Edition Colour Index International online)*

Tracer	Solution	CI generic name or CI number
Photine CU	2 l of 20% solution in 5 l of water	Fluorescent Brightener 15
DY-96	220.5 g in 5 l of water	Direct Yellow 96
Sodium fluorescein	200 g powder in 2 l of water	Acid yellow 73 no. 45350
Rhodamine-WT	200 g of 20% solution in 2 l of water	Acid Red 388

and marked. For each test, dye was injected below the soil 'C' horizon into a temporary pit (approximately 30 × 30 cm) dug to bedrock. Dye solutions injected for each experiment are listed in Table 1, following injection a further 20–30 l of water was introduced to the pit. chalk at the base of the pits always had visible fractures and the dye solution typically took 5–50 min to drain before the remaining water could be applied. Locations of injection points were accurately surveyed relative to fixed surface features to enable relocation. Flow rates of the seeps sampled were monitored at each sampling (time taken for a measured mass of water to accumulate) to permit estimation of tracer recoveries. Rainfall was monitored using tipping-bucket rain gauges connected to data loggers at two locations in the study area.

Qualitative tests

Tracers used for qualitative tests were the optical brightener Photine-CU and the dye Direct Yellow 96 (DY-96) to establish surface to subsurface hydraulic connections. Detectors for these two solutes were hanks of cotton wool, previously tested for fluorescence under UV light, placed in a lined bucket to intercept flow entering the underground installations. All inlets had detectors placed and checked prior to any tracer injections to ensure that there were no other tracers present in the fracture system. Detectors were placed before injection to avoid contamination, and were replaced and checked at intervals of 3–7 weeks thereafter. Detection of dye was made by simply illuminating the cotton wool under UV light, the dye-contaminated detectors showing an obvious blue-white (Photine) or yellow (DY-96) fluorescence.

Quantatitive tests

During the first sodium fluorescein and Rhodamine-WT test, water samples were collected from two sites (DW4 and DW8) infrequently. In the further tests 13 inlets were sampled every 12 h after an initial period of 48 h. Quantitative analyses of tracer concentrations of the water samples were made using a Perkin-Elmer LS-3 Fluorescence Spectrometer using constant offset scans to determine the position and size of tracer fluorescence peaks. Tracer concentrations were determined by comparison of sample fluorescence intensity above background to that of standard solutions. For each sample site, background fluorescence was determined in samples taken prior to tracer injections. Detection limits for sodium fluorescein above the variability in background fluorescence for these samples were approximately 0.1 $\mu g\, l^{-1}$ and Rhodamine-WT 1 $\mu g\, l^{-1}$.

Modelling of tracer transport and breakthrough curves

Modelling was carried out using the Barker Code (Barker & Foster 1981), which is a model for solute behaviour in a dual porosity aquifer, simulating fracture flow and matrix diffusion. It is a one-dimensional (1D) dual porosity model used to simulate flow and tracer transport through fractured chalk (Barker & Foster 1981). In this model, flow and transport of solute through a fractured porous medium is simulated and transport by both advection and diffusion are incorporated. The fractures are assumed to be parallel, of constant aperture and equally spaced. Solute of known concentration is introduced at the inflow and travels at constant velocity along the fractures. Diffusion perpendicular to the fractures is included allowing exchange of solute between groundwater in the porous rock matrix and fractures. Thus, the net transport of solute between fracture and matrix water depends on the solute concentration gradients. Dispersion within the fracture, simulating the effect of variable aperture and channelling within fractures, is also included. The solute concentration at the inflow is input as a user-defined source function and may be variable with time. The model output gives the variation with time of the breakthrough concentration at the fracture outflow and the mass of solute residing in the matrix. More details on the model and the mathematical background are given in Barker & Foster (1981).

Parameters were fixed using initial breakthrough data, thus fixing the velocity and first arrival time.

The assumptions that were made using the model are listed below. Parameter values used in the modelling will tend to maximize the predicted diffusional loss from fracture flow, whilst remaining in a range believed to be representative of the chalk.

Model assumptions. Three assumptions were made.

Saturated matrix. The narrow pore throats that exist in chalk will prevent drainage except at exceptional high suctions of approximately 3 atm (Pruess 1999). Thus, matrix pore water is assumed to be immobile.

Fissures are assumed to be always saturated. There is thus no net drainage of the fissure network. The periods during which the tracer tests were carried out, October–April, coincide with the period of highest recharge and thus it is assumed that the fissures are saturated (Barker & Foster 1981; Downing *et al.* 1993). Although in reality the fissures will not remain saturated, the presence of film flow may allow continuous fissure flow even in the presence of an air phase, so the model is not unrealistic. However, the velocity may differ from that for saturated flow, thus affecting the characteristic time for diffusion into matrix pore water.

Uniform parallel fissures are assumed. Variation in aperture width, roughness and degree of wetting will affect the velocity of flow and the characteristic time for diffusion (Doe 2001). At faults there is an increase in the density of fissures. The increase in intensity of fissures could lead to changes in the tortuosity, and the properties of the matrix may vary. However, the purpose of the model here is to estimate 'average' properties by fitting model curves to the data.

Modelling was carried out using the Barker Code with a single parameter being varied for each model run. The parameters that were not varied were the matrix block size, matrix porosity, advection time and dispersivity (summarized in Table 2). The model was run in an Excel spreadsheet written by John Barker (University College London).

Results

An overview of all the tracer experiments performed appears in Table 3. Quantitative tests were performed along hydraulic connections previously established in qualitative tests. Some nearby additional sites were monitored in addition to the 'proven' connection as the qualitative methodology may not have been sensitive enough to identify all linkages.

Qualitative traces

The pattern of observed recoveries from qualitative traces is presented graphically in Figure 4. Recovery of tracer during a 'positive' interval could have occurred at any time during that interval, so tracer travel times are not closely constrained. There may be a single 'spike' arrival of dye that causes a positive or the integration of a low tracer concentration over any part of the period of the positive interval. Tracer injections were made vertically above the locations of sampled seeps. Tracer injection number 2 gave positive recoveries at DW4 and DW8; DW8 is vertically below the injection point and DW4 11 m displaced horizontally. Intervening seeps at sites DW5, DW6 and DW7 showed no tracer recovery.

Quantitative traces

Constant offset scans were carried out to detect the presence of sodium fluorescein and Rhodamine-WT (none of the Rhodamine-WT experiments showed

Table 2. *Summary of input parameters used in the Barker Code models described in the text. Base case values were used, except in Model Run 1 (where aperture was varied) and Model Run 2 (where route length was varied for a range of fracture apertures). A value for* D_e *is based on the ranges reported by Hartman (2005)*

Parameter	Symbol	SI units	Base case value	Minimum	Maximum
Diffusion coefficient	D_e	$m^2\ s^{-1}$	5×10^{-11}		
Matrix porosity	Φ_m		0.35		
Advection time	t_a	s	Velocity is fixed by arrival time of 48 h through a 40 m unsaturated zone, thus a velocity of $0.79\ m\ h^{-1}$ is used		
Fracture spacing (width of matrix block between fractures)	b	m	0.1		
Dispersivity	α	m	Fixed as 10% of route length (= 4 m)		
Aperture	a	m	0.0005	0.0003	0.005
Route length	R	m	40	5	60

Table 3. *Summary of tracer experiments and the results obtained from the qualitative (to establish hydraulic connections) and quantitative tests. The distance between injection sites is at least 500 m horizontally compared to vertical travel distances <50 m. It is therefore assumed that there is no crossover between injections at different sites*

Substance	Surface location	Unsaturated thickness (m)	Date of injection	Date and location of positive
Qualitative tests				
Photine CU	1	*c.* 35	5 February 2004	No positive response to date
Direct Yellow	2	*c.* 38	5 February 2004	24 March 2004 DW 8
Photine CU	3	*c.* 38	10 January 2005	On 5 February 2005 there were weak positives at DW20 and DW21. DW21 showed no further positives. 21 February 2005 DW20 showed strong positive
Photine CU	4	*c.* 15	10 January 2005	Weak positive DW34 5 February 2005 only
Quantitative tests				
Sodium fluorescein	2	*c.* 38	10 January 2005	0.019% from DW8 and 0.0035% from DW4
Rhodamine-WT	3	*c.* 38	1 April 2005	No recovery
Sodium fluorescein	3	*c.* 38	1 April 2005	0.004% from DW21
Rhodamine-WT	2	*c.* 38	1 April 2005	No recovery

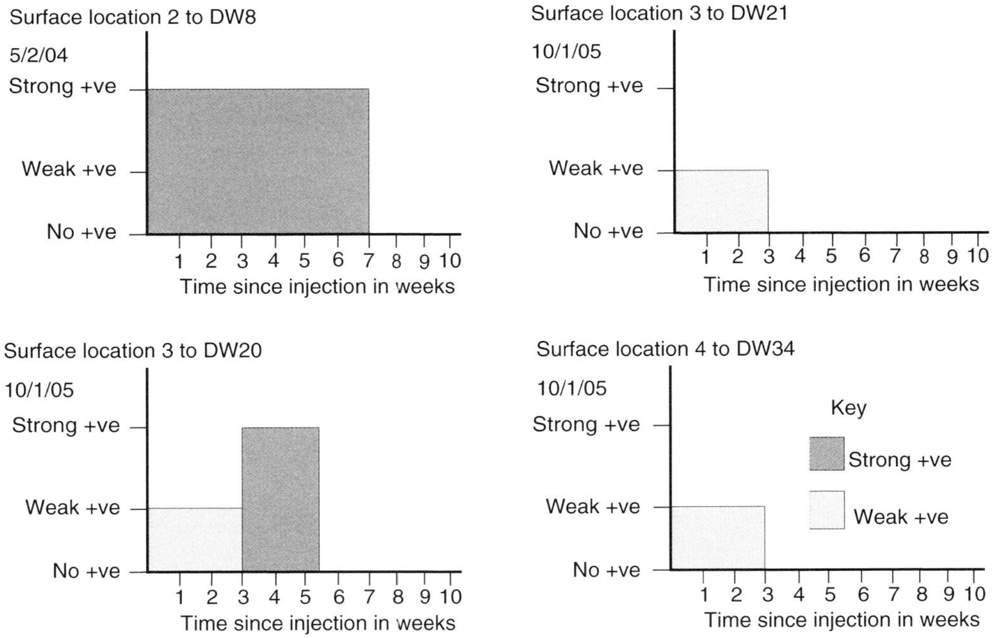

Fig. 4. Schematic representation of the results obtained from all of the qualitative tests. N. B. 1 week = 168 h. All tests continued with detectors in place for a further 10 weeks longer than the 10 weeks depicted with no further positives. Detectors were placed at all 37 monitored inlets during all tests and only positives are shown. Tracer injection at surface location 1 produced no detectable positive traces.

any positive result). Owing to access restriction the temporal resolution of the tracer tests was limited. There are no very short time period samples in the tracer tests reported here, so we cannot totally exclude the possibility of very rapid early arrival of tracer. Repeat traces at the same sites with continuous monitoring of tracer arrival in this early phase are currently underway and indicate that very rapid arrivals do not occur.

Sodium fluorescein injected at surface location 2 was recovered in samples from inlet sites DW4 and DW8, the only two sites monitored during this test. The two tracer breakthrough curves are presented in Figure 5. These are plotted as sodium fluorescein concentrations determined from the excess fluorescence above the background fluorescence of three samples taken at the same site prior to tracer injection. Tracer breakthrough at low concentration (*c.* 1 $\mu g\,l^{-1}$) was observed at DW8 in the first sample at 96 h and at similar low concentrations in the subsequent two samples. No tracer was detected at DW4 at these times. The samples at 1008 h show a significant increase in tracer concentration at DW8 (20 $\mu g\,l^{-1}$) and tracer was then detected at DW4 (3.7 $\mu g\,l^{-1}$). These are the maximum concentrations observed at these inlets and concentrations subsequently fell. However, the long time interval between sampling means that it is unlikely that the actual peak concentration had been sampled at either inlet. The breakthrough curve at DW4 is thus essentially a muted version of the response seen at DW8. The observed peaks at DW4 and DW8 at 1008 h occur following 2 days of rainfall (totalling 11 mm) that was the equivalent of two-thirds of the rainfall of the preceding 2 weeks.

Inlet DW21 has a very different tracer breakthrough curve to that at DW4 and DW8 (see Fig. 6) showing a distinctive spike and falling limb. Tracer was present at 1.4 $\mu g\,l^{-1}$ in the first sample taken 48 h after injection and the maximum observed concentration is 8.1 $\mu g\,l^{-1}$ at 60 h. Tracer concentration fell to 1.1 $\mu g\,l^{-1}$ at 72 h and remained present at less than 1 $\mu g\,l^{-1}$ up to the final sample at 372 h.

The average flow rates for DW4 and DW8 was 0.16 and 0.17 ml s^{-1}, respectively, with DW6 and DW7 both being lower at 0.017 ml s^{-1} but DW5 being 0.18 ml s^{-1}. The average flow rate of DW21 was 0.0025 ml s^{-1}, whereas DW20 was 3.7 ml s^{-1} and DW19 0.012 ml s^{-1}.

Modelling results

Results from the modelling are presented in Figures 7 and 8. Figure 7 depicts the results of a model investigation of tracer breakthrough for various assumed fracture apertures. Travel time was set to 48 h, based on initial breakthrough observed at DW21 (see 'Quantitative tests' above) and a vertical

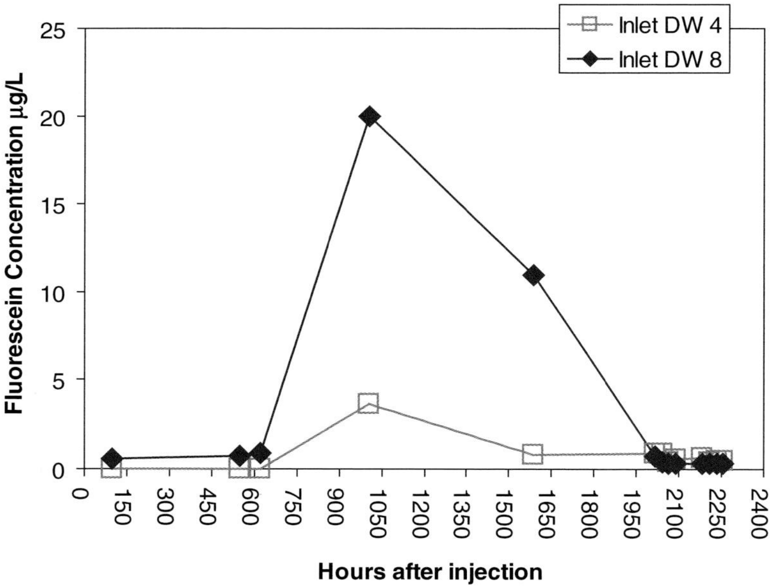

Fig. 5. Breakthrough curves of sodium fluorescein concentration against time after injection. At inlet DW8 initial breakthrough had occurred by 96 h after injection and total tracer recovery was 0.019% of the tracer injected. At DW4 initial breakthrough occurred between 624 and 1008 h, and total tracer recovery was 0.0035%.

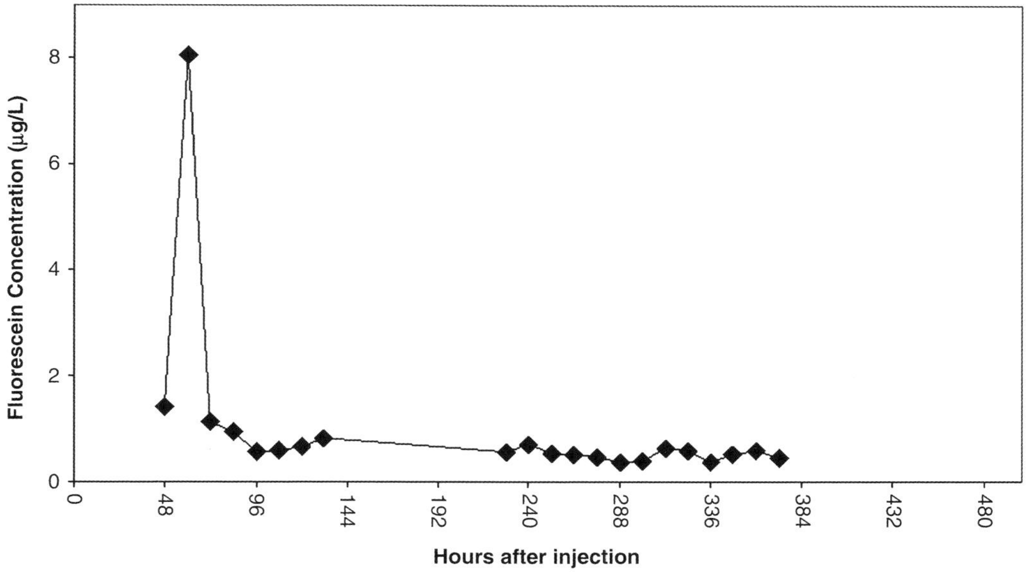

Fig. 6. Breakthrough curve of sodium fluorescein concentration at inlet DW21 against time after injection. A total of 0.004% of tracer injected was recovered. The gap in sampling between 132 and 228 h was caused by mechanical failure.

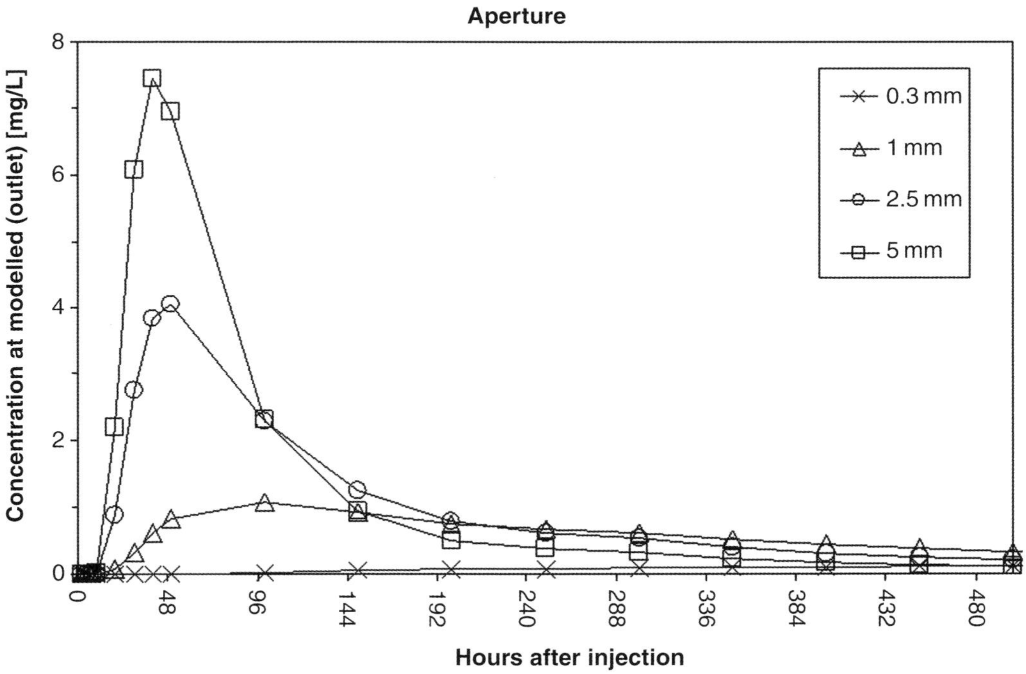

Fig. 7. Plot of the modelled output concentration of a single fracture against time since tracer injection (Model Run 1, parameters as in Table 2). Aperture size has the greatest significance of all of the parameters; an aperture of 0.3 mm has a very flat profile, whilst the 5 mm aperture has a distinctive peak. Total model duration is 2000 h, only the first 500 h are shown.

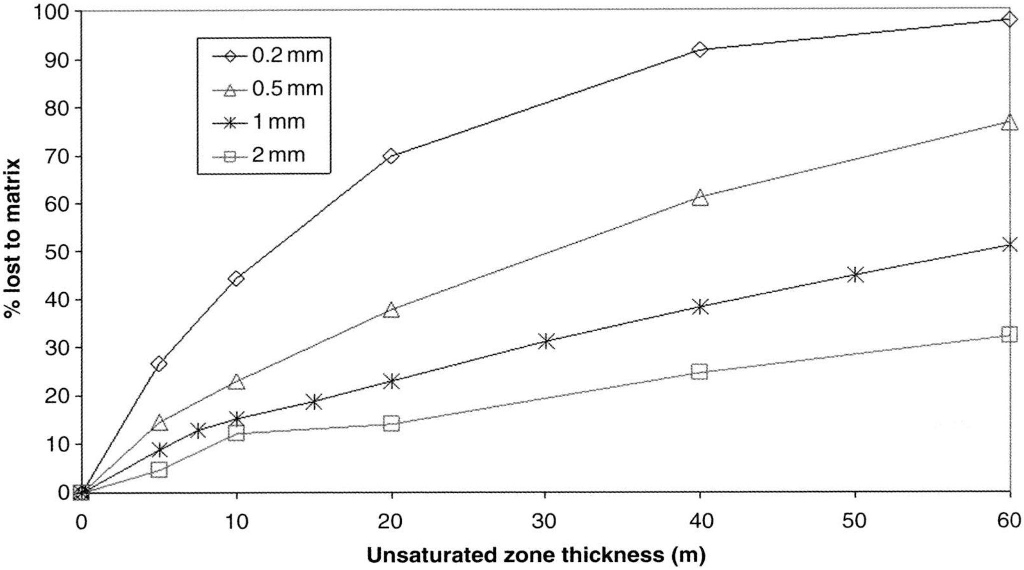

Fig. 8. Plot of output from Model Run 2 (see Table 2). Loss of tracer from the fracture water to the matrix is plotted as a function of increasing unsaturated zone thickness for different fracture apertures. The total model duration is 2000 h (approximately 12 weeks) to simulate the observed timescale of rapid bypass flow.

distance of 40 m. The relatively large fracture apertures (5 and 2.5 mm) produced predicted curves that have marked concentration peaks and tails. The smaller apertures modelled (0.3 and 1 mm) produced flat, highly attenuated breakthrough curves that may not exceed detection limits for actual tracer tests ($<0.1\ \mu g\ l^{-1}$). Model runs using longer travel times (e.g. 1200 h) resulted in highly attenuated breakthrough curves for all fracture apertures used.

Figure 8 plots the proportion of tracer (or contaminant) that migrates from fracture water into the chalk matrix by diffusion against unsaturated zone thickness in the first 2000 h of each of the model runs. There is a clear relationship between aperture size and attenuation of tracer by diffusive loss to the matrix. Smaller fracture apertures resulted in a high degree of attenuation, whereas the largest fracture apertures produced far lower degrees of attenuation even at unsaturated zone thicknesses up to 60 m.

Discussion

Tracer travel times

The rapid arrival times seen in both qualitative (Fig. 4) and quantitative tests (Figs 5 & 6) represent evidence for bypass flow. Although the travel times in qualitative tests are poorly constrained, they do indicate arrivals on timescales of hundreds of hours or less (Fig. 4). Quantitative traces demonstrate initial tracer breakthrough at DW8 and DW21 by 96 and 48 h, respectively. These results lead to minimum velocities of 9.5–19 m day^{-1} through unsaturated chalk (Table 4). Minimum vertical tracer velocities calculated from qualitative tracer tests are in the range 0.23–19 m day^{-1} (Table 4). Such high velocities cannot be the result of piston flow through the chalk matrix (typically 0.5 m $year^{-1}$: Barraclough *et al.* 1994) and all positive traces to date indicate that rapid bypass flow must occur in the chalk unsaturated zone.

Tracer breakthrough curves

The tracer breakthrough curve observed for the test at DW21 (Fig. 6) is similar in form to the modelled results for a 2.5 or 5 mm aperture fracture (Fig. 7). There are, however, differences between the observed peak concentrations and those predicted from the model, the observed concentrations being 1000 times lower. Price *et al.* (1992) observed very low recoveries with similar tracer tests, with travel times of 2 km day^{-1}. The inlet at DW21 is not necessarily the unique emergence point of a single fracture transporting tracer and may be the combined flow of many fractures, only one or some of which became labelled by the injected tracer. The close correspondence of the form of modelled and observed breakthrough curves indicates that (at least part of)

Table 4. *Summary of the minimum velocities for tracer tests. Surface locations 2 and 3 have an unsaturated thickness of 38 m above monitored inlet, surface location 4 is only 15 m above DW34*

Quantitative traces				
From surface location to inlet	Initial breakthrough (h)	Peak arrival (h)	Velocity (m day^{-1})	
			Breakthrough	Peak
2 to DW4	504	1008	1.8*	0.94*
2 to DW8	96	1008	9.5	0.91
3 to DW21	48	60	19	15

Qualitative traces		
	Maximum travel for time breakthrough (h)	Minimum velocities (m day^{-1})
2 to DW8	4032	0.23
3 to DW20	504 initially then strong positive by 1080 h	0.84–1.8
3 to DW21	504	1.8
4 to DW34	504	0.72

*Horizontal displacement included.

the observed rapid bypass flow takes place through relatively wide fractures. The prediction of such large fracture apertures for these flows may imply that some proportion of unsaturated zone transport is by pipe flow (perhaps in solutionally enlarged conduits) or in solutionally widened channels in fractures rather than by flow in plane fractures.

The tracer breakthrough curves observed at DW4 and DW8 (Fig. 5), where peak concentrations arrive at around 1000 h, do not conform to the predictions of the model used. The pattern of tracer arrival in the qualitative trace to DW20 (Fig. 4) may be similar in form, as only a weak positive resulted during the period 0–504 h, whilst a strong positive in the 504–1008 h interval may indicate the later arrival of a tracer 'peak'. Assuming continuous and constant flow, such long travel times should result in high degrees of attenuation of tracer by diffusion into the chalk matrix. The presence of significant peaks in the breakthrough concentration curve after such periods (as observed at DW8 and, to a lesser extent, DW4, see Fig. 5) does not therefore fit with model predictions. This may indicate that in these cases one or more of the model assumptions is incorrect. The most likely candidate assumption is that of continuous and uniform saturated flow in fractures. Flow may be discontinuous because the input is not sufficient to maintain flow or a physical barrier causes ponding. Ponding could occur on marl or flint horizons, known to be present in the chalk at this site. Either cause would lead to tracer becoming immobilized until flushed out during a later increase in flow. Provided that diffusive loss of tracer into the chalk matrix is also restricted during cessation of flow, a distinct tracer peak with a delayed arrival time could be observed. The fact that the peaks in tracer concentration at DW4 and DW8 followed a significant rainfall event suggests that their long delay could be owing to cessation of fracture flow earlier in the test period. Bottrell & Atkinson (1992) observed similar storage and flushing of tracer after rainfall in the unsaturated zone of the UK Carboniferous Great Scar Limestone aquifer.

The very low concentrations of tracer that are detectable for long periods after peak arrival (Figs 5 and 6) are consistent with the predictions from the model calculations (Fig. 7).

Implications for contaminant transport

The tracer tests described have shown clear evidence of rapid 'karstic' fracture bypass flow in the chalk unsaturated zone. This has clear implications for contaminant transport as it provides a pathway for rapid migration of contaminants from surface to the water table. Modelling of flow in fractures of different apertures shows that delivery of high concentrations of contaminant to the unsaturated zone requires flow in fractures with relatively large aperture (e.g. 2.5–5 mm aperture, Fig. 7). The tracer breakthrough curves observed are consistent with flow in such fractures. Importantly, flows through such fractures can be sufficiently rapid that relatively little attenuation of solutes by diffusional loss into matrix pore water can take place. In Figure 8 model results for integrated diffusional loss from such fracture flows are plotted against unsaturated zone thickness. Even for an unsaturated zone thickness of 60 m, flows in fractures of 2.5–5 mm aperture are so rapid that only 30–50% of contaminant undergoes attenuation by this mechanism.

Observed tracer concentrations are lower than the model predictions, despite the rapid tracer travel times. This could result from the nature of point injections used in tests, with subsequent tracer dilution by mixing with unlabelled recharge water.

Conclusions

- Tracer methodology has been successfully utilized to investigate solute transport in a thick unsaturated zone on the chalk aquifer.
- Rapid tracer breakthrough times in all positive tests confirm the hypothesis that karstic bypass flow takes place in unsaturated zone of the chalk aquifer.
- Shape of quantitative tracer test breakthrough curves are consistent with flow through large-aperture (>1 mm) fracture networks.
- Hydrological transport processes in the unsaturated zone of the chalk aquifer thus have similarities to those in the vadose zone of karstic limestone aquifers. The rapid contaminant transport permitted by fracture transport limits the attenuation capacity of the unsaturated zone of the chalk aquifer.

We thank all the landowners who granted access permission for this work and NERC (grant number NER/D/S/2003/00694), the Environment Agency and Yorkshire Water for supporting this project.

References

BARKER, J. A. & FOSTER, S. S. D. 1981. A diffusion exchange model for solute movement in fissured porous rock. *Quarterly Journal of Engineering Geology*, **14**, 17–24.

BARRACLOUGH, D., GARDNER, C. M. K. & WELLINGS, S. R. & COOPER, J. D. 1994. A tracer investigation into the importance of fissure flow in the unsaturated zone of the British Upper Chalk. *Journal of Hydrology*, **156**, 459–469.

BOTTRELL, S. H. & ATKINSON, T. C. 1992. Tracer study of flow and storage in the unsaturated zone of a karstic limestone aquifer. *In*: HOTZL, H. & WERNER, A. (eds) *Tracer Hydrology*. Balkema, Rotterdam, 207–211.

CHILTON, P. J. *ET AL.* 1997. Trends in nitrate concentrations in the Yorkshire chalk aquifer. Yorkshire Water and BGS.

DOE, T. W. 2001. What do drops do? Surface wetting and network geometry effects on vadose-zone fracture flow. *In*: *U.S.N.C.f.R.M. Panel on Conceptual Models of Flow and Transport in the Fractured Vadose Zone, Board on Earth Sciences and Resources, Commission on Geosciences, Environment, and Resources, National Research Council (Editor), Conceptual Models of Flow and transport in the Fractured Vadose Zone*. National Academy Press, Washington, D.C., 243–270.

DOWNING, R. A., PRICE, M. & JONES, G. P. 1993. The making of an aquifer. *In*: DOWNING, R. A., PRICE, M. & JONES, G. P. (eds) *The Hydrogeology of the Chalk of North-west Europe*. Clarendon Press, Oxford, 1–13.

FOSTER, S. S. D. & SMITH-CARINGTON, A. 1980. The interpretation of tritium in the Chalk unsaturated zone. *Journal of Hydrology*, **46**, 343.

HARTMANN, S. 2005. *Flow and Transport in the Confined Chalk Aquifer in East Yorkshire*. Ph.D. Thesis, University of Leeds, Leeds.

KNAPP, M. F. 2005. Diffuse pollution threats to groundwater: a UK water company perspective. *Quarterly Journal of Engineering Geology and Hydrogeology*, **38**, 39–51.

PRICE, M., ATKINSON, T. C., BARKER, J. A., WHEELER, D. & MONKHOUSE, R. A. 1992. A tracer study of the danger posed to a chalk aquifer by contaminated highway run-off. *Proceedings of the Institution of Civil Engineers-Water Maritime and Energy*, **96**, 9–18.

PRICE, M., LOW, R. G. & MCCANN, C. 2000. Mechanisms of water storage and flow in the unsaturated zone of the chalk aquifer. *Journal of Hydrology*, **233**, 54–71.

PRUESS, K. 1999. A mechanistic model for water seepage through thick unsaturated zones in fractured rocks of low matrix permeability. *Water Resources Research*, **35**, 1039–1051.

Stable isotopes in aqueous sulphate as tracers of natural and contaminant sulphate sources: a reconnaissance study of the Xingwen karst aquifer, Sichuan, China

S. H. BOTTRELL

Institute of Geological Sciences, School of Earth and Environment, University of Leeds, Leeds LS2 9JT, UK (e-mail: ear6shb@earth.leeds.ac.uk)

Abstract: Isotopic compositions of sulphate ($\delta^{18}O$ and $\delta^{34}S$) have been analysed in groundwaters from a karst aquifer in Xingwen, China to assess their use as indigenous tracers of different pollutant sulphate sources. Sulphate $\delta^{18}O$ is highly effective at distinguishing sulphate from atmospheric 'acid rain' sources (higher $\delta^{18}O$ values) from sulphate produced by aqueous pyrite oxidation (natural or acid mine drainage), which always has lower $\delta^{18}O$. The range of sulphate $\delta^{34}S$ produced by aqueous oxidation of different pyrite sources is sufficiently wide to enable different natural and pollutant sulphate sources to be distinguished. Despite the fact that streams containing processing fines and pyrite mine drainage both derive sulphate from oxidation of ore materials, there is still a clear distinction in their sulphate $\delta^{34}S$. A combination of sulphur and oxygen isotopic measurements is thus highly effective at discriminating between all the sulphate sources to the karst aquifer and this indigenous tracer provides a powerful tool for assessing the impact of acid mine drainage on karst groundwater.

Increasing anthropogenic sulphur emissions since the industrial revolution have increased sulphur fluxes to surface water and groundwater (see Ivanov 1983 for estimates of anthropogenic contributions to fluxes). This has occurred through both diffuse source and point source contamination of the natural environment. Diffuse sulphur pollution sources are principally acid rain (SO_2 from combustion of fossil fuels and from ore smelting) and application of sulphate-containing agrichemicals. Point source sulphur pollution arises from a very wide range of industrial and urban sources, major contributors being acid mine drainage (AMD), discharge and spillage of industrial chemicals, and breakdown of sulphur-containing detergents and other cleaning products discharged to the environment in sewage.

In September 1992 the British China Caves Project organized an expedition to the karst region of the Xingwen National Park, Sichuan, China. During this expedition sampling was undertaken for a reconnaissance study of the isotopic compositions of sulphate sources to the karst aquifer. Natural sources, including acidic high sulphate runoff from shale catchments, as well as mining and processing of pyrite ore contribute to the aquifer sulphate budget (Bottrell 1993). High sulphate concentrations alone therefore are not a unique indicator of AMD pollution of groundwater in the Xingwen aquifer, as high sulphate concentrations occur naturally in some catchments. The sampling strategy for this reconnaissance attempts to characterize the variability in isotopic composition between the different natural and anthropogenic sulphate sources, and thereby assess whether sulphate isotopic compositions can be of use in constraining the impact of AMD on aquifer sulphate budgets.

Sulphate is a ubiquitous component of natural waters and a common constituent of many anthropogenic effluents to the hydrosphere. Both $^{34}S/^{32}S$ and $^{18}O/^{16}O$ ratios of sulphate may be measured routinely and the variation of these isotopic ratios therefore potentially provides a powerful indigenous discriminant of sulphate sources. Except at very low pH or under extremely reducing conditions, sulphate is chemically stable and resistant to isotopic exchange or fractionation processes in the hydrosphere, so sulphate isotopic compositions should behave as conservative tracers of sulphate source in many surface water and groundwaters. Sulphate isotopic compositions therefore provide a potential tool for the delineation of the impact of anthropogenic sulphur sources on natural waters. In this study the isotopic compositions of sulphate from different inputs to the Xingwen aquifer were measured and the mechanisms by which different sulphate sources inherit their isotopic compositional 'fingerprints' investigated. The ability of sulphate stable isotopes to discriminate between different natural and anthropogenic sources of sulphur, and thus their utility as tracers of anthropogenic sulphur in karst groundwaters, are then assessed.

From: PARISE, M. & GUNN, J. (eds) *Natural and Anthropogenic Hazards in Karst Areas: Recognition, Analysis and Mitigation*. Geological Society, London, Special Publications, **279**, 123–135. DOI: 10.1144/SP279.11
0305-8719/07/$15.00

Pyrite oxidation: natural and anthropogenic

Pyrite (FeS_2) is a common constituent of many sedimentary rocks, usually formed as a result of bacterial processes during diagenesis of the sediment after burial (Berner 1970). Pyrite and other sulphide minerals are also major constituents of many ore bodies. When exposed to an oxidizing, near-surface environment, pyrite and other sulphides become unstable and undergo oxidation reactions. This may involve oxidation by molecular oxygen derived from air (often via oxygenated shallow groundwaters)

$$FeS_2 + \tfrac{7}{2}O_2 + H_2O = 2SO_4^{2-} + 2H^+ + Fe^{2+} \quad (1)$$

or may involve another oxidizing agent, e.g. Fe^{3+},

$$FeS_2 + 14Fe^{3+} + 8H_2O = 2SO_4^{2-} + 16H^+ + 15Fe^{2+}. \quad (2)$$

These reactions are often microbially mediated, as energy is released in the transformation of sulphide to sulphate. The stable end products of these reactions are sulphate and protons; essentially sulphuric acid is being produced and the waters become highly acidic.

The process of pyrite oxidation takes place entirely naturally, as part of the normal surface weathering process. However, human activities may accelerate the process beyond its natural rate. This is often the result of mining activities; these modify groundwater flow paths and introduce oxygenated surface waters into previously reducing environments. In addition, mine spoil dumped at the surface may contain pyrite and/or other sulphide minerals that then become weathered by percolating rainfall. Both of these scenarios give rise to the phenomenon of acid mine drainage (AMD).

Previous work and context of this study

Natural variation in the abundance of stable isotopes of the light elements (particularly C, O and H) have been used as tracers of provenance and of geochemical processes for over half a century. Sulphur isotope studies have traditionally been applied to the study of ore deposit formation (e.g. Ohmoto & Rye 1979; Bierlein *et al.* 1996) and the redox cycling of sulphur in marine sediments (see review by Bottrell & Raiswell 2000). More recently, sulphur isotope studies have been applied to tracing the fate and impact of pollutant sulphur in the natural environment (e.g. Krouse 1977; Novak *et al.* 1996; Mayer *et al.* 2001; Kirste *et al.* 2003). These studies showed that, provided the pollutant sulphur introduced into the environment has an isotopic composition distinct from that naturally present, the movement of sulphur between different environmental compartments can be traced through changes in their isotopic compositions. Similar studies have been applied to the tracing of natural and pollutant sulphur sources in natural waters, both rivers and lakes (e.g. Hitchon & Krouse 1972; Robinson & Bottrell 1997; Li *et al.* 2006) – including those impacted by AMD (e.g. Edraki *et al.* 2003) – and groundwaters (e.g. Kimblin 1995; Moncaster *et al.* 2000; de Caritat *et al.* 2003). However, few such studies have been applied to groundwaters in carbonate karst aquifers. Two sulphur isotopic studies have identified significant contributions of evaporite sulphate to karst spring waters (Krothe & Libra 1983; Gunn *et al.* 2006). Bottrell *et al.* (2000) identified a number of sulphur sources and factors affecting the sulphur isotopic composition of sulphate in a karst aquifer at Castleton, central England. Significant sulphur sources were found to be: sulphate in rainwater; application of agricultural slurries; weathering of pyrite present in some limestone lithologies; and oxidation of sulphide ore minerals present in vein systems. Bottrell *et al.* (2000) used mass balance considerations to show that the input of sulphate from the ore mineral source must have been enhanced by historical mining activities.

Isotopic tracers have distinct advantages over artificially introduced tracers in groundwater systems: the tracer is indigenous to the sulphur source(s) and the tracer signal is present throughout any contamination episode. Furthermore, sulphate ions can also acquire an oxygen isotopic composition characteristic of their source. Different sulphur oxidation mechanisms (e.g. Reactions 1 and 2 earlier) incorporate oxygen atoms from atmospheric O_2 and from environmental water in different proportions. These two sources of oxygen atoms have widely different isotopic compositions (atmospheric O_2 strongly enriched in ^{18}O and environmental water weakly to strongly depleted in ^{18}O), so sulphate produced by different mechanisms can have a wide range of different oxygen isotopic compositions (e.g. Holt & Kumar 1991; McCarthy *et al.* 1998).

Study area

Xingwen county lies in the south of China's Sichuan Province, approximately 80 km SE of Yibin, a city on the Chiang Jiang (Yangtze River). The limestone escarpment at Xingwen is spectacularly karstified (Waltham *et al.* 1993) and hosts many large caves that were explored and surveyed during the 1992 China Caves Project (Waltham & Willis 1993). The Xingwen karst is formed in a 350 m-thick middle Permian limestone sequence (the Maokou and Qixia formations) which have an elliptical outcrop of approximately 60 × 20 km around a major pericline. Xingwen lies on the south side of this pericline, where the limestone dips at around 5°–25° and hence there is a wide limestone outcrop on the dip slope (Fig. 1). The limestones unconformably overlie Silurian sandstones and siltstones, which form the core of the pericline. Upper Permian shales and sandstones overlie the limestones, and these contain beds of coal and, almost immediately above the limestone, a 1.5 m-thick bed of pyrite ore.

The pyrite and coal deposits form the basis of much of the local industry; coal is mined for domestic and industrial fuel as well as for use in conjunction with pyrite in sulphur production and to fuel lime kilns (Waltham & Willis 1993). The mining of coal and pyrite ore and the mineral processing, by which pyrite is beneficiated from the ore, lead to pollution of some waters. In addition, the smelting of pyrite to produce sulphur leads to the release of sulphur dioxide to the atmosphere (with localized smog and 'acid rain') and the creation of large slag heaps containing unused pyrite.

The geomorphology and hydrology of the karstic aquifer developed in the limestone are described in detail by Waltham & Willis (1993) and Waltham *et al.* (1993) and the salient features summarized in Figure 1. Allogenic recharge comes from sinking streams with catchments on both Silurian and Permian non-carbonate rocks. Most of this water follows three major cave conduit systems that unite within the aquifer and discharge at a single spring. These flows are augmented by autogenic recharge: percolation inlets fed by rainfall onto the limestone outcrop, which supports no surface drainage. The chemistry of the surface waters and groundwaters of the area are described in detail by Bottrell (1993); here we consider the aquifer sulphate budget and the isotopic variations within the sulphate. The only identifiable natural sulphate source is from weathering of sulphides in the bedrocks of catchments on the Silurian and Upper Permian strata. Anthropogenic sulphate sources are derived from the various operations associated with the extraction and processing of the pyrite ore (Waltham & Willis 1993; Bottrell 1993), these are:

- direct inputs of mine drainage at the limestone–shale contact;
- inputs of fine-grained, pyrite-bearing sediment from ore-processing plant into rivers sinking in the limestone;
- atmospheric contributions of sulphate (as dry deposition or rainfall) from SO_2 released during conversion of pyrite to sulphur.

Catchments feeding the Xingwen karst aquifer

A detailed description of the aquifer and the caves is given by Waltham *et al.* (1993). The catchments discussed below are marked on Figure 1.

Autogenic recharge on Permian limestone

All rainfall onto the limestone outcrop evaporates or sinks; the limestone outcrop supports no surface drainage. This water collects into percolation flows underground and these percolation inlets augment the allogenic flows from the stream sinks. Where soil is present in fractures, valleys and on doline floors it is intensively but traditionally farmed. The major sulphur works in the area are built on the limestone outcrop. Samples were taken from small percolation inlets into the caves.

Silurian sandstone catchments

Rivers drain from these catchments into sink caves at the base of the limestone escarpment. All the catchments are fertile farmland and are intensively but traditionally farmed, with rice, maize and tobacco being the main crops. There is no industrialization of these catchments. Samples were taken at the Xia Dong sink and in the Tiencuan Dong cave river.

Permian shale catchments

These catchments are underlain by a pyritic shale–coal sequence and are farmed with varying degrees of intensity. Drainage from these areas feeds into gorge and sinkhole systems in the upper part of the limestone. Samples were taken from streams above the pyrite mines where there was no coal-mining activity higher in their catchments.

Mine drainage

The pyrite ore bed at the base of the Permian shale sequence (and immediately above the limestone) is worked in a series of mines of various sizes. These work down the dip of the bed into the hillside and are drained through their floors into the underlying limestone aquifer. There are also numerous small coal mines in the sequence above emitting drainage from their entrances and into sinking streams.

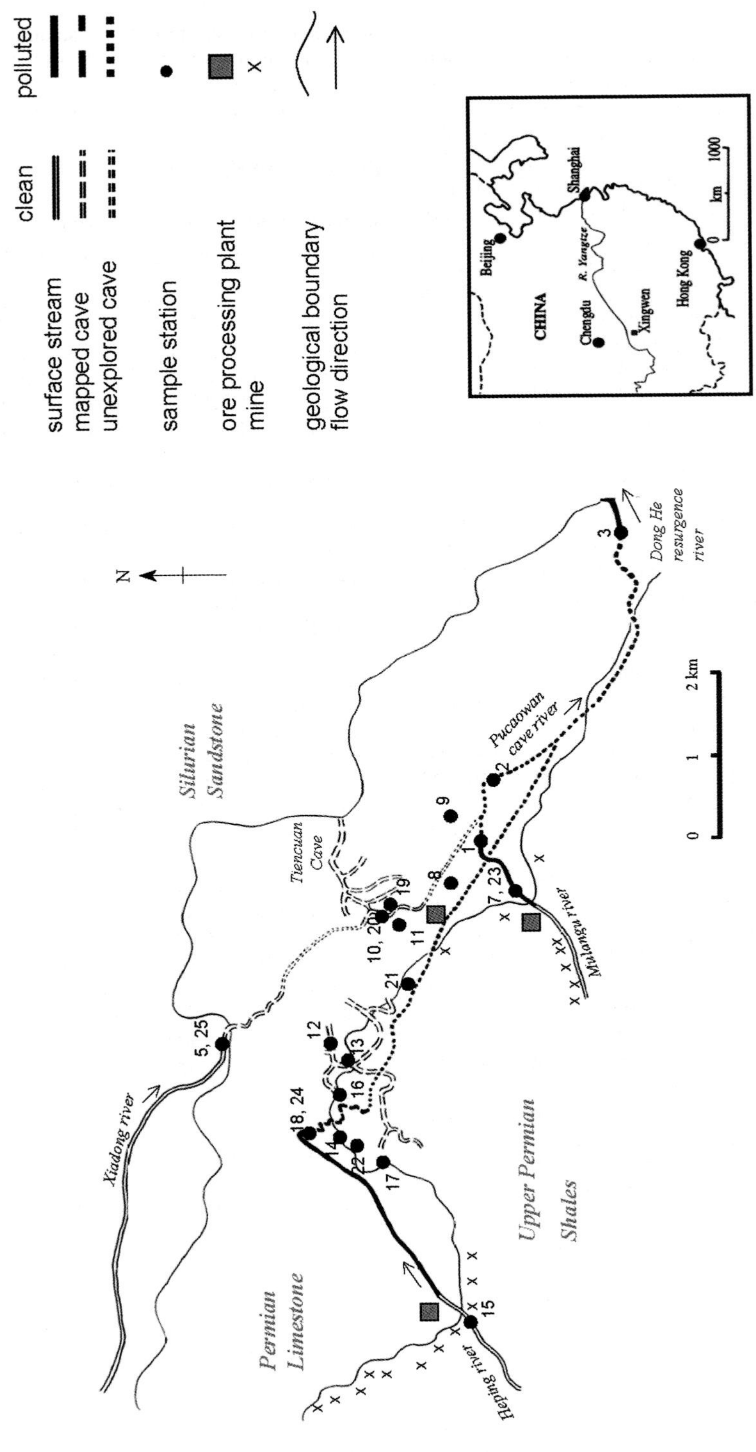

Fig. 1. Surface geology and hydrology of the Xingwen aquifer with locations of sampling points and major pyrite-processing plants.

Streams contaminated with processing tailings

The Mulangu River and the Heping River are both heavily contaminated by tailings from modern ore-processing plants, rendering them highly turbid. Both rivers sink into caves and carry a considerable load of suspended sediment. There are smaller traditional ore-washing plants at many of the mines, but these discharge very much smaller amounts of effluent into sinking streams.

All of the water recharging the Xingwen karst drains out at a single resurgence river at Dong He Dong. This is in the Daba River valley, immediately SE of the Dong He resurgence in Figure 1; the Daba River crosses the limestone outcrop without sinking and the valley acts as a discharge boundary controlling the base level of groundwater in the aquifer.

Sampling and analytical details

The sampling locations are marked on Figure 1. As far as possible, three or four samples were taken of each of the catchment water types listed above. A further two samples were taken, one at the Pucaowan cave in the centre of the aquifer (location 2, Fig. 1) and one at the Dong He resurgence (location 3, Fig. 1), the single resurgence for the Xingwen karst; these will represent mixtures of waters derived from the catchments described above. All of the samples were collected at base-flow conditions, prior to the rainfall and flooding that affected the area later in the expedition.

Water pH and conductivity were measured *in situ* at the sample site or, in the case of some samples collected underground, immediately on being brought to surface. All samples were filtered in the field onto preweighed filters and the sediment dried for weighing and analysis. A measured volume of the filtrate was acidified to pH 2.5 with HCl and treated with $BaCl_2$ solution to recover sulphate as $BaSO_4$ for gravimetric determination of SO_4 and isotopic analysis. For a subset of the samples, two further aliquots of the filtrate were retained; one untreated for anion analysis by ion chromatography; a second acidified with HNO_3 for cation analysis by inductively coupled plasma-optical emission spectrometer (ICP-OES). The pyrite content of dried sediment samples was analysed using the chromous chloride reduction technique and the sulphide trapped as CuS for isotopic analysis (Newton *et al.* 1995). Sulphate-O isotopic measurements were made on CO_2 gas prepared from $BaSO_4$ by reduction with graphite (McCarthy *et al.* 1998) and water-O on equilibrated CO_2 (Epstein & Mayeda 1953). Sulphur isotopic measurements were made on SO_2 gas prepared by oxidation of Ag_2S (prepared from BaS residues from sulphate-O analyses above) or CuS using the Cu_2O technique of Robinson & Kusakabe (1975). Isotope analyses are reported in standard delta notation relative to the V-SMOW (Vienna Standard Mean Ocean Water) oxygen and V-CDT (Vienna Cañon Diablo Triolite) sulphur standards.

Results and discussion

Sulphur content and isotopic composition of rocks, ores and manufactured sulphur

Summary data on concentrations of disseminated pyrite in rocks and isotopic compositions of the recovered sulphide and of physically separated macroscopic pyrite crystals from rocks and pyrite ore are given in Table 1. Also presented are analyses of sulphur produced by smelting of the pyrite ore.

In the Xingwen region, pyrite oxidation reactions take place in three situations.

- As part of natural weathering reactions. The upper Permian shale sequence is pyrite rich and natural weathering by shallow groundwaters involves pyrite oxidation and releases acid sulphate to streams sinking at the limestone contact. This process is less important in the Silurian catchment, as the lithologies here contain less pyrite.
- As a result of mining operations for pyrite and coal AMD is produced. The mines drain through their floors into the limestone or discharge them from adit entrances into sinking streams.
- Effluent from ore-washing plants is discharged into sinking streams and contains significant amounts of fine-grained pyrite, which oxidizes in the surface water and groundwater. Pyrite oxidation also takes place where the sediment accumulates in groundwater conduits and cave passages in the limestone aquifer. At the same plants, pyrite oxidation will also take place in the slag heaps and contaminate the drainage from these.

A further source of acidity and sulphate in the area is 'acid rain' resulting from sulphur dioxide released during the pyrite smelting process. This is a combination of 'rain-out' and dry deposition, and will be washed into soil and percolation waters in the vicinity of the sulphur production plants.

Sulphate geochemistry of Xingwen surface waters and groundwaters

All of the field and isotopic data are reported in Table 2 and are discussed below in groups

Table 1. *Concentration and isotopic composition of sulphur in rocks and other materials*

Sample type	Sulphide concentration (wt %)				Isotopic composition (‰ V-CDT)			
	No.	Mean	SD	Range	No.	Mean	SD	Range
Silurian sandstone	4	0.04	0.04	<0.02–0.10	0*	–	–	–
Silurian siltstone	5	0.14	0.07	0.06–0.24	4*	−2.5	1.6	−1.0 to −4.1
Permian shale	6	0.55	0.20	0.31–0.81	7†	1.2	4.7	−6.0 to 5.8
Permian coal:								
Macroscopic pyrite	–	–	–	–	4	5.4	0.9	4.1–6.3
Disseminated pyrite	3	0.52	0.18	0.35–0.71	3	−5.4	3.0	−3.0 to −7.2
Pyrite ore‡	–	–	–	–	3	−10.7	4.7	−6.1 to −15.5
Product sulphur	–	–	–	–	3	−6.4	0.8	−5.7 to −7.2

*Too little sulphide for isotopic analysis in some or all samples.
†Includes one additional sample of macroscopic pyrite.
‡Separated macroscopic pyrite.

according to the type of catchment. Sample locations are shown on Figure 1.

Group I – rain and percolation waters (five samples). Rainwater sample 9A was collected in a cleaned plastic bowl during a heavy rain storm, whereas 9B was a larger sample collected from a plastic roof (after rain washing) for $BaSO_4$ precipitation for isotopic analysis. Their closely similar values of pH (6.24 and 6.31), conductivity (0.07 and 0.08 mS cm^{-1}) and sulphate (18 and 16 mg l^{-1}) confirm that 9B is uncontaminated with respect to sulphate.The pH is within the normal range for rainwater, but the conductivity and SO_4^{2-} concentration are rather high. The percolation waters are slightly alkaline (pH 7.2–7.8) and contain substantial concentrations of SO_4^{2-} (45–190 mg l^{-1}). The rain and percolation sulphate is of very uniform isotopic composition ($\delta^{34}S = +0.4$ to +1.6‰; $\delta^{18}O = +5.8$ to +7.1‰) suggesting a single source of SO_4^{2-}, almost certainly an atmospheric source from smelting operations (see below), the increased sulphate concentrations in the percolation waters over rainwater resulting from evapotranspiration effects and/or additional SO_4^{2-} derived from atmospheric dry deposition.

Group II – Silurian catchment (four samples). Waters draining the Silurian catchment were sampled at the major stream sink at Xiadong and in Tiencuan cave (Fig. 1). All have alkaline pH (7.6–8.1 at surface and 7.3 after transit through the aquifer), similar conductivities (0.18–0.23 mS cm^{-1}) and low SO_4^{2-} concentrations (36–48 mg l^{-1}) with similar isotopic compositions ($\delta^{34}S = -4.4$ to −4.9‰; $\delta^{18}O = +1.0$ to +1.6‰).

Group III – Permian shale catchments (four samples). These samples have low pH (3.0–4.4), high conductivity (1.2–1.5 mS cm^{-1}) and high SO_4^{2-} concentration (760–1600 mg l^{-1}). Their sulphate isotopic compositions are similar with $\delta^{34}S = +5.4$ to 6.9‰ and $\delta^{18}O$ between −1.4 and −2.2‰.

Group IV – mine drainage (three samples). Three samples (14 and 21) were collected from pyrite mines; they have low pH (2.9 and 3.3), very high conductivity (7.9 and 4.8 mS cm^{-1}), and 12 800 and 6400 mg l^{-1} SO_4^{2-}. The other sample (16) was collected issuing from fissures in limestone immediately below a disused mine and is therefore probably partially neutralized, having pH = 4.6, conductivity of 0.7 mS cm^{-1} and 320mg l^{-1} SO_4^{2-}. Their sulphate isotopic compositions are all similar, with $\delta^{34}S = +1.6$ to +3.4‰ and $\delta^{18}O = -1.8$ to −3.0‰.

Group V – waters contaminated with pyrite fines (five samples). Three samples (1, 7 and 23) were taken from the Mulangu River, which receives effluent fines from mineral-processing plants prior to sinking at Mulangu Dong cave (Fig. 2). The fines render the water turbid and have concentrations of 2600–10 200 mg l^{-1}, of which 177–240 mg is pyrite (the rest being mainly clay minerals). Dissolved SO_4^{2-} concentrations are 680–780 mg l^{-1} associated with low pH (3.8–4.0) and high conductivity (1.1–1.2 mS cm^{-1}). Two samples (18 and 24) are from the Heping Dong stream, which sinks at Heping Dong (Fig. 2) and again receives mineral-processing fines. These contained high levels of suspended solids (8600 and 9400 mg l^{-1}), of which 540 and 860 mg l^{-1}, respectively, was pyrite. These waters have very low pH (2.4 and 2.6), high conductivity (4.7 and 4.8 mS cm^{-1}) and high sulphate (6750 and 6800 mg l^{-1}). The suspended pyrite and aqueous sulphate isotope compositions of these

Table 2. *Analyses of dissolved and suspended components of water samples*

Sample No.	Description	pH	Conducting (mS cm^{-1})	Sample type	Concentration (mg l^{-1})	δ^{34}S V-CDT	δ^{18}O V-SMOW
9A	Rainwater	6.24	0.07	Dissolved SO_4^{2-}	18	–	–
9B	Rainwater	6.31	0.08	Dissolved SO_4^{2-}	16	+1.0	+6.4
8	Percolation inlet	7.78	0.59	Dissolved SO_4^{2-}	190	+1.2	+5.8
				Water	–	–	−7.37
11	Percolation inlet	7.23	0.39	Dissolved SO_4^{2-}	45	+0.4	+7.1
12	Percolation inlet	7.57	0.41	Dissolved SO_4^{2-}	100	+1.6	+5.9
				Water	–	–	−7.89
19	Percolation inlet	7.47	0.43	Dissolved SO_4^{2-}	120	+0.9	+6.1
5	Silurian catchment	8.13	0.20	Dissolved SO_4^{2-}	40	−4.4	+1.3
10	Silurian catchment	7.30	0.23	Dissolved SO_4^{2-}	48	−4.8	+1.6
20	Silurian catchment	7.35	0.21	Dissolved SO_4^{2-}	42	−4.6	+1.0
25	Silurian catchment	7.62	0.18	Dissolved SO_4^{2-}	36	−4.9	+1.5
13	Permian shale catchment	4.09	1.40	Dissolved SO_4^{2-}	1400	+5.9	−1.4
15	Permian shale catchment	2.95	1.40	Dissolved SO_4^{2-}	760	+6.9	−2.0
17	Permian shale catchment	4.37	1.23	Dissolved SO_4^{2-}	1150	+5.4	−2.2
				Water	–	–	−7.94
22	Permian shale catchment	3.65	1.51	Dissolved SO_4^{2-}	1600	+6.2	−1.7
14	Pyrite mine drainage	2.91	7.92	Dissolved SO_4^{2-}	12 800	+1.6	−3.0
16	Pyrite mine drainage	4.56	0.73	Dissolved SO_4^{2-}	320	+3.4	−1.8
21	Pyrite mine drainage	3.29	4.79	Dissolved SO_4^{2-}	6400	+2.1	−2.0
1	Contaminated stream	3.75	1.17	Dissolved SO_4^{2-}	680	−1.1	−0.2
				Suspended pyrite	204	−6.4	–
7	Contaminated stream	3.98	1.13	Dissolved SO_4^{2-}	780	−0.3	−0.9
				Suspended pyrite	241	−7.1	–
18	Contaminated stream	2.60	4.80	Dissolved SO_4^{2-}	7000	−1.3	−0.7
				Suspended pyrite	860	−8.1	–
23	Contaminated stream	3.81	1.22	Dissolved SO_4^{2-}	810	+0.1	−1.0
				Suspended pyrite	177	−6.2	–
24	Contaminated stream	2.42	4.67	Dissolved SO_4^{2-}	6750	−1.0	−0.1
				Suspended pyrite	540	−8.3	–

(*Continued*)

Table 2. *Continued*

Sample No.	Description	pH	Conducting ($mS\ cm^{-1}$)	Sample type	Concentration ($mg\ l^{-1}$)	$\delta^{34}S$ V-CDT	$\delta^{18}O$ V-SMOW
2	Pucaowan cave river	3.96	0.88	Dissolved SO_4^{2-}	520	+0.7	−1.3
3	Dong He resurgence river	4.67	0.73	Dissolved SO_4^{2-}	450	−0.6	−0.9
				Water	–	–	−8.04

samples all exhibit very narrow ranges. Pyrite ranges from −6.2 to −8.3‰, while aqueous sulphate is more enriched in ^{34}S (+0.1 to −1.3‰) and has $\delta^{18}O$ levels from −0.1 to −1.0‰.

Group VI – mixed aquifer waters. These were collected at Pucaowan cave, in the central part of the aquifer, and at the Dong He resurgence cave, where all of the aquifer water issues (Fig. 2). Both have pH values (4.0 and 4.7) far lower than normal karst waters and arise by mixing of karst waters with one or more of the acid sulphate waters discussed above (Bottrell 1993).

Summary – oxygen isotopic data. Four water samples were analysed for oxygen isotope composition and fall in a tight group between −7.37 and −8.04‰V-SMOW (Table 2, Fig. 2a). This is close to the mode of data from the IAEA Kunming station (−8.2‰V-SMOW), which lies at similar altitude and 350 km to the SW.

The sulphate oxygen isotopic data are plotted as a histogram in Figure 2a and show two major groupings: rainwater and percolation water sulphates group between +5.8 and +7.1‰; while the majority of other waters are ^{18}O-depleted, lying between −0.2 and −3.0‰. The Group II samples from the Silurian catchment lie at intermediate values at +1.3 and +1.6‰.

Summary – sulphur isotopic data. The sulphur isotopic data show a wide spread of values but the pairs or groups of samples from each catchment type have characteristic compositions (Fig. 2b). The Group III shale slope drainage waters are relatively ^{34}S-enriched at around +6‰, while mine drainage waters (Group IV) and processing-fines-contaminated waters (Group V) are lighter at around +2 to +3‰ and 0 to −1‰, respectively. It is noteworthy that Groups IV and V both have SO_4^{2-}-S significantly enriched in ^{34}S relative to pyrite in the fines, at around −6 to −7‰ (Table 2) and ores at −6 to −16‰ (Table 2). The Silurian catchment

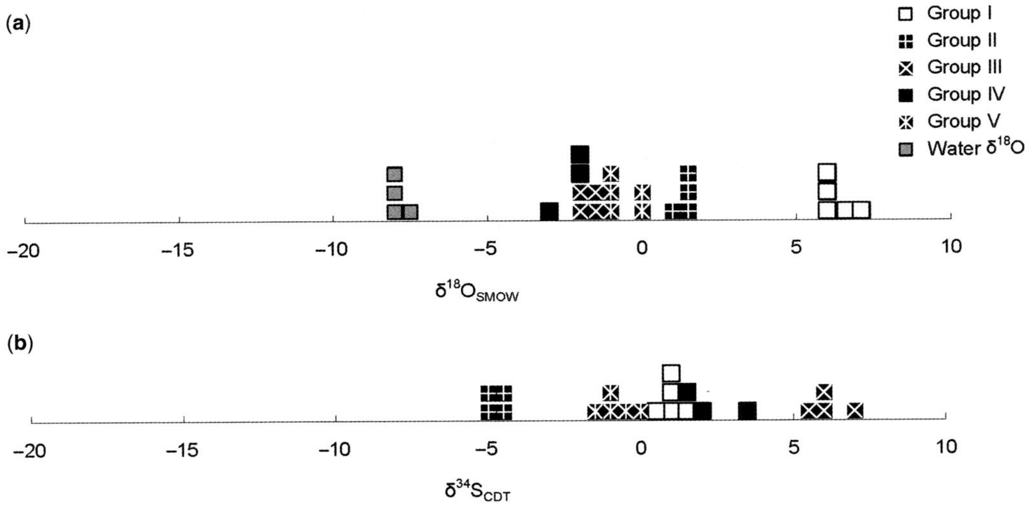

Fig. 2. Distributions of (**a**) oxygen and (**b**) sulphur isotopic compositions of aqueous sulphate in different water types distinguished in the text. The $\delta^{18}O$ of water samples are also plotted.

waters (Group II) are ^{34}S-depleted relative to all the others at −4 to −5‰.

Sulphur in the atmospheric-derived SO_4^{2-} (Group I) has an isotopic composition around +1‰; this is significantly ^{34}S-enriched relative to both the pyrite ore (Table 1) and the sulphur produced by smelting (at around −6 to −7‰, Table 1).

Processes affecting the isotopic composition of sulphate: how have different sulphate sources acquired their isotopic 'fingerprint'?

Oxygen isotopes. The O isotopic composition of aqueous SO_4^{2-} can be described in terms of two major factors. The source of O, either from ^{18}O-enriched atmospheric oxygen at +23‰ (Horibe *et al.* 1973) or water-derived O that is usually ^{18}O-depleted at 0‰ or lighter, exerts a primary control. However, isotopic enrichment effects may be associated with incorporation of O into SO_4^{2-}, and the overall situation is represented by a general equation:

$$\delta^{18}O_{sulphate} = (\delta^{18}O_{atm} + \varepsilon_{atm})F_a + (\delta^{18}O_{water} + \varepsilon_{water})(1 - F_a) \quad (3)$$

where $\delta^{18}O_{atm}$ and $\delta^{18}O_{water}$ are the O isotopic compositions of atmospheric and water-derived O, respectively; ε_{atm} and ε_{water} are the respective isotope enrichment factors associated with incorporation of O from that source into SO_4^{2-}; and F_a is the fraction of SO_4^{2-}-O derived from atmospheric O_2. This equation simplifies to:

$$\delta^{18}O_{sulphate} = 14.3F_a + \delta^{18}O_{water}(1 - F_a) \quad (4)$$

since the isotopic composition of O_{atm} varies little from +23‰ (Horibe *et al.* 1973) and ε_{atm} is −8.7‰ (Lloyd 1967). Thus, if $\delta^{18}O$ is known for both SO_4^{2-} and local environmental water, the fractions of atmospheric and water-derived oxygen in the SO_4^{2-} may be calculated. It must be noted, however, that this equation holds true only for the final O isotope composition of the SO_4^{2-} and cannot be used to directly infer mechanisms of reaction. As reaction intermediates must be involved in oxidation from S^{2-} to SO_4^{2-} (this involves the loss of eight electrons and no more than two may be lost in any one reaction step) and the possibility of O isotopic exchange with environmental water exists at each intermediate stage, the final SO_4^{2-}-O isotopic composition will represent an integration of signals due to both O source and O isotopic exchange.

The use of SO_4^{2-}-O isotopic composition has been proposed to distinguish between sulphide oxidation reactions utilizing O_2 and Fe^{3+} as oxidizing agents (Reactions 1 and 2 earlier) by Taylor *et al.* (1984). However, they assume that the O isotope composition of the SO_4^{2-} product is controlled simply by the stoichiometry of incorporation of atmospheric v. water-derived O according to these reactions. Should any intermediate sulphoxy anion be open to O isotopic exchange with water this assumption breaks down and the proportion of water-O atoms incorporated into sulphate will not reflect the importance of Reaction (2) v. Reaction (1).

Fractions of atmospheric-derived O, were calculated using Equation (4), for the samples of this study using a mean of the measured groundwater $\delta^{18}O$ values. Sulphate in rain and percolation water has 60–68% atmospheric O, while the main group of waters have SO_4^{2-} with only 22–35% atmospheric O. The consistency of both S and O isotope compositions between percolation and rainwater suggests that a single atmospheric source followed by increase in concentration by evapotranspiration or addition of a dry deposition sulphate component produced the percolation waters (without isotopic modification). The smelting of pyrite ore to produce S is the dominant local source of SO_2 that would initially be formed with atmospheric O. However, SO_2 undergoes rapid O isotope exchange with atmospheric water vapour and, since even in polluted atmospheres H_2O/SO_2 is approximately 10^6, the SO_2 acquires an O isotope composition controlled by meteoric water (Holt *et al.* 1981). The subsequent oxidation reactions of SO_2 to form SO_4^{2-} are complex. Addition of an atmospheric O_2 molecule to a fully water-equilibrated SO_2 molecule would form SO_4^{2-} with approximately a 50% atmospheric-O signature. However, if this oxidation reaction occurred sufficiently rapidly that the SO_2 was not fully equilibrated with water vapour, SO_4^{2-} with greater fractions of atmospheric O would result, as are observed here (see also Holt & Kumar 1991).

The main group of aqueous SO_4^{2-} averages 29% atmospheric O. A theoretical value of 25% atmospheric O would be obtained if the S^{IV} intermediate (dominantly HSO_3^- in this case, Hocking & Lee 1977) were to be fully isotopically water-equilibrated and the final oxidation step were addition of an O atom derived from an atmospheric source, i.e.:

$$HSO_3^- + \tfrac{1}{2}O_2 = H^+ + SO_4^{2-} \quad (5)$$

$SO_{2(aq)}$ becomes a minor but significant species at the low pH of many of these waters (Hocking & Lee 1977). Since O isotopic exchange between SO_2 and H_2O is rapid (Holt *et al.* 1981), O isotopic equilibration of HSO_3^- via $SO_{2(aq)}$ is likely to occur.

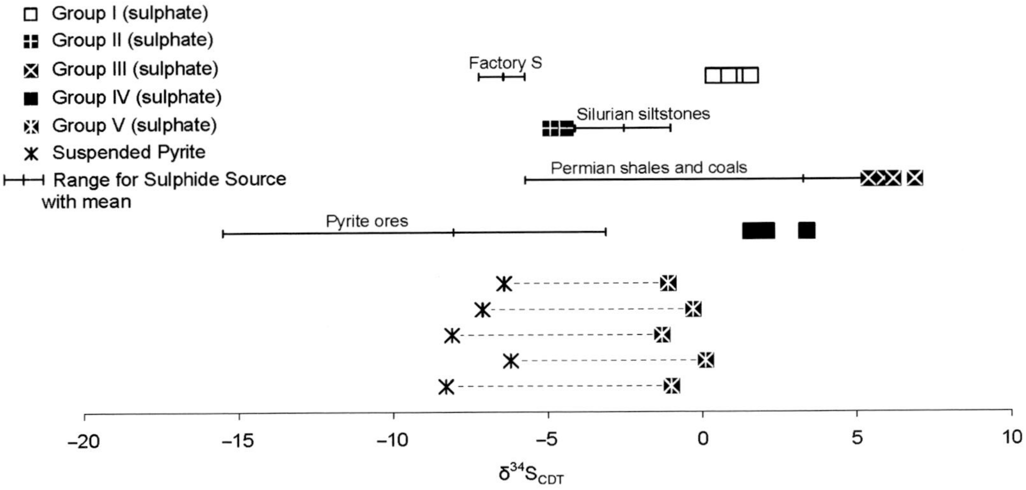

Fig. 3. $\delta^{34}S$ of sulphate in different water types compared to the range of $\delta^{34}S$ of their respective sulphur sources. For Group V the compositions of sulphate and suspended pyrite in individual samples are plotted. Note that for groups I, IV and V aqueous sulphate has $\delta^{34}S$ significantly higher than the sulphur source range.

Higher proportions of atmospheric-derived O will result in the final SO_4^{2-} if atmospheric O is utilized in earlier oxidation steps but the intermediates are oxidized to SO_4^{2-} before equilibration with water is complete.

The two Group II waters from the Silurian catchment have SO_4^{2-}-O isotopic compositions intermediate between those of SO_4^{2-} produced by the atmospheric oxidation routes discussed above, with 42 and 43% atmospheric-derived O. These intermediate compositions could be derived in one of two ways.

- These waters are low in SO_4^{2-} and could be a mixture of atmospheric SO_4^{2-} from S smelting and SO_4^{2-} derived by natural weathering of bedrock pyrite (with an O isotope composition similar to that of the other sulphates produced by aqueous oxidation reactions). A mixture of 40% atmospheric SO_4^{2-} and 60% aqueous SO_4^{2-} would yield the appropriate oxygen isotopic composition.
- The SO_4^{2-} in these waters originates entirely by natural weathering reactions of bedrock pyrite but that, because of the different conditions under which the reaction takes place (higher pH, low SO_4^{2-}), O isotopic exchange between S^{IV} species and water is restricted because of the lower concentrations of the required intermediates.

Sulphur isotopes. The $\delta^{34}S$ of each of the water types described earlier are compared to the source of pyrite in each catchment in Figure 3. Sulphate in drainage from the Silurian and Permian shale catchments (groups II and III) has $\delta^{34}S$ close to the range of substrate pyrite compositions. This is consistent with a majority of studies where sulphate produced by pyrite oxidation is found to be isotopically similar to the initial pyrite (Nakai & Jensen 1964; Toran & Harris 1989 and references therein).

In the case of the atmospheric SO_4^{2-} (Group I) the displacement in isotopic composition probably arises from isotopic equilibration between the small amount of SO_2 produced and the major S product, whilst both are vapours during smelting. The equilibrium isotopic fractionation between $S_{(g)}$ and $SO_{2(g)}$ is between 9 and 11‰ at temperatures between 350 and 450°C, with ^{34}S enrichment in the SO_2 (Grinenko & Thode 1970). The exact pyrite smelting temperature is uncertain, but this process would introduce a fractionation of the correct sense and magnitude to explain the observed approximately 8‰ enrichment in ^{34}S in the atmospheric SO_4^{2-} over the S product (Fig. 3).

The Group V waters contaminated by pyrite fines show ^{34}S enrichment in aqueous SO_4^{2-} of 5.3 and 6.8‰ relative to the suspended pyrite. In order to achieve such a net fractionation, some S species enriched in ^{32}S relative to SO_4^{2-} must be lost from the system. It is unlikely that this is a solid SO_4-bearing mineral, since alunite (which might form on neutralization of some of the Al-rich acidic waters: Bottrell 1993) does not exhibit significant S isotope fractionation relative to aqueous SO_4^{2-} (see the review by Rye *et al.* 1992). More likely is loss of S by degassing

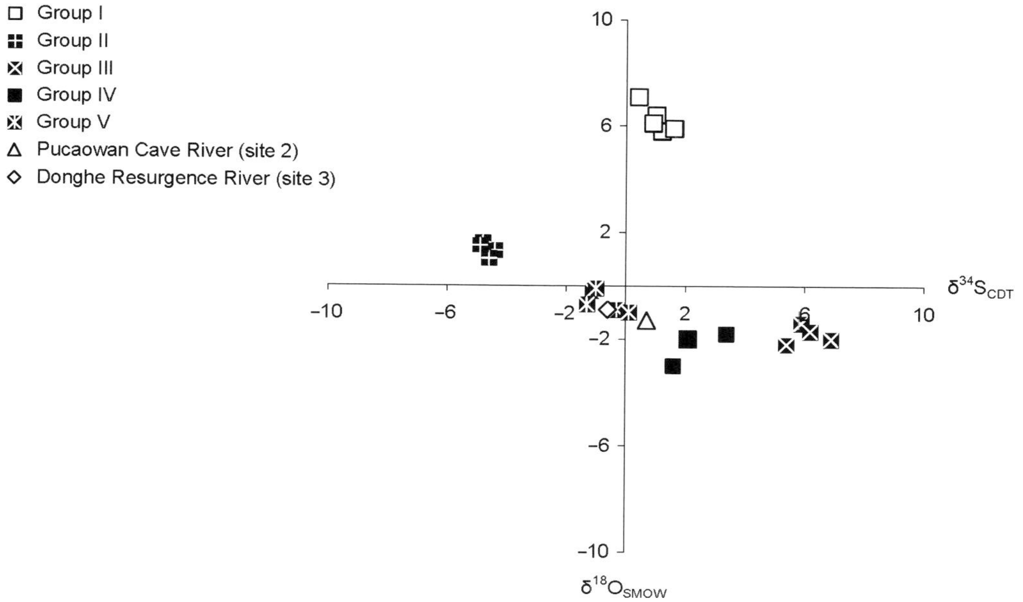

Fig. 4. Plot of $\delta^{18}O$ v. $\delta^{34}S$ for all aqueous sulphate samples (groups I–V and two mixed aquifer waters). Note the clear distinction in $\delta^{18}O$ between Group I samples and all others. Groups II–V all show clear distinction in terms of their $\delta^{34}S$ values.

of SO_2 from the waters. $SO_{2(aq)}$ is a significant species in these low pH waters (Hocking & Lee 1977) and evidence for SO_2 degassing comes from two sources. First, there is a strong sulphurous smell associated with waterfalls in the Mulangu River and an acid sulphurous haze over pyrite-contaminated rivers in the caves (Waltham & Willis 1993; Bottrell 1993). Secondly, a series of pairs of pH measurements taken above and below a 20 m waterfall in the Mulangu River show an increase in pH below the falls (of 0.15–0.66 pH units), which is consistent with loss of significant amounts of SO_2. If $SO_{2(aq)}$ were in S isotopic equilibrium with SO_4^{2-} (which might be possible with HSO_3^- and HSO_4^- acting as exchange intermediates in these low pH waters) an isotopic fractionation of 30‰ would result at 20 °C (the temperature of the surface waters; calculation based on Sakai 1968). If this fractionation mechanism pertains in this case then loss of 20% of aqueous S as $SO_{2(g)}$ with oxidation of the remainder to SO_4^{2-} would produce the observed ^{34}S enrichment of approximately 6‰ between suspended pyrite and aqueous SO_4^{2-}.

The mine drainage waters (Group IV) show even larger ^{34}S enrichment in aqueous SO_4^{2-} relative to pyrite ore (*c.* 9‰, Fig. 3). However, in these cases fractionation processes may not be the only factor as waters entering the mines through the overlying shales may already contain SO_4^{2-} enriched in ^{34}S.

Sulphate isotopes as tracers and the aquifer sulphate budget

The SO_4^{2-} – $\delta^{18}O$ and –$\delta^{34}S$ data are combined in Figure 4. This highlights the effectiveness of SO_4^{2-}-^{18}O in distinguishing sulphate from atmospheric (Group I, $\delta^{18}O$ around +6 to +7‰) and hydrospheric oxidation pathways (groups II–V, $\delta^{18}O$ from +1.5 to −3‰). The hydrospheric-derived SO_4^{2-} shows a spread of S isotope compositions allowing the different groupings to be distinguished. Sulphate in the mixed waters from within the aquifer are both relatively ^{18}O depleted, falling amongst the hydrospherically-derived group, and neither therefore contains any significant component of atmospheric-derived SO_4^{2-}. The low SO_4^{2-} concentration of the ^{34}S-depleted Group II waters precludes them contributing any significant component to either of the mixed aquifer waters, as the latter have high SO_4^{2-} concentration (Table 2) and S isotope compositions closer to the ore-derived sulphates. Indeed, the discharge from the aquifer at the Donghe spring has SO_4^{2-}-S isotope compositions almost identical to the Group V (pyrite-fines-contaminated) waters. The high sulphate in the discharge from the aquifer as a whole is thus dominantly the result of contamination by processing fines and not natural inputs of high sulphate waters from the Permian shale catchments or mine drainage. In the water sampled at Puccaowan cave, however, sulphate from either or both Group IV (mine drainage) or Group III (shale catchment) waters are present at significant concentrations relative to the processing fines contribution. The very large component of contamination by processing fines compared to mine drainage in the aquifer as a whole indicates that remediation efforts would probably be most beneficial if focused on processing-plant effluent.

This study of the Xingwen aquifer demonstrates the effectiveness of sulphate stable isotopes in distinguishing between different natural and anthropogenic sources of sulphate. This indigenous isotopic tracer provides a powerful tool for assessing the impact of acid mine drainage on karst groundwater at Xingwen. Sulphate stable isotopes thus provide a powerful indigenous tracer of solute and contaminant sources that may be of potential use in a variety of surface water and groundwater studies.

Conclusions

- The stable isotope composition of aqueous SO_4^{2-} is shown to be effective at discrimination of different SO_4^{2-} sources in the Xingwen karst. Isotopic composition proves to be a powerful indigenous tracer that enables the relative contributions of different aquifer sulphate sources to be assessed and the impact of pollutant sulphate on groundwaters to be quantified.
- Oxygen isotopic compositions of SO_4^{2-} are effective at distinguishing atmospheric from aqueous oxidation pathways owing to the different oxidation mechanisms characterizing these environments. The final O isotope composition is controlled by the nature of the S^{IV} species (SO_2 in air, HSO_3^- in water) that approaches isotopic equilibrium with water (vapour or liquid) prior to addition of either one or two atmospheric O atoms to form SO_4^{2-}.
- Sulphate-S isotopic compositions are related to the isotopic composition of the sulphide source. However, in low pH waters there is evidence of fractionation of S isotopes during oxidation, apparently related to loss of ^{34}S depleted SO_2.

Fieldwork for this project was undertaken as part of the 1992 British China Caves Project expedition to Xingwen organized by A. Eavis and Prof. Z. Xuewen. I thank all the expedition's sponsors, the team members and our Chinese hosts for making the work possible and enjoyable. The early parts of this work were completed at the Institute of Geological and Nuclear Sciences, Lower Hutt, New Zealand as part of a Nuffield Fellowship. I thank B. Robinson and his colleagues for their help and discussion. Later analyses at Leeds were supported by NERC grant GR3/8134 when D. Hatfield and M. McCarthy assisted with many of the analyses. I. Fairchild and M. Petitta provided constructive reviews that helped to improve this paper.

References

BERNER, R. A. 1970. Sedimentary pyrite formation. *American Journal of Science*, **268**, 1–23.

BIERLEIN, F. P., ASHLEY, P. M. & SECCOMBE, P. K. 1996. Origin of hydrothermal Cu–Zn–Pb mineralization in the Olary Block, South Australia: evidence from fluid inclusions and sulphur isotopes. *Precambrian Research*, **79**, 281–305.

BOTTRELL, S. H. 1993. Water chemistry in the Xingwen caves, China. *Cave Science*, **20**, 87–92.

BOTTRELL, S. H. & RAISWELL, R. 2000. Sulphur isotopes and microbial sulphur cycling in sediments. *In*: RIDING, R. E. & AWRAMIK, S. M. (eds) *Microbial Sediments*. Springer, Heidelberg, 96–104.

BOTTRELL, S. H., WEBBER, N., GUNN, J. & WORTHINGTON, S. R. H. 2000. The geochemistry of sulphur in a mixed allogenic-autogenic karst catchment, Castleton, Derbyshire, UK. *Earth Surface Processes and Landforms*, **25**, 155–165.

DE CARITAT, P., KIRSTE, D., CARR, G. & MCCULLOCH, M. 2003. Groundwater in the Broken Hill region, Australia: recognising interaction with bedrock and mineralisation using S, Sr and Pb isotopes. *Applied Geochemistry*, **20**, 767–787.

EDRAKI, M., GOLDING, S. D., BAUBLYS, K. A. & LAWRENCE, M. G. 2003. Hydrochemistry, mineralogy and sulfur isotope geochemistry of acid mine drainage at the Mt. Morgan mine environment, Queensland, Australia. *Applied Geochemistry*, **20**, 789–805.

EPSTEIN, S. & MAYEDA, T. 1953. Variation of ^{18}O content of waters from natural sources. *Geochimica et Cosmochimica Acta*, **4**, 213–224.

GRINENKO, V. A. & THODE, H. G. 1970. Sulphur isotope effects in volcanic gas mixtures. *Canadian Journal of Earth Sciences*, **7**, 1402–1409.

GUNN, J., BOTTRELL, S. H., LOWE, D. J. & WORTHINGTON, S. R. H. 2006. Deep groundwater flow and geochemical processes in limestone aquifers: evidence from thermal waters in Derbyshire, England, UK. *Hydrogeology Journal*, **14**, 868–881.

HITCHON, B. & KROUSE, H. R. 1972. Hydrogeochemistry of the surface waters of the Mackenzie River drainage basin, Canada: III. Stable isotopes of oxygen, carbon and sulphur. *Geochimica et Cosmochimica Acta*, **36**, 1337–1357.

HOCKING, M. B. & LEE, G. W. 1977. Calculated sulfur dioxide equilibria at low concentrations between air and water. *Water, Air and Soil Pollution*, **8**, 255–262.

HOLT, B. D. & KUMAR, R. 1991. Oxygen isotope fractionation for understanding the sulphur cycle. *In*: KROUSE, H. R. & GRINENKO, V. A. (eds) *Stable Isotopes: Natural and Anthropogeniuc Sulphur in the Environment*. Wiley, Chichester, 27–41.

HOLT, B. D., KUMAR, R. & CUNNINGHAM, P. T. 1981. Oxygen-18 study of the aqueous-phase oxidation of sulphur dioxide. *Atmospheric Environment*, **15**, 557–566.

HORIBE, Y., SHIGEHARA, K. & TAKAKUWA, Y. 1973. Isotope separation factors of carbon dioxide–water system and isotopic composition of atmospheric oxygen. *Journal of Geophysical Research*, **78**, 2625–2629.

IVANOV, M. V. 1983. Major fluxes of the global biogeochemical cycle of sulphur. *In*: IVANOV, M. V. & FRENEY, J. R. (eds) *The Global Biogeochemical Sulphur Cycle*. Wiley, Chichester, 449–463.

KIMBLIN, R. T. 1995. The chemistry and origin of groundwater in Triassic sandstone and Quaternary deposits, northwest England and some UK comparisons. *Journal of Hydrology*, **172**, 293–311.

KIRSTE, D., DE CARITAT, P. & DANN, R. 2003. The application of the stable isotopes of sulfur and oxygen in groundwater sulfate to mineral exploration in the Broken Hill region of Australia. *Journal of Geochemical Exploration*, **78**, 81–84.

KROTHE, N. C. & LIBRA, R. D. 1983. Sulfur isotopes and hydrochemical variations in spring waters of southern Indiana, USA. *Journal of Hydrology*, **61**, 267–283.

KROUSE, H. R. 1977. Sulphur isotope abundances elucidate uptake of atmospheric sulphur emissions by vegetation. *Nature*, **265**, 45–46.

LI, X., MASUDA, H., KUSAKABE, M., YANAGISAWA, F. & ZENG, H. 2006. Degradation of groundwater quality due to anthropogenic sulfur and nitrogen contamination in the Sichuan Basin, China. *Geochemical Journal*, **40**, 309–332.

LLOYD, R. M. 1967. Oxygen-18 composition of oceanic sulphate. *Science*, **156**, 1228–1231.

MAYER, B., PRIETZEL, J. & KROUSE, H. R. 2001. The influence of sulfur deposition rates on sulfate retention patterns and mechanisms in aerated forest soils. *Applied Geochemistry*, **16**, 1003–1020.

MCCARTHY, M. D. B., NEWTON, R. J. & BOTTRELL, S. H. 1998. Oxygen isotopic compositions of sulphate from coals: implications for primary sulphate sources and secondary weathering processes. *Fuel*, **77**, 677–682.

MONCASTER, S. J., BOTTRELL, S. H., TELLAM, J. H., LLOYD, J. W. & KONHAUSER, K. O. 2000. Migration and attenuation of agrochemical pollutants: insights from isotopic analysis of groundwater sulphate. *Journal of Contaminant Hydrology*, **43**, 147–163.

NAKAI, N. & JENSEN, M. L. 1964. The kinetic isotope effect in the bacterial reduction and oxidation of sulfur. *Geochimica et Cosmochimica Acta*, **28**, 1893–1912.

NEWTON, R. J., BOTTRELL, S. H., DEAN, S. P., HATFIELD, D. & RAISWELL, R. 1995. An evaluation of the use of the chromous chloride reduction method for isotopic analyses of pyrite in rocks and sediment. *Chemical Geology*, **125**, 317–320.

NOVAK, M., BOTTRELL, S. H., FOTTOVA, D., BUZEK, F., GROSCHEOVA, H. & ZAK, K. 1996. Sulphur isotope signals in forest soils of Central Europe along an air-pollution gradient. *Environmental Science and Technology*, **30**, 3473–3476.

OHMOTO, H. & RYE, R. O. 1979. Isotopes of sulfur and carbon. *In*: BARNES, H. L. (ed.) *Geochemistry of Hydrothermal Ore Deposits*. Wiley, New York, 509–567.

ROBINSON, B. W. & BOTTRELL, S. H. 1997. Discrimination of sulphate sources in pristine and polluted New Zealand river catchments using stable isotopes. *Applied Geochemistry*, **12**, 305–319.

ROBINSON, B. W. & KUSAKABE, M. 1975. Quantitative preparation of sulphur dioxide, for $^{34}S/^{32}S$ analysis, from sulphides by combustion with cuprous oxide. *Analytical Chemistry*, **47**, 1179–1181.

RYE, R. O., BETHKE, P. M. & WASSERMAN, M. D. 1992. The stable isotope geochemistry of acid sulfate alteration. *Economic Geology*, **87**, 225–262.

SAKAI, H. 1968. Isotopic properties of sulphur compounds in hydrothermal processes. *Geochemical Journal*, **2**, 29–41.

TAYLOR, B. E., WHEELER, M. C. & NORDSTROM, D. K. 1984. Isotope composition of sulphate in acid mine drainage as measure of bacterial oxidation. *Nature*, **308**, 538–541.

TORAN, L. & HARRIS, R. F. 1989. Interpretation of sulfur and oxygen isotopes in biological and abiological sulfide oxidation. *Geochimica et Cosmochimica Acta*, **53**, 2341–2348.

WALTHAM, A. C. & WILLIS, R. G. 1993. *Xingwen.China Caves Project 1989–1992*. British Cave Research Association, Bridgwater.

WALTHAM, A. C., BROOK, D. & BOTTRELL, S. H. 1993. The caves and karst of Xingwen, China. *Cave Science*, **20**, 75–86.

Intrinsic vulnerability of the Alburni karst system (southern Italy)

D. DUCCI

Department of Geotechnical Engineering, 'Federico II' University, Piazzale Tecchio, 80, 80125 Naples, Italy (e-mail: daniela@unina.it)

Abstract: In order to, respectively, assess the resource vulnerability and the source vulnerability of the Alburni karst system (southern Italy), the COP and VULK methods have been applied in the framework of the 'European Approach' as proposed in 2003 by the COST (Co-Operation in Science and Technology) 620 task group – Action: 'Vulnerability and Risk Mapping for the Protection of Carbonate – Karst – Aquifers'. The Alburni massif, with more than 500 caves, is the most important karst area of southern Italy. The drainage network is not continuous; there are several endoreic basins and large zones without superficial drainage but with many sinking streams (streams disappearing into swallow holes) and dolines. The caves (more than 100), well explored by speleologists, directly communicate with the above mentioned infiltration areas. The most important ones are hierarchically well organized and extend over a wide area. Some of them recharge important springs with very short transit times. The final vulnerability map shows that the prevalent vulnerabilities are high and very high owing to the widespread karstification and the presence of large vegetated areas with gentle slopes that favour fast infiltration on the plateau.

Groundwater protection is a major environmental concern in many countries. For example, in Europe more than 50% of water supplies are obtained from groundwater. Carbonate rocks, which often contain karst aquifers that are extremely vulnerable to contamination, crop out in 35% of the European territory (Daly *et al.* 2002).

In the last 30 years, several methods of mapping the intrinsic vulnerability of groundwater to contamination have been developed, based on different approaches: hydrogeological setting, parametric system models, index methods, and mathematical and statistical models. At present, the parametric system models like DRASTIC (Aller *et al.* 1987) and SINTACS (Civita & De Maio 2000) are widely used and favoured by the advent of geographical information systems (GIS).

Intrinsic vulnerability only depends on the natural properties of the aquifer (soil, lithology, hydraulic properties, recharge, etc.) and is independent of the nature of the contaminant. In karstic environments, the properties of the aquifer are highly anisotropic with concentrated recharge via dolines and swallow holes, and generally very high velocities because of conduit porosities (Gunn 1986).

Some methods for vulnerability mapping provide only adaptations for application in karst environment, as the less important role of the 'depth to water' (e.g. SINTACS: see applications in Corniello & Ducci 2000; Corniello *et al.* 2004; Ducci & Rossi 2007). Other methods are especially designed for karst environment, taking into account the function of the epikarst and of the karstic network (e.g. EPIK: Doerfliger & Zwahlen 1998). A comparison between SINTACS and EPIK – methods, carried out in Corniello & Ducci 2000, using qualitative and statistic approaches – showed marked differences. The result obtained using the EPIK method was much more severe, attaching great importance to the karstic landform, while the SINTACS method indicated a lower degree of groundwater resource vulnerability.

In this study, a new approach, proposed by the COST 620 task group Action: 'Vulnerability and Risk Mapping for the Protection of Carbonate – Karst – Aquifers', has been applied to the highly karstified area of the Alburni Massif, located in the southern part of the Italian peninsula.

Hydrogeological setting of the Alburni karst system

The carbonate massif of the Alburni Mountains is one of the more important hydrogeological structures of the Campania region in southern Italy. It stretches for about 23 km along a NW–SE axis, is 9–10 km wide, covers 246 km^2 (Fig. 1), has a mean elevation of about 790 m above sea level (a.s.l.) and a maximum elevation of 1742 m. The main areas of concentrated infiltration are located at altitudes ranging between 1000 and 1600 m, while much of the water table and the most important springs range in elevation between 60 and

From: PARISE, M. & GUNN, J. (eds) *Natural and Anthropogenic Hazards in Karst Areas: Recognition, Analysis and Mitigation*. Geological Society, London, Special Publications, **279**, 137–151. DOI: 10.1144/SP279.12
0305-8719/07/$15.00

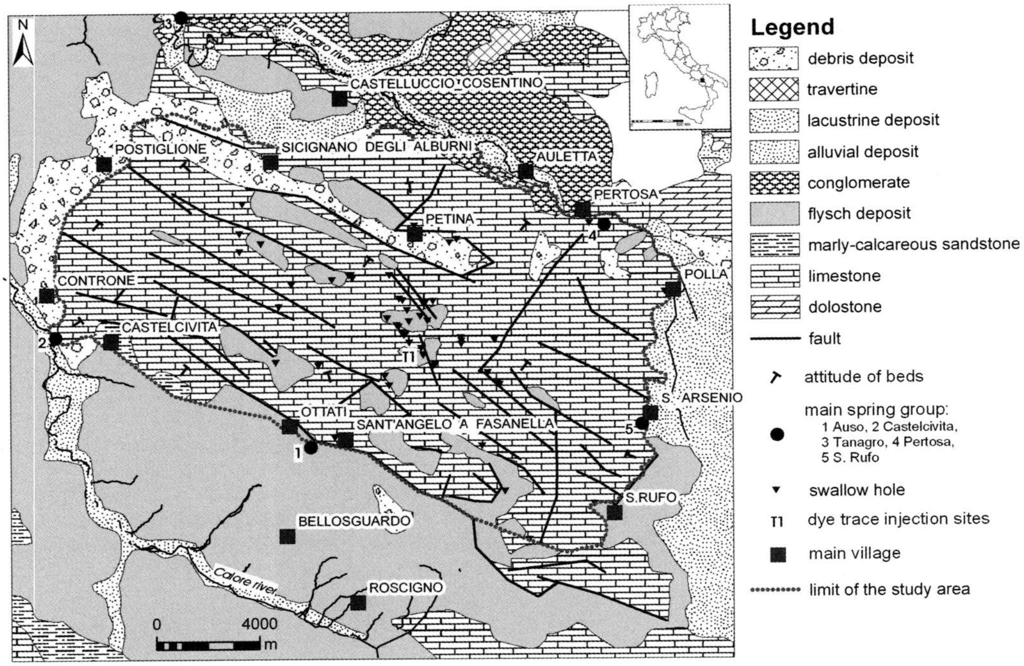

Fig. 1. Hydrogeological map of the Alburni karst system.

150 m. The structure of the Alburni Mountains is monoclinal, inclined to the SW and is well delimited by fault systems. Pliocene–Quaternary extensional tectonics gave a strong contribution to the present morphological and structural setting of the chain. In particular, east of the Alburni ridge, a complex E-dipping normal fault system has been identified by Berardi *et al.* (1996) and Brozzetti & Salvatore (2004).

The massif is composed predominantly of thick Mesozoic limestones on which Miocene flysch has been preserved in small structural depressions; lesser amounts of Quaternary clastic deposits outcrop mostly along the northern and western borders (Fig. 1). The presence of several karst forms influences groundwater flow, which is very complex (Bellucci *et al.* 1991, 1995; Santo 1993; Celico *et al.* 1994). The drainage network is not continuous. There are several endoreic basins and large zones without superficial drainage but with many swallow holes and dolines. The areas of concentrated infiltration are represented by dolines located on the plateau, especially in the central sector, that recharges large volumes into some swallow holes following heavy rainfall. The caves (more than 100), well explored by speleologists in the past, directly communicate with the abovementioned infiltration areas. The most important ones are hierarchically well organized and extend over a wide area.

Four groundwater basins are known to feed the main springs (Celico *et al.* 1994).

Along the NE boundary of the mountains is the Pertosa group, one of the main basal springs of the massif. It has a mean discharge of 1.1 m^3 s^{-1} and flows partly inside the karstic unit and partly along the Tanagro River course. Along the NW side of the mountains is the Low Tanagro River group, a set of springs located along a 3 km-long segment of the river causing stream discharge to increase to about 3.5 m^3 s^{-1} (Celico *et al.* 1990).

In the SW sector, the main spring group is Castelcivita, flowing partly inside the karstic unit and partly along the course of the Calore River. The mean discharge is 1.8 m^3 s^{-1} and the springs are connected directly with the infiltration area by means of different conduits, as indicated from their quick increase in discharge and variations in physical and chemical parameters during heavy rainfall events (Santo 1993).

Similarly, the Auso spring group, located in the central part of the southern boundary, has a highly variable discharge ranging from l s^{-1} to m^3 s^{-1} and is connected with the infiltration area by means of karstic conduits as demonstrated by fluorescein tracer tests (Bellucci *et al.* 1991).

Pollution vulnerability assessment of the Alburni karst system in the framework of the 'European Approach'

In the year 2000, the European Water Directive stated 'water is not a commercial product like any other but, rather, a heritage which must be protected, defended and treated as such'. Following this statement, in the year 2003, a task group of COST 620 (Action: 'Vulnerability and Risk Mapping for the Protection of Carbonate – Karst – Aquifers') proposed the European Approach in the COST 620 Final Report (Zwahlen 2003).

The European Approach to groundwater vulnerability mapping is based on an origin–pathway–target conceptual model (Fig. 2). The possible contamination event is assumed to originate at the land surface. For resource protection, the groundwater surface in the aquifer is the target; for source protection, the spring or well is the target. For resource protection, the pathway extends through the vadose zone (also referred to as the overlying layers) and for source protection, it includes passage through the aquifer's phreatic zone (Goldscheider 2002).

The European Approach adopted the K (karst network), P (protective cover) and I (infiltration) factors from the EPIK method (Doerfliger & Zwahlen 1998), although much modified. Epikarst, although considered an important factor, was incorporated into the P factor. The European Approach uses four factors in assessing intrinsic vulnerability: Overlying layers (O), Concentration of flow (C), Precipitation regime (P) and Karst network development (K) (Fig. 2). The factors O, C and K represent the internal characteristics of the system, while the P factor is an external stress applied to the system.

For resource vulnerability mapping, the factors O, C and P should be taken into consideration, while the factor K should be additionally taken into account for source vulnerability mapping. The European Approach does not specify how the factors should be measured or categorized. It is not a methodology. However, the approach could be applied, and has been applied at a variety of test sites, using locally developed methods: e.g. PI (Goldscheider *et al.* 2000; Goldscheider 2002); COP (Vías *et al.* 2002); VULK (Jeannin *et al.* 2001); Time-Input (Kralik 2001). In this study, the COP method (Vías *et al.* 2002) was applied to map the resource contamination vulnerability of the Alburni karst system area. To assess the source vulnerability of the Auso springs, the additional factor K of the European Approach has been considered using the VULK method (Jeannin *et al.* 2001).

Input data

The topographical maps used for this study are 1:25 000–50 000 in scale and are overlain by a 30 m grid. For the calculation of a sinking stream's distance, discussed below, a 5 m resolution has been used. The slope inclination has been calculated using a GIS digital elevation model partially supplied from 'Comunità Montana Alburni' and partially from new digitalization.

Geological information for the model was provided by a 1:100 000 geological map, 1:50 000 hydrogeological maps and, partially, by 1:10 000 hydrogeological maps. These also have provided, together with published hydrogeological cross-sections (Bellucci *et al.* 1991, 1995; Santangelo & Santo 1997), the basis to estimate the thickness of the hydrogeological units. The karst forms (dolines, swallow holes, caves and sinking streams) have been recognized from topographic maps, from black and white aerial photographs (1:15 000) and, especially, from published speleological data derived from field investigation (Bellucci *et al.* 1995).

Areas without of permanent vegetation are categorized as land-use classes 3.3 (sparsely vegetated areas) according to Corine Land Cover classification (Büttner *et al.* 2002). The Corine Map has been partially supplied from Comunità Montana Alburni and partially extracted from 'Regione Campania' digital data. Soil types and thicknesses were obtained from pedological maps and published data derived from field investigations (di Gennaro 2002).

To evaluate the yearly rainfall amount and the number of rainy days for wet years, the rainfall rate and the number of days were calculated for each year by interpolating precipitation data from 22 rain gauging stations using kriging interpolation techniques. The process to construct the annual rainfall maps was carried out in a GIS environment. The wet years were identified by comparing the 47 annual rainfall raster maps (1951–1997 published data from the Italian Hydrographic Survey) and selecting the 20 years with highest values of rainfall.

The flow velocity in the groundwater basin feeding the Auso springs required by the VULK method for the calculation of the K factor, has been derived from direct measurements by means of field tracer tests using fluorescein (Bellucci *et al.* 1991) during maximum velocity flood flow conditions.

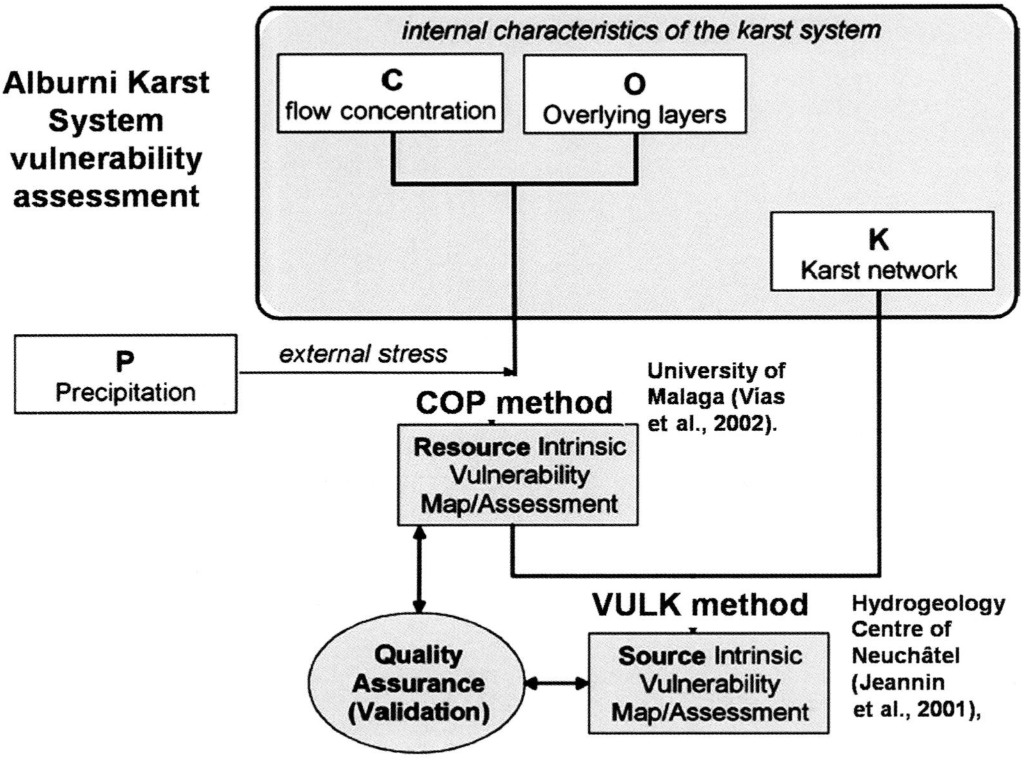

Fig. 2. The application of the 'European Approach' to assess groundwater vulnerability (modified from Zwahlen 2003): creation of resource and source vulnerability maps by a combination of the factors C, O and P evaluated by the COP method (Vías *et al.* 2002) and K evaluated by the VULK method (Jeannin *et al.* 2001).

The resource vulnerability map evaluated by the COP method

The COP method (Vìas *et al.* 2002) is based on three factors (flow Concentration, Overlying layers and Precipitation), as is the European Approach for assessing the intrinsic vulnerability of groundwater resources (Daly *et al.* 2002; Zwahlen 2003). Two conditions are necessary to assess intrinsic vulnerability by the COP method. First, the contaminant depends on the characteristics of the water to move through the aquifer. Second, the contaminant infiltrates from the surface by means of rainfall.

The O factor refers to the protection of the vadose zone against a contaminant event. The C and P factors are used as modifiers that correct the degree of protection provided by the overlying layers (O factor). The COP method is summarized in Figure 3, which shows the different values assigned to the variables evaluated for each factor, the procedure used to calculate the C, O and P factors, and the final vulnerability index.

The 'O' factor

The O factor indicates the capability of the vadose zone, by means of various processes, to filter or attenuate contaminants and thus reduce their adverse effects.

Two data layers are used in order to evaluate the O factor, each related to layers in the vadose zone with a different hydrogeological behaviour: the soil [O_S] and the lithology of the vadose zone [O_L]. The soil subfactor [O_S] deals with the biologically active zone of weathering, composed of minerals, organic substance, water, air and living matter. It comprises the A and B pedological horizons, and is determined by their texture (clayey, silty, loam or sandy) and thickness (>1 m, 0.5–1 m, <0.5 m) (Figs 3 & 4).

The lithology subfactor [O_L] reflects the attenuation capacity of each rock type of the vadose zone. For its quantification, three parameters were adopted (Figs 3 & 5):

- lithology and fractures (ly in Fig. 3);
- thickness of each rock type (m in Fig. 3);
- degree to which an aquifer is confined (cn in Fig. 3).

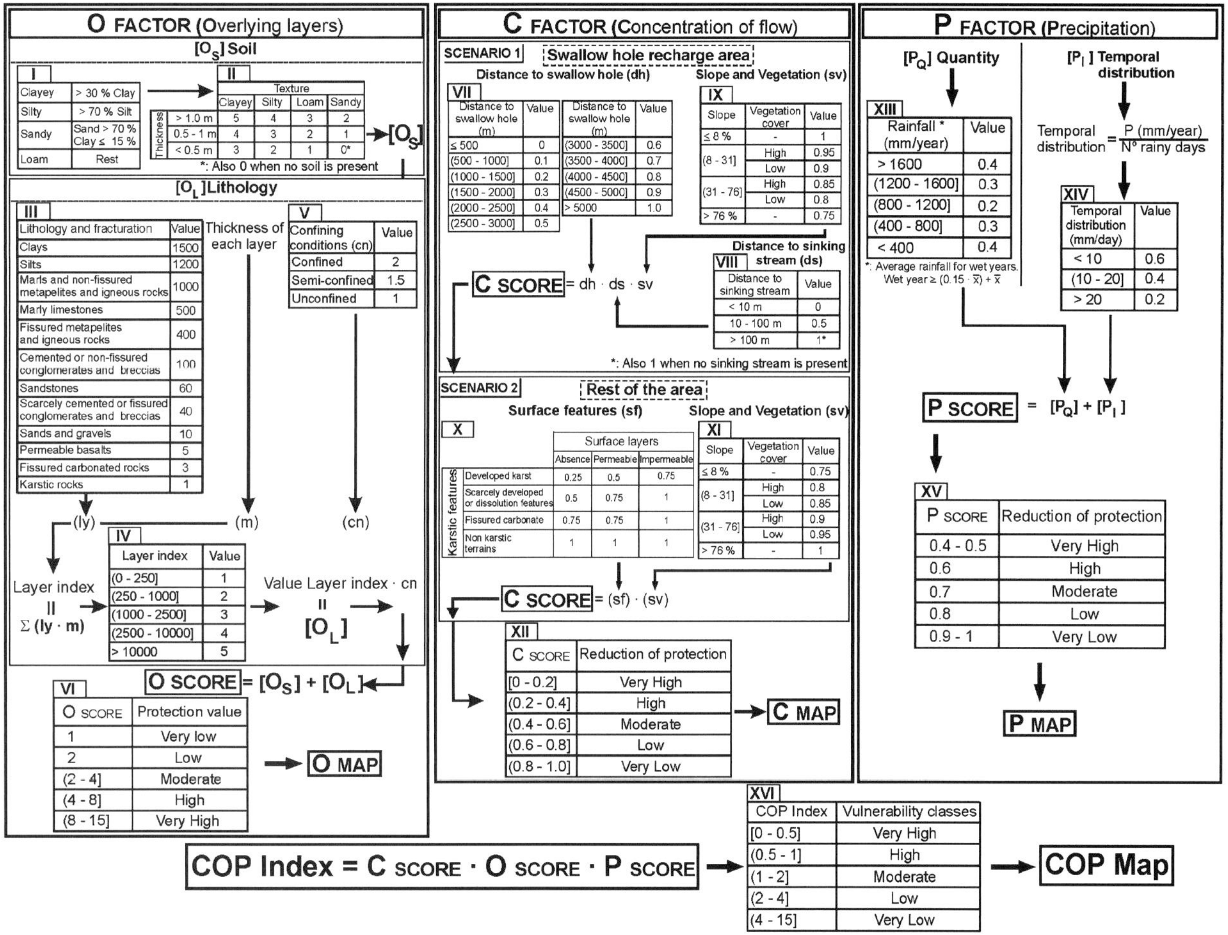

Fig. 3. Diagram of the COP method, showing the differentiation of the C, O and P factors (from Zwahlen 2003).

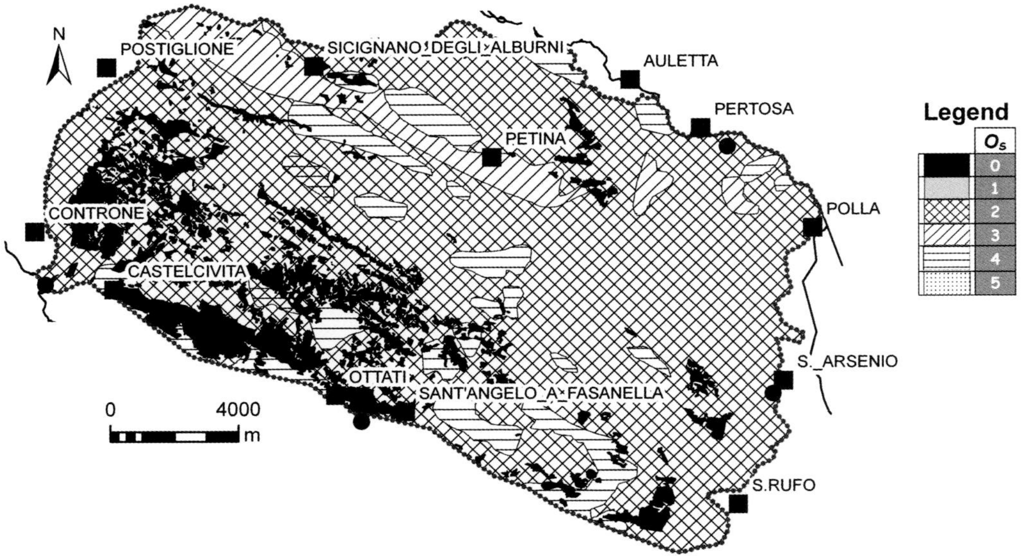

Fig. 4. The soil subfactor [O_S] of the COP method for the Alburni karst system.

In the study area, the lithology consists of limestone, marly-calcareous sandstone, flysch deposits and debris deposits. The aquifer is unconfined and the thickness of the vadose karst ranges from few metres to approximately 1300 m.

The resulting O factor ($O_{SCORE} = O_S + O_L$) is predominantly in the moderate protection class (Figs 3 & 6).

The 'C' factor

The C factor takes into account the surface conditions that control water flowing towards zones of rapid infiltration and then represents the degree of concentration of flow towards karstic features that are directly connected with the phreatic zone.

Two conditions are modelled: (1) the catchment area of a stream sinking through a swallow hole;

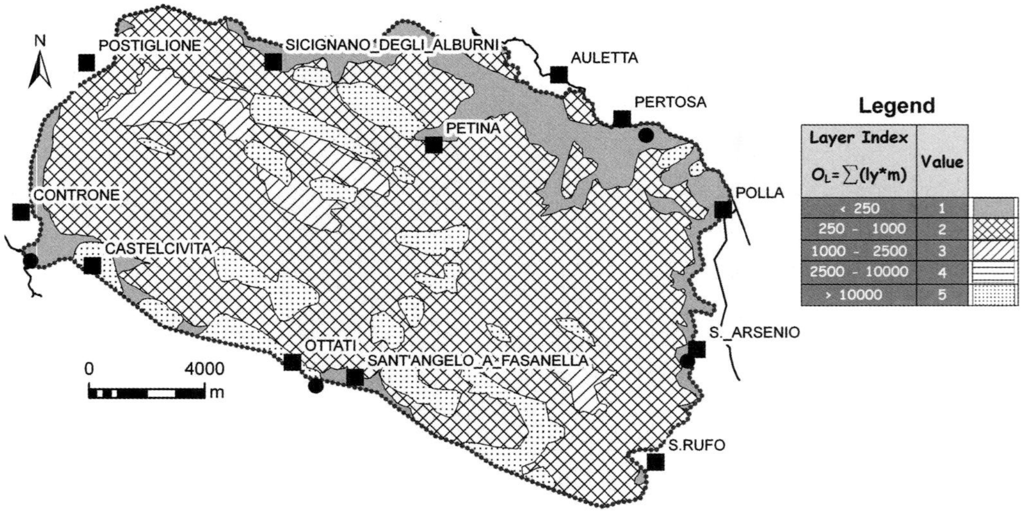

Fig. 5. The lithology subfactor [O_L] of the COP method for the Alburni karst system.

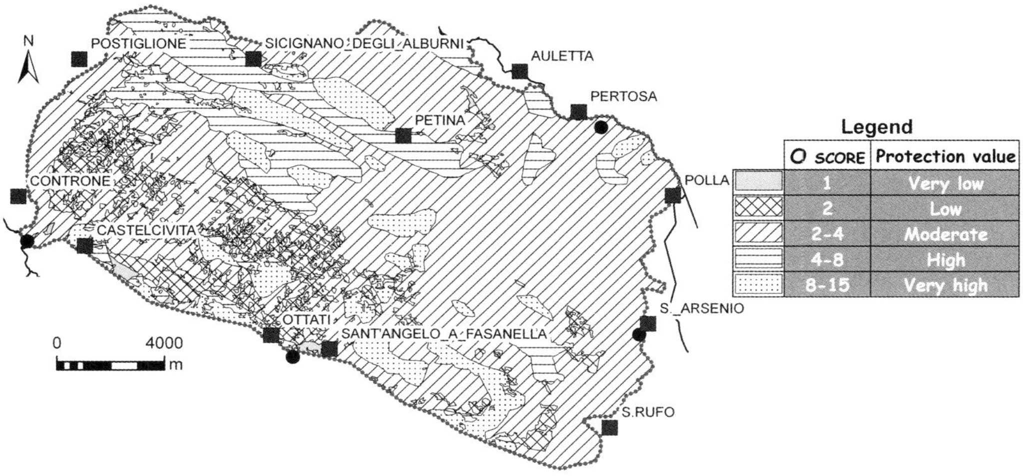

Fig. 6. The overlying layers of the COP method [$O_{SCORE} = O_S + O_L$] for the Alburni karst system.

and (2) the rest of the area. In the first condition, in which almost all of the study area is included, the C factor requires accounting for four variables:

- distance to swallow hole (dh in Fig. 3) (Fig. 7);
- distance to sinking stream (ds in Fig. 3) (Fig. 8);
- slope (s in Fig. 3) (Fig. 9);
- vegetation (v in Fig. 3) (Figs 10 & 11).

The classification of the variables, and the corresponding values in terms of protection or vulnerability, are assigned according to the COP method requirements shown in Figure 3. Nevertheless, in some cases the values proposed seem unjustified, as for example the distance to sinking stream (as explained below) and the slope classification (Veni 1997 found that in the Edwards Plateau area of the United States, the majority of vadose cave development occurs on slopes of less than 5%).

In the study area, where there are many swallow holes and sinking streams, the C factor ($C = dh \times ds \times sv$) is prevalently in a very high and high vulnerability class (called very high and high reduction of protection in the COP method) (Figs 3 & 12).

It is crucial to make a criticism about the classes of distance to sinking stream (ds) provided by the COP method (and also by the PI method – see

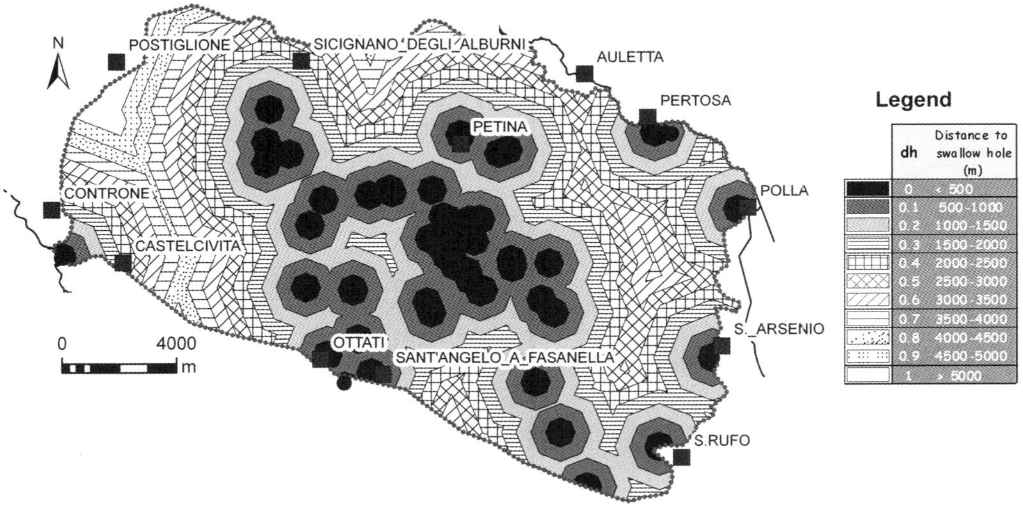

Fig. 7. The distance to swallow holes [dh] classified by the COP method.

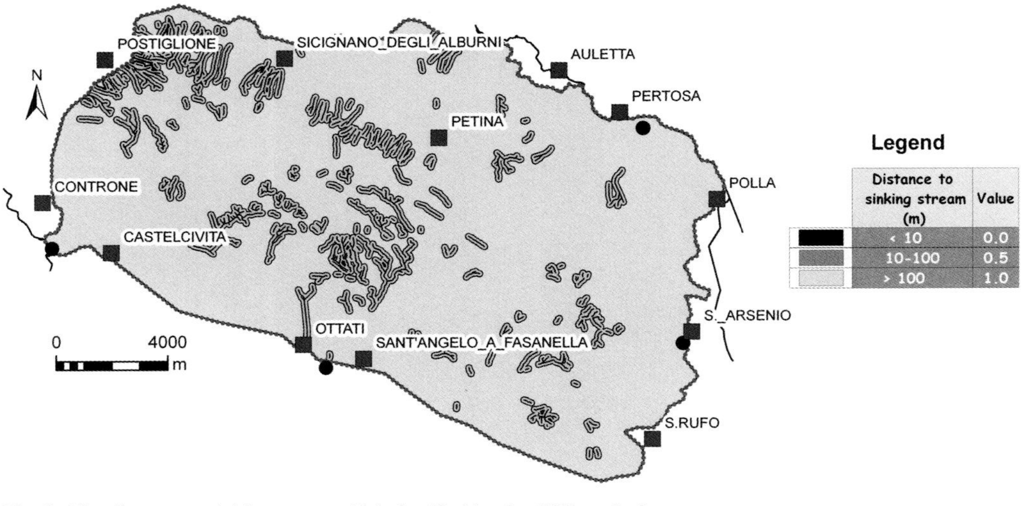

Fig. 8. The distance to sinking streams [ds] classified by the COP method.

Goldscheider 2002) and shown in Figures 3 & 8. In particular, the class '<10 m' at the 1:50 000 scale is a map distance of 0.2 mm, and 1 mm on the 1:10 000 scale. Such distances are difficult to accurately draw and represent from and on hard-copy maps.

The 'P' factor

The P factor considers the characteristics of the water that transports contaminants through the vadose zone. The influence of precipitation on vulnerability is not as great as that of the flow concentration, and was developed to differentiate zones with widely varying rates of rainfall, as occurs in Europe.

The P factor concerns the temporal distribution of precipitation in a certain period of time. To estimate this subfactor, two variables were considered: (1) mean annual precipitation for the wet years [P_Q], shown in Figure 13; and (2) average number of rainy days in a wet year, shown in Figure 14. The intensity [P_I] is their ratio (Fig. 15). The intensity of rainfall in karst media facilitates the development of high and

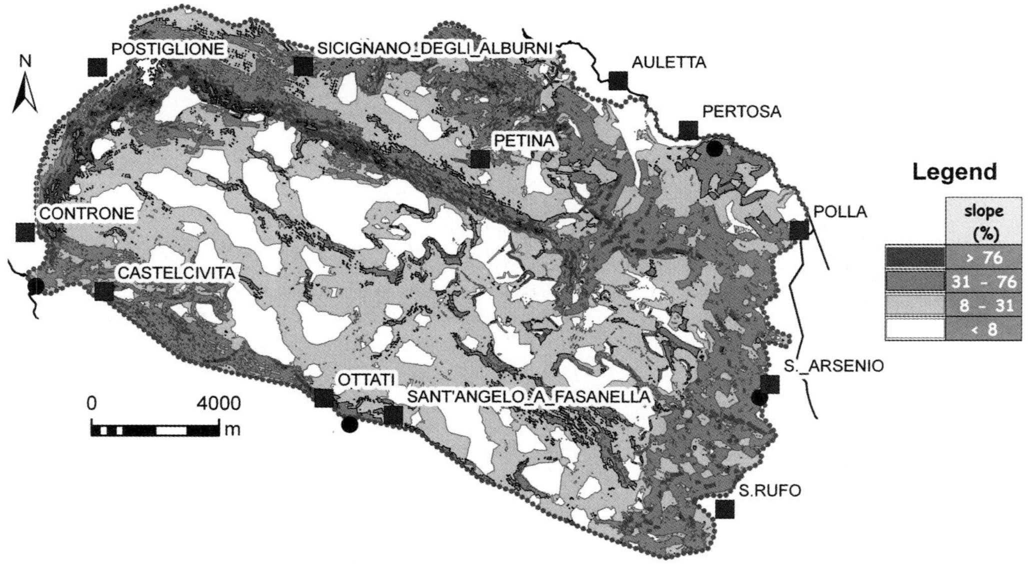

Fig. 9. The slope [S] classified by the COP method.

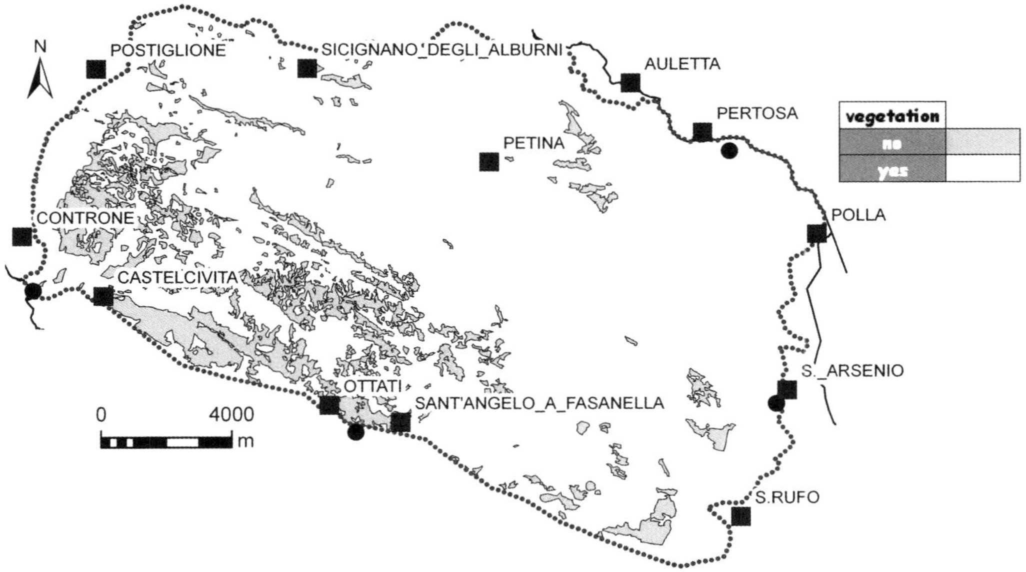

Fig. 10. Areas with absence of permanent vegetation (3.3.3 CORINE class: 'sparsely vegetated areas'). Areas with vegetation are prevalently 'broad-leaved forests' (3.1.1 CORINE class).

fast infiltration volumes through fissures and karst conduits.

The final value of the P factor is the sum of the two subfactors [P_Q] and [P_I]. The values of P range between 0.4, the value for minimum protection, and 1, the value for the greatest protection. Values closer to 1 indicate that precipitation has little influence on the protection.

In the Alburni karst system, where the P factor was evaluated for the years 1951–1997, almost the whole area is in a moderate vulnerability class (called moderate reduction of protection in the COP method) (value 0.7), and a very small area near Sicignano degli Alburni (<1%) is in high vulnerability class (called high reduction of protection in the COP method) (value 0.6) (Fig. 16).

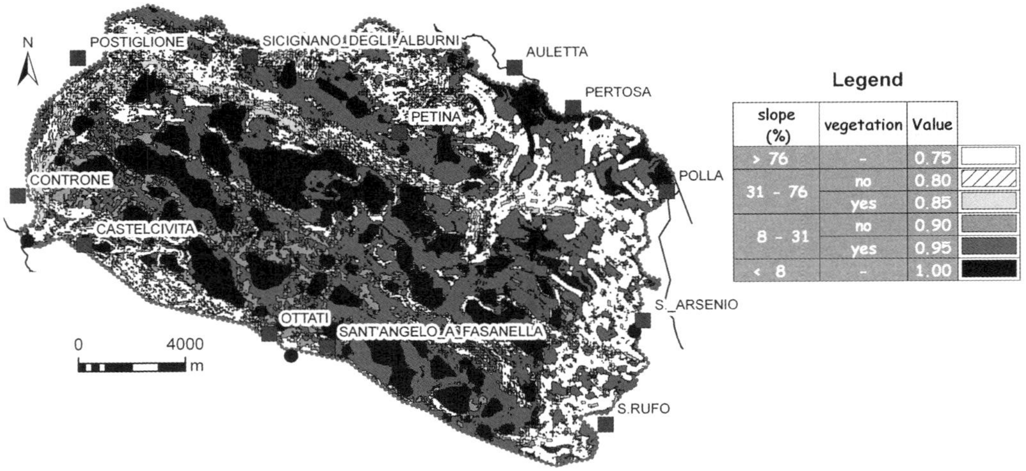

Fig. 11. The slope-vegetation map [sv].

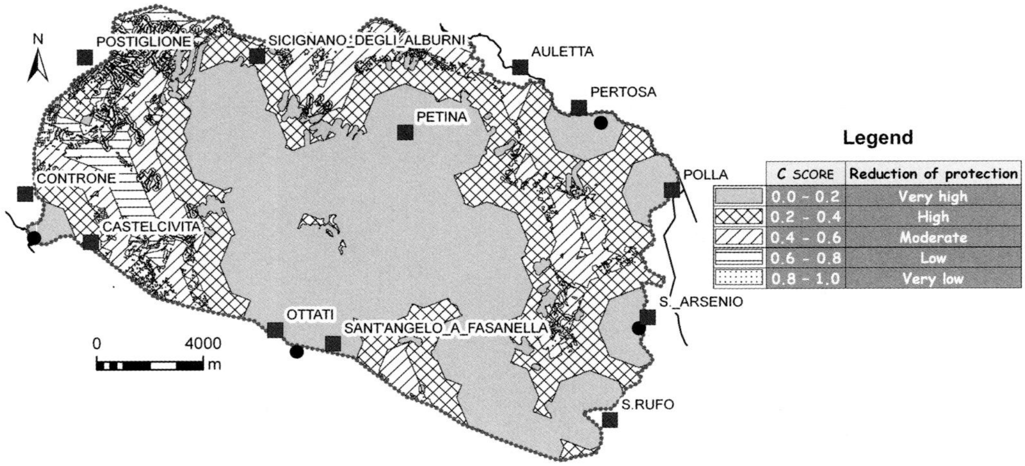

Fig. 12. The concentration of flow of the COP method [C_{SCORE}] for the Alburni karst system.

Finally, the COP Index is derived by the multiplying the C_{SCORE}, the O_{SCORE} and the P_{SCORE} (Fig. 3). Values are between 0 and 15, with the lower values representing the higher degrees of vulnerability.

The COP vulnerability map (Fig. 17) shows that the prevalent vulnerability degrees are high and very high owing to widespread karstification and to gentle slopes favouring fast infiltration (factor C). Swallow holes and sinking streams bypass the protective capacity (factor O) of the impervious sediments (flysch deposits – see Fig. 1) in terms of thickness and permeability, reducing the transit time of the infiltration water.

The influence of precipitation (factor P) on vulnerability is very low because of the distribution of rainfall in terms of amount (Fig. 13), number of rainy days (Fig. 14) and intensity (Fig. 15) is similar. In fact, the range of variation in P scores (Fig. 16) does not allow discrimination of areas.

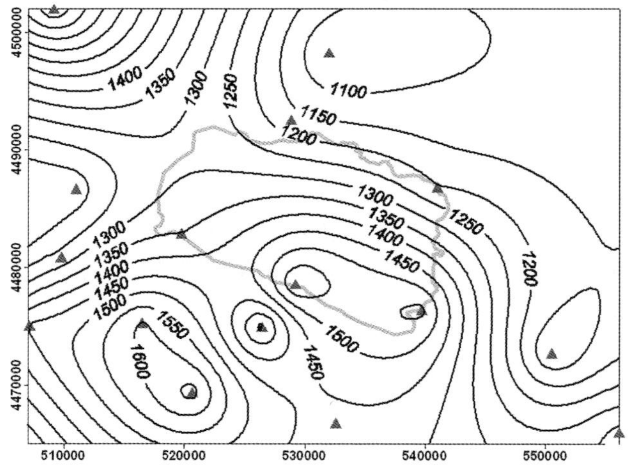

Fig. 13. Rainfall map (mm $year^{-1}$) of the Alburni karst system: 1951–1997 average for wet years on the basis of 22 rain gauge stations (dark triangles).

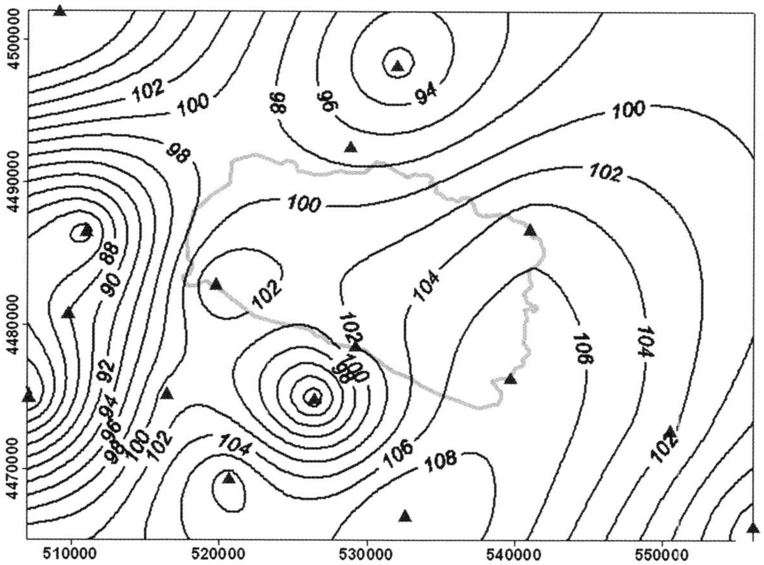

Fig. 14. Number of rainy days of the Alburni karst system: 1951–1997 average for wet years on the basis of 22 rain gauge stations (grey triangles).

The source vulnerability map evaluated using the VULK method

To assess the vulnerability of a groundwater source, the additional factor K (karst network development) of the European Approach must be considered. To assess the factor K in this study, the VULK method (Jeannin *et al.* 2001) has been used. The source vulnerability map has been evaluated only for the Auso spring, which has a highly variable discharge

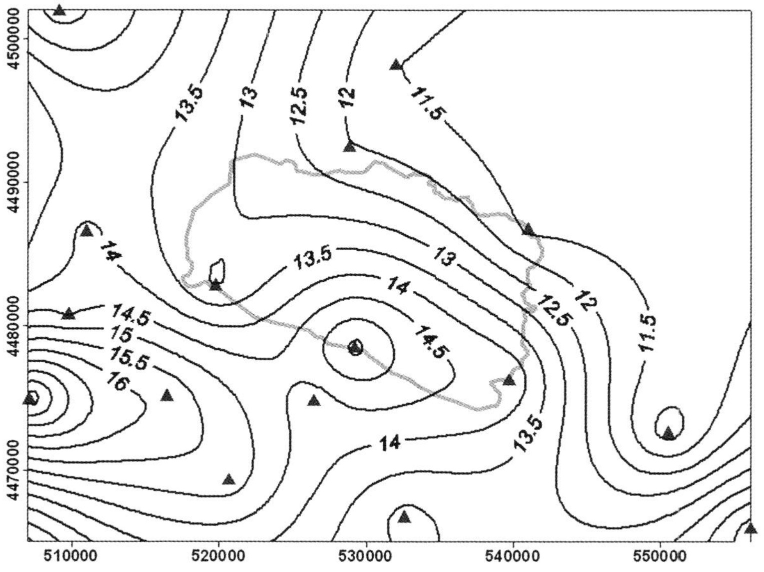

Fig. 15. Rainfall intensity (P (mm year^{-1})/Number of rainy days) for the Alburni karst system: 1951–1997 average for wet years on the basis of 22 rain gauge stations (grey triangles).

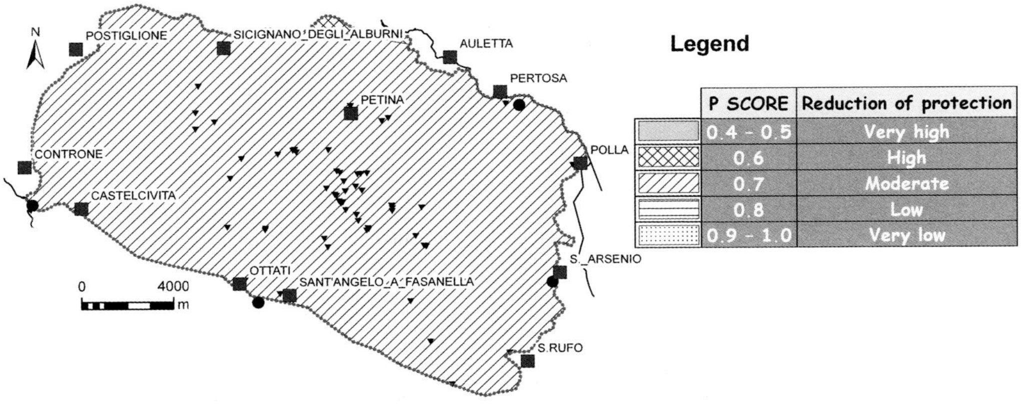

Fig. 16. The precipitation factor of the COP method [P_{SCORE}] for the Alburni karst system.

(Fig. 18), and where the groundwater basin and velocities feeding the spring have been at least partly delineated based on tracer testing between some swallow holes and the springs (Bellucci *et al.* 1991; Celico *et al.* 1994).

VULK is a simple analytical computer program to model the transport of contaminants through the different subsystems of a karst aquifer. VULK uses the factors O (overlying layers) and K (karst network development) of the European Approach. The K factor derives from the flow distance and the flow velocity, often measured directly by means of field experiments such as tracer tests.

The fluorescein tracer tests (Bellucci *et al.* 1991) revealed rapid groundwater flow (about 120 m h^{-1}) between the Piani di S. Maria swallow hole and the springs; flow in this direction is assumed to reach the maximum velocities for the area. Along the boundary of the basin the velocity is assumed to be zero. Intermediate flow velocities were calculated using ILWIS 3.3 GIS (http://www.itc.nl/ilwis/). The delimitation of the groundwater basin feeding the Auso springs has been copied from Celico *et al.* (1994), who drew it on the basis of a hydrogeological budget calculation coupled to the dye traces.

The flow distance is the horizontal distance from the swallow hole to the spring, and for this study it is assumed to occur almost entirely in the phreatic part of the aquifer (Jeannin *et al.* 2001). The horizontal distance from the Piani di S. Maria swallow hole to the spring was measured with ILWIS 3.3 GIS on the 1:50 000 and 1:25 000 topographic maps.

The required distance [m] and the mean flow velocity [m s^{-1}] are shown, respectively, in

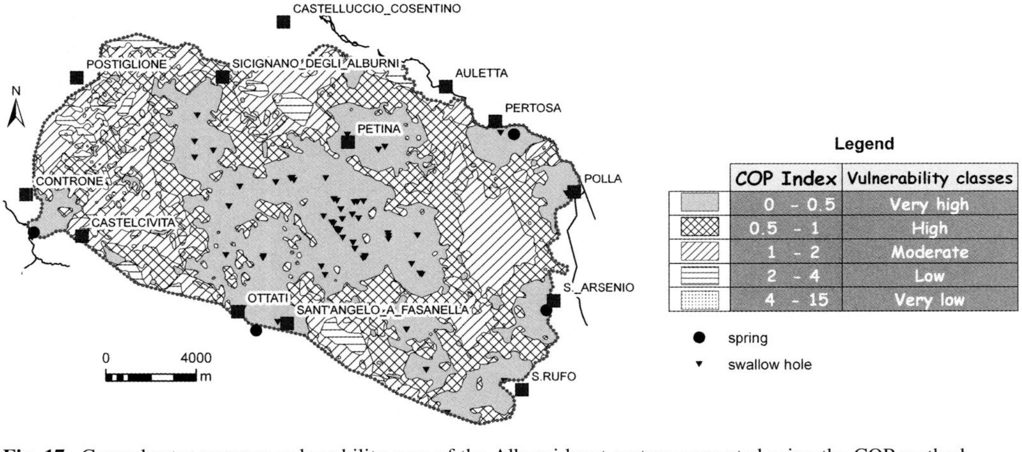

Fig. 17. Groundwater resource vulnerability map of the Alburni karst system computed using the COP method.

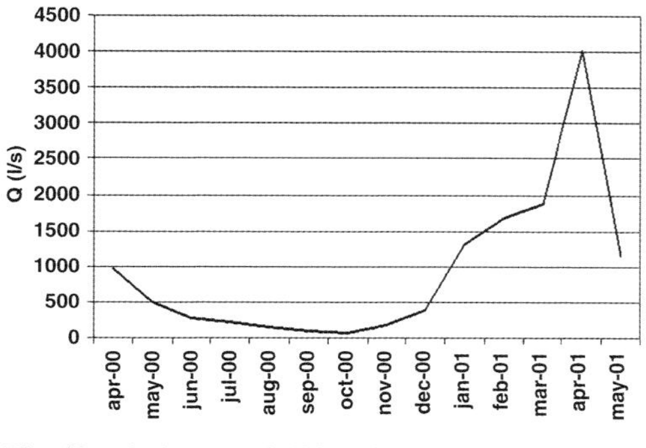

Fig. 18. Discharge variability of the Auso spring.

Figures 19 and 20. Dividing distance by velocity for each hydrogeological units (one, in the case of the Auso spring) gives the transfer time K (Fig. 21).

The K VULK map (Fig. 21), even though using strongly simplified assumptions, emphasizes that an accidental contamination in the Piani di S. Maria area could reach the spring in 2 days and highlights the very high vulnerability of the Alburni karst system.

Conclusions

The resource vulnerability map (Fig. 12) shows that the prevalent vulnerability degrees are high and very high owing to the widespread karstification and to the presence, on the plateau, of large areas with gentle slopes favouring fast infiltration. Moreover, in the event of a contamination occurrence in some areas, like the groundwater basin of the Auso spring, the springs will be quickly polluted, as testified by the source vulnerability map (Fig. 21). Interpreting the maps' information in the light of land use, development planning and groundwater protection, the Alburni karst area requires great care regarding the location of the potential contamination sources.

The European Approach to evaluate contamination vulnerability by the COP method is useful and easy enough to use, but it needs a robust database. In the Alburni karst system, where the surface karst landforms (e.g. the dolines) are very copious draining the groundwater flow, the COP method, like the first EPIK method (Doerfliger & Zwahlen 1998; Corniello & Ducci 2000), correctly gives a strong emphasis to the karstic landform.

To evaluate the K factor it is necessary to consider the flow velocity using dye tracing. At present, the VULK method seems to be suitable, but must be implemented in order to avoid oversimplification and is currently being further developed at the Hydrogeology Centre of Neuchâtel (Zwahlen 2003).

The methods have to take more into account the 'scale problem', establishing a clear relationship between input scale, working scale and output scale. It is not possible to apply these methods to all map scales (state, regional, local) because some elements (e.g. buffers of 10 or 100 m) may not be precisely mappable, and using GIS the

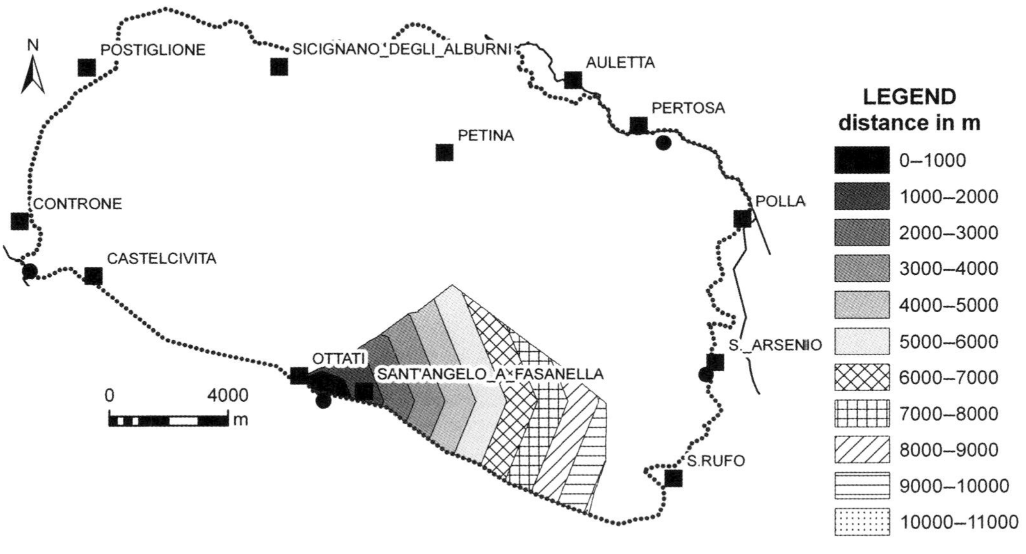

Fig. 19. Classified distance from the Auso spring for the area feeding the spring.

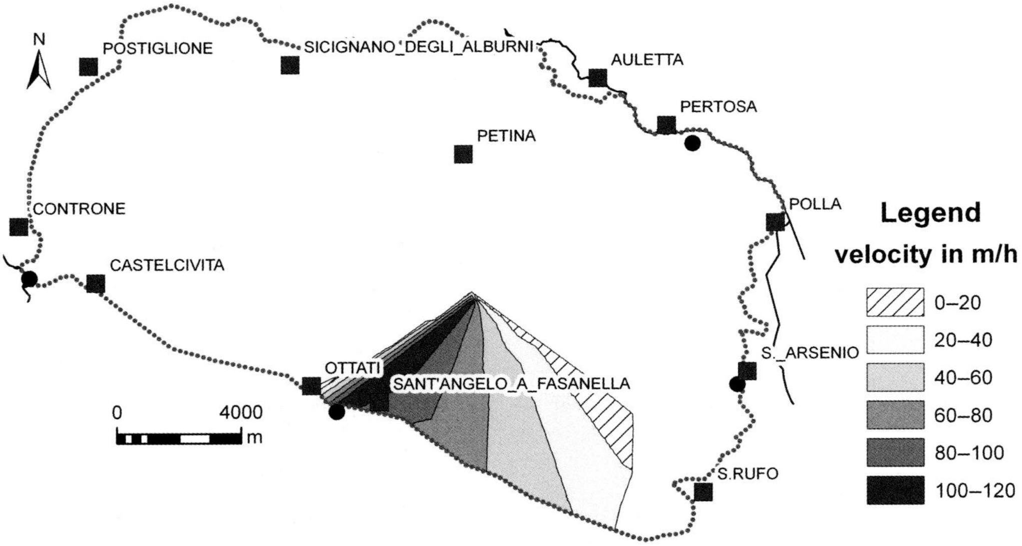

Fig. 20. Classified groundwater flow velocity for the area feeding the Auso spring.

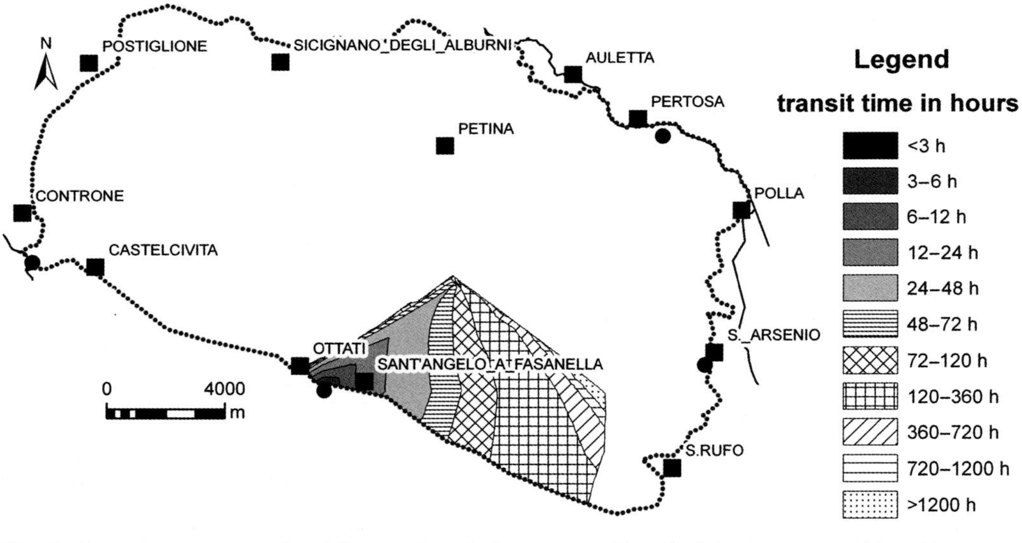

Fig. 21. Groundwater source vulnerability map (transit time source – K map) of the Auso spring (Alburni karst system) computed using the VULK method.

accuracy required, and then the resolution, is completely different.

I would like to thank the Comunità Montana Alburni who allowed the use of digital morphological and land-use data, and Prof. I. Bernetti of the University of Florence for the kind and helpful explanations and suggestions for this project. Finally, I would like to thank the reviewers for their comments, and especially Dr G. Veni for his careful revisions and suggestions that enabled me to improve this paper.

References

Aller, L., Bennett, T., Leher, J. H., Petty, R. J. & Hackett, G. 1987. *DRASTIC: A standardized system for evaluating ground water pollution potential using hydrogeologic settings*. US Environmental Protection Agency, Ada, OK, **600/2-87-035**.

BELLUCCI, F., GIULIVO, I., PELELLA, L. & SANTO, A. 1991. Carsismo ed Idrogeologia dei Monti Alburni. *Geologia Tecnica & Ambientale*, **3**, 5–12.

BELLUCCI, F., GIULIVO, I., PELELLA, L. & SANTO, A. 1995. *Monti Alburni. Ricerche Speleologiche.* De Angelis, Avellino.

BERARDI, F., DE ROSA, G. & TOZZI, M. 1996. Vincoli strutturali di superficie per una ricostruzione geometrica del massiccio dei Monti Alburni (Appennino meridionale). *Memorie della Società Geologica Italiana*, **51**, 13–21.

BROZZETTI, F. & SALVATORE, A. 2004. Active extension along two Apennines transects crossed by the crop03 (Tuscany–Umbria) and crop04 (Campania–Lucania) profiles: a comparison. *In*: *Reunion de Sciences de la Terre*, Strasbourg 20–25 September 2004.

BÜTTNER, G., FERANEC, J. & JAFFRAIN, G. 2002. *Corine Land Cover Update 2000: Technical Guidelines.* Technical Report, **89**. EEA (European Environment Agency, Copenhagen, Denmark).

CELICO, P., SALZANO, G. & VALLETTA, M. 1990. Rapporti idrogeologici tra il massiccio carbonatico degli Alburni e la bassa valle del Tanagro (Campania). *Memorie Descrittive della Carta Geologica d'Italia*, **38**, 283–298.

CELICO, P., PELELLA, L., STANZIONE, D. & AQUINO, S. 1994. Sull'idrogeologia e l'idrogeochimica dei Monti Alburni (SA). *Geologica Romana*, **30**, 687–698.

CIVITA, M. & DE MAIO, M. 2000. *Valutazione e cartografia automatica della vulnerabilità degli acquiferi all'inquinamento con il sistema parametrico SINTACS R5.* Pitagora, Bologna.

CORNIELLO, A. & DUCCI, D. 2000. Pollution vulnerability assessment in karstic aquifers. A case study of the Matese Mountains (southern Italy). *In*: SILILO, O. (ed.) *Groundwater: Past Achievements and Future Challenges.* Balkema, Rotterdam, 725–730.

CORNIELLO, A., DUCCI, D. & MONTI, G. M. 2004. Aquifer Pollution Vulnerability in the Sorrento Peninsula, Southern Italy, Evaluated by the SINTACS Method. *Geofisica Internacional*, **43**, 575–581.

DALY, D., & DASSARGUES, A. *ET AL.* 2002. Main concepts of the European Approach for (karst) groundwater vulnerability assessment and mapping. *Hydrogeology Journal*, **10**, 340–345.

DI GENNARO, A. 2002. *I sistemi di terre della Campania. Regione Campania.* Risorsa srl, Assessorato alla Ricerca scientifica. Selca, Firenze.

DOERFLIGER, N. & ZWAHLEN, F. 1998. *Practical Guide, Groundwater Vulnerability Mapping in Karstic Regions (EPIK).* Swiss Agency for the Environment, Forests and Landscape (SAEFL), Bern.

DUCCI, D. & ROSSI, S. A. 2007. Non-independent parameters in the aquifer vulnerability assessment methods: GIS application in three sample areas. *Environmental Geology*, in press.

EUROPEAN WATER DIRECTIVE. 2000. *Directive 2000/60/EC of the European Parliament and of the Council of 23 October 2000 Establishing a Framework for Community Action in the Field of Water Policy.* EC, Brussels.

GOLDSCHEIDER, N. 2002. *Hydrogeology and vulnerability of karst systems – examples from the Northern Alps and Swabian Alb.* Dissertation, University of Karlsruhe wwww.ubka.uni-karlsruhe.de/vvv/2002/biogeo/3/3.pdf

GOLDSCHEIDER, N., KLUTE, M., STURM, S. & HÖTZL, H. 2000. The PI method – a GIS-based approach to mapping groundwater vulnerability with special consideration of karst aquifers. *Zeitschrift für Angewandte Geologie*, **46**, 157–166.

GUNN, J. 1986. Modelling of conduit flow dominated karst aquifers. *In*: GÜNAY, G. & JOHNSON, A. I. (eds) *Karst Water Resources.* IAHS, Publications, **161**, 587–596. IAHS, Wallington.

ITALIAN HYDROGRAPHIC SURVEY. 1948–1999. *Annali Idrologici, Parte I e II, Compartimento di Napoli.* Naples.

JEANNIN, P.-Y., CORNATON, F., ZWAHLEN, F. & PERROCHET, P. 2001. VULK: a tool for intrinsic vulnerability assessment and validation. *In*: ZWAHLEN, F. & MUDRY, J. (eds) *7th Conference on Limestone Hydrology and Fissured Media.* Université de Franche-Comté, Besançon. Mémoires hors Série, **13**, 185–188.

KRALIK, M. 2001. *Strategie zum Schutz der Karstwassergebiete in Österreich.* [*A Strategy for Protecting Karst Groundwater in Austria*]. Report, **BE-189**. Environment Agency, Vienna.

SANTANGELO, N. & SANTO, A. 1997. Endokarst processes in the Alburni massif (Campania, southern Italy): evolution of ponors and hydrogeological implications. *Zeitschrift für Geomorphologie, N. F.* **41**, 229–246.

SANTO, A. 1993. Idrogeologia dell'area carsica di Castelcivita (M.ti Alburni, SA). *Geologia Applicata ed Idrogeologia*, **28**, 666–673.

VENI, G. 1997. The effects of aridity and topography on limestone cave development. *In*: *Proceedings of the 12th International Congress of Speleology.* International Union of Speleology, La Chaux-de-Fonds, Switzerland, 373–376.

VÍAS, J. M., ANDREO, B., PERLES, M. J., CARRASCO, F., VADILLO, I. & JIMÉNEZ, P. 2002. Preliminary proposal of a method for vulnerability mapping in carbonate aquifers. *In*: CARRASCO, F., DURÁN, J. J. & ANDREO, B. (eds) *Karst and Environment.* 2nd Nerja Cave Geological Symposium and 2002 meeting of IGCP Project 'World Correlation on Karst Ecosystems', Nerja, Malaga, 61–67.

ZWAHLEN, F. (ed.). 2003. *COST Action 620. Vulnerability and risk mapping for the protection of carbonate (Karst) Aquifers.* Final Report. http://capella.unine.ch/chyn/pdf

Evaluating the impact of quarrying on karst aquifers of Salento (southern Italy)

M. DELLE ROSE[1], M. PARISE[1] & G. F. ANDRIANI[2]

[1]*National Research Council, IRPI, Bari, Italy (e-mail: m.parise@ba.irpi.cnr.it)*

[2]*Department of Geology and Geophysics, University of Bari, Bari, Italy*

Abstract: This paper describes a case study in the Salento karst (Apulia, southern Italy) in a site that has been intensively used to quarry limestones in the last 30 years. After quarrying activity had stopped, the site was transformed into legal and illegal landfills where solid and liquid wastes have been repeatedly dumped, with serious consequences for the groundwater resources. In this paper, through a geological, petrographical and hydrogeological approach, we attempt to assess the consequences of the anthropogenic activities on the local hydrogeology, with particular regard to the surficial aquifer that is contained in the Plio-Quaternary calcarenites cropping out in the area. Application of some well-known methods to assess the vulnerability of aquifer systems to contamination by human activities (DRASTIC, SINTACS, LeGrand and GOD) highlights the limits of such an approach in karst environment, and the necessity to include in the methods data strictly related to the peculiarity of karst. This is further evidenced by application of the EPIK method, specifically designed for karst areas. The final part of the paper focuses on the need of a thorough understanding of the hydrogeological setting for a better management and policy action of karst environments.

Owing to a number of geological, morphological and hydrogeological features, karst is one of the most susceptible environments to negative impacts from human activities (Nicod 1972; White 1988; Ford & Williams 1989; Parise & Pascali 2003). As regards groundwater protection, the intimate connection between surface and underground drainage and the rapidity of the surficial water to enter and percolate through the carbonate rocks result in a high intrinsic vulnerability of the karst aquifers. This may favour fast movement of contaminants towards the water table and the resulting deterioration in groundwater quality, sometimes with unrecoverable effects (Memon & Prohic 1989; Kacaroglu 1999; Escolero *et al.* 2002).

Environmental damage and pollution in karst often derives from anthropogenic activities such as uncontrolled quarrying (Langer 2001). Accompanying the growth of urbanization during the 20th century, an accelerated pace of limestone quarrying has been registered in several areas of the Mediterranean basin, including Italy (Bondesan & Meneghel 1990; Gams *et al.* 1993; Gunn 1993). The most direct consequence of quarrying activity is the deterioration of the original karst landscape, accompanied by the partial or total destruction of surface (dolines) and subsurface (caves) karst landforms, with loss of the geological, palaeontological and archaeological evidence therein present (Gillieson 1996; Gunn 2004). In addition, the quarry sites may later be used (most of the times illegally) as landfills to dump wastes of different kinds.

Laws and management procedures generally do not take into account the peculiarity of karst environments (Ekmecki & Gunay 1997; LaMoreaux *et al.* 1997; Daly *et al.* 2002), and this results in the absence of adequate policy and management tools as well as in the uncertainty of the water authority to face an emergency. The extent and complexity, often not fully understood, of catchment basins in karst represent the main problems in defining the areas of protection, and frequently causes conflict between the rural environment and urban consumers (Nicod & Salomon 1999; Bonacci 2004).

In southern Italy, the Salento peninsula (Fig. 1), mostly made of carbonate rocks, is dissected by an extensive network of quarries. Land-use transformation and degradation of the karst environment have produced heavy modification in the natural landscape, leading in many cases to pollution, or to deterioration and partial destruction of natural sites, including several caves. This paper, through description of the human impact in a sample site of Salento and an evaluation of the consequences derived from the quarrying activity, emphasizes the need to correctly manage such environment in order to preserve the natural landscape and water quality in karst aquifers.

From: PARISE, M. & GUNN, J. (eds) *Natural and Anthropogenic Hazards in Karst Areas: Recognition, Analysis and Mitigation*. Geological Society, London, Special Publications, **279**, 153–171. DOI: 10.1144/SP279.13
0305-8719/07/$15.00

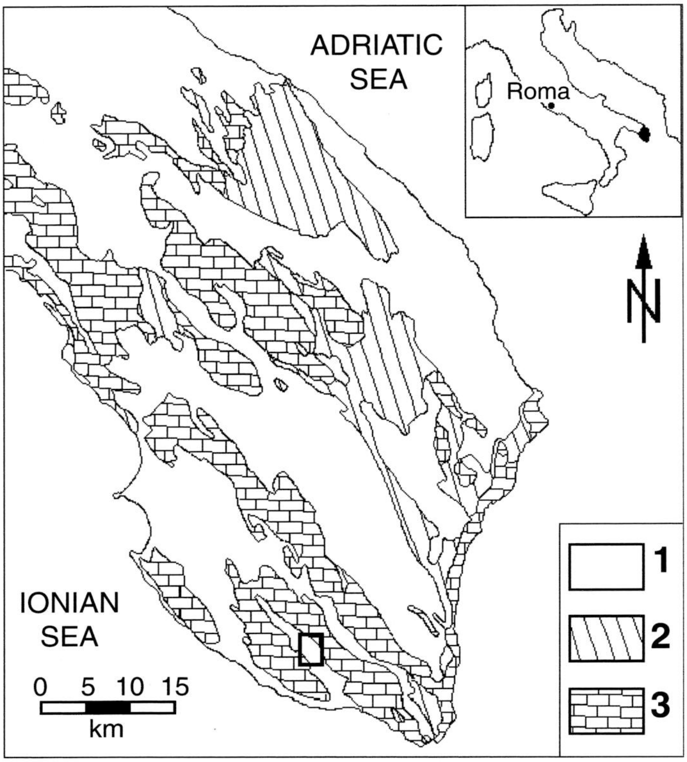

Fig. 1. Geology of Salento Peninsula. Key: 1, Clayey deposits and calcarenites (Pliocene–Pleistocene); 2, bioclastic carbonate rocks (Paleogene) and calcarenites (Miocene); and 3, carbonate platform rocks (Cretaceous). The rectangle shows location of the Burgesi area.

Study area

The Burgesi graben is situated in the southern part of Salento peninsula, within the NNW–SSE-striking horst and graben structure of this sector of Apulia (Martinis 1970; Ricchetti *et al.* 1988; Tozzi 1993). At Burgesi, a Plio-Quaternary sequence overlies Cretaceous limestone bedrock with a transgressive and unconformable contact (D'Alessandro & Massari 1997), whereas the adjacent horsts are exclusively Cretaceous limestone (Fig. 2). The Plio-Quaternary sequence, with an overall maximum thickness of some 35 m, consists of calcarenites with intervening clayey deposits (Fig. 3). The bottom calcarenites, which can be correlated to the upper Pliocene–early Pleistocene Calcarenite di Gravina Formation (Robba 1969), are massive, medium- to fine-grained and 0–10 m thick. The intermediate unit (maximum thickness 10 m) is made up of argillaceous silts with sandy intercalations, that can be correlated to the early Pleistocene Argille Sub-Appennine Formation (Dell'Anna *et al.* 1985). The top units are formed by middle Pleistocene medium- to coarse-grained massive calcarenites and subordinate calcirudites, bounded to the SE by bedded and laminated calcarenites ('Calcarenite della Casarana': D'Alessandro & Massari 1997); thickness of the latter unit ranges from a few to about 15 m.

The contact between the Cretaceous limestone and the Plio-Quaternary sequence occurs through high-angle normal faults, whose presence is marked at the surface by bands of diffuse deformation some tens of metres wide, evidenced by well-developed joints in both the limestones and the calcarenites. Structural surveys identified the main discontinuity systems in the Cretaceous, NE-dipping, carbonate rock mass. The limestones are characterized by secondary permeability that develops essentially through the main discontinuity systems in the rock mass, striking NE–SW and, subordinately, NW–SE (Fig. 4). In contrast, the Plio-Quaternary sequence does not generally show evident discontinuities, with the exception of the aforementioned bands at the contact with the Cretaceous limestones; there

Fig. 2. Geological and morphological sketch of the study area (simplified after Martinis 1970). Key: a, top calcarenites; b, argillaceous silts with sandy intercalations; c, bottom calcarenites; d, limestone bedrock; e, fault; f, strata attitude; g, solution doline; and h, collapse dolines.

a limited development of WNW–ESE and, subordinately, NW–SE fractures have been observed.

As generally is typical of karst, the Burgesi graben hosts a dry valley without any hierarchically organized stream system. The surface runoff is aggregated into tiny rivulets and, in very few cases, into small gullies. The drainage at the surface is also characterized by several closed depressions filled by several metres of reddish-brown residual soils. The area is mostly used for agriculture, with the presence of olive trees and cultivated fields.

Hydrogeologically, two aquifers are present (Fig. 5): the shallow aquifer, that is within the top calcarenites and the sandy levels of the argillaceous silts, is extensively used for irrigation. Its lower limit is the impervious layers in the argillaceous silts that act as a low permeability barrier. The water table is 10–15 m below the surface (Calò *et al.* 1992). The deep and basal aquifer, the only local supply of

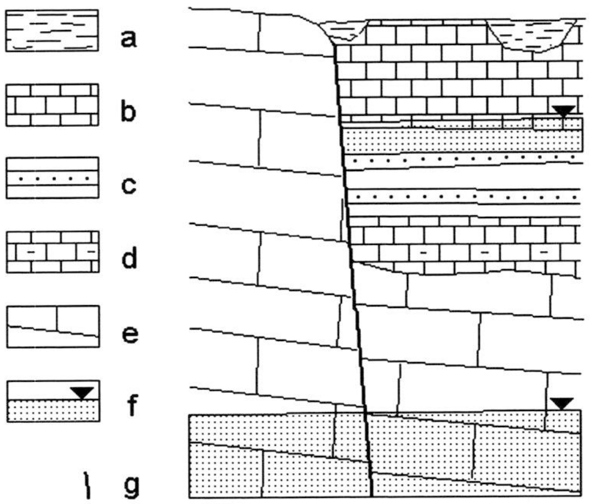

Fig. 3. Sketch of the stratigraphy in the study area. Key: a, terre rosse and clay deposits; b, top calcarenites; c, argillaceous silts with sandy intercalations; d, bottom calcarenites; e, limestones; f, water tables; and g, fault.

drinkable water, is an unconfined aquifer within the Cretaceous limestone (Calcari di Melissano Formation). Its lower boundary is the zone of transition with salt waters of continental intrusion (Tadolini & Tulipano 1981). In the study area, the water table is about 100 m below the ground surface.

Since the beginning of the 20th century the Burgesi graben has been impacted by anthropogenic activities such as stone clearing, removal of the soil cover and intense quarrying in the top calcarenites aimed at extraction of materials for construction purposes. Interpretation of multi- temporal aerial photographs, dating back to 1972, integrated with field surveys allowed the distribution of the quarrying activity during the last 30 years to be assessed.

In 1972 only a few quarries of very limited extent were present in the area. Since then, however, quarrying activity has proceeded rapidly, and Figure 6 shows how the quarries have enlarged. The average depth of the quarries is presently around 10 m, with a maximum at an original doline (quarry A in Fig. 6) where a depth of 21 m is reached. There, anthropogenic activity has resulted in severe transformation of the original landscape, and proceeded through: (i) removal of the terre rosse deposits; and (ii) quarrying of the calcarenites. Consequently, quarrying has heavily modified the thickness and characteristics of both the soil cover and the unsaturated zone, significantly reducing the depth of the water table of the shallow aquifer.

Besides quarrying activity, further serious environmental problems occurred when the sites began to be used to dump solid and liquid wastes. At least since 1980 several episodes of uncontrolled and illegal waste disposal have been registered, the most dangerous of which occurred in September 2000 and consisted of many hundreds of drums containing waste industrial oils being dumped in three quarries in the middle Pleistocene calcarenites. The fluids in the drums were

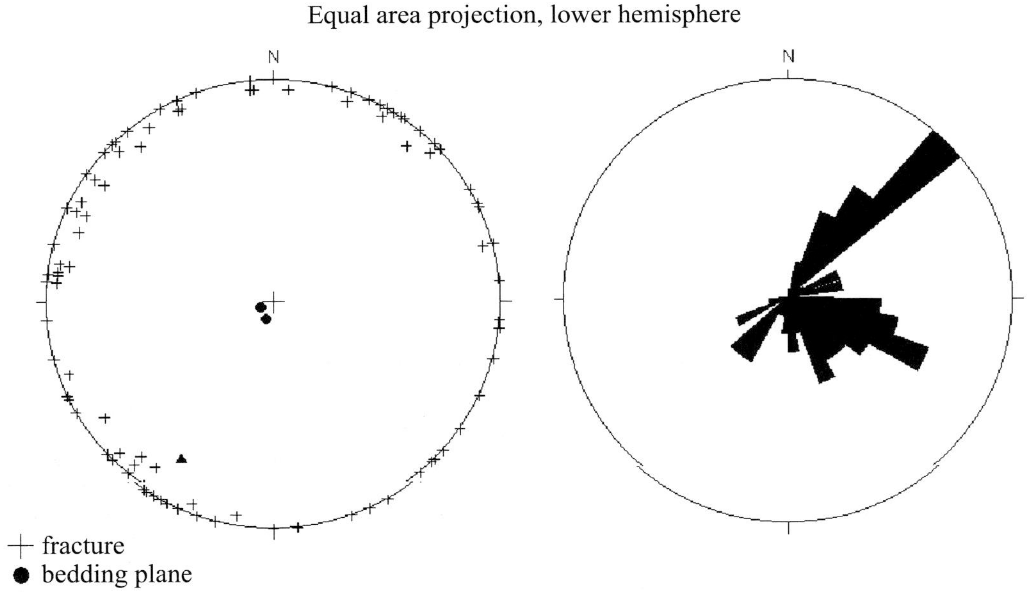

Fig. 4. Representation of discontinuity systems in the Cretaceous limestones at Burgesi ($n = 92$).

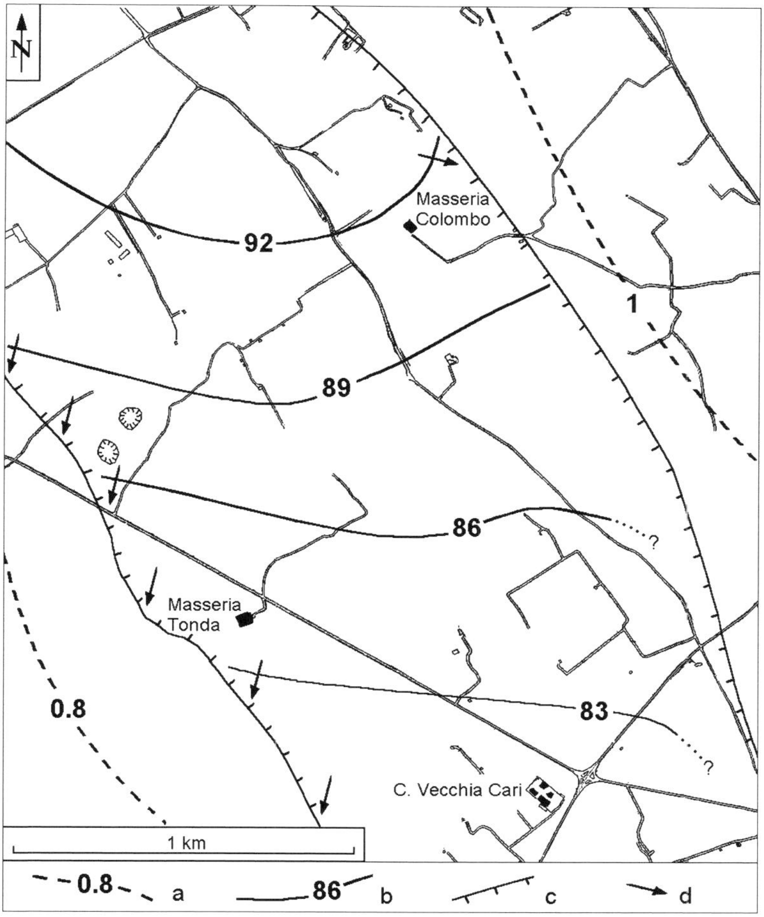

Fig. 5. Isophreatic map (modified after Calò *et al.* 1990). Key: a, isophreatic lines of the deep water table (elevations in metres above sea level); b, isophreatic lines of the shallow water table (elevations in metres above sea level); c, fault; and d, likely paths of water infiltration from the shallow to the deep water table.

highly contaminated with phenols, polychlorinated biphenyls and polyaromatic hydrocarbons (Table 1). At quarry A, the oil from tens of broken drums flowed out and filtered into the subsoil over about 24 h (Figs 7 & 8). At other sites, large amounts of solid wastes and hundreds of drums are still present (Figs 9 & 10), the only protection being a plastic cover. Again, the possibility of leaching of contaminants through the discontinuity systems in the rock mass is very high. At quarry C (Fig. 6), some tens of drums were removed in a few days in October 2000 when the illegal landfills were brought to the attention of the public through newspaper articles.

Owing to the geological and hydrogeological setting, the main problem resulting from human activities has been the reduction, locally the almost complete elimination, of the top calcarenites that originally acted as a protection to the infiltration of pollutants. For these reasons, the geology and petrography of the top calcarenites in the Plio-Quaternary sequence of the area were assessed to highlight the filtering role that this material originally had.

As regards management of the area, the only policy actions carried out in the past by the local authorities consisted of the prohibition of water pumping from those wells located close to the polluted sites. The prohibition covered an area with a radius of 1 km, centred at the sites of quarries A and B, and lasted for 4 months. Sampling and analysis of water from four wells continued for 1 year, and the drums were covered by waterproof plastic to limit as much as possible leaching of pollutants underground. Recently (June 2005) a programme of environmental restoration started in a sector of the Burgesi graben, the aim being to clear the wastes, revegetate the area and use it for recreational activities, including nature trails.

Methods of study

Many methods are available to assess the likelihood of groundwater contamination at a specific site. Although they have slightly different approaches, and are sometimes adapted to the local geological settings, all these methods consider a number of factors concerning three main environments: the landscape surface, the unsaturated (vadose) zone and the aquifer (or saturated zone). As regards the landscape surface, the local topography and the definition of the catchment boundaries are probably the most representative factors. The unsaturated zone requires very detailed analyses and modelling as it depends on a number of factors, including the sorption capacity of the rock and soil mass above the aquifer, and a definition of the physical and chemical characteristics of the soil. The aquifer, or saturated zone, has to be defined in terms of type of aquifer (unconfined, confined, partially confined), depth of the water table, and those characteristics that determine the possibility of dilution of the likely pollutants and their residence time. Moreover, as infiltration water is the main vector to transport pollutants, groundwater recharge also has to be taken into account.

Some methodologies have been implemented to evaluate the likelihood of groundwater contamination at regional scale, while others at a point centre (LeGrand 1964, 1983; Aller *et al.* 1985; Davidson *et al.* 1988). To provide some examples,

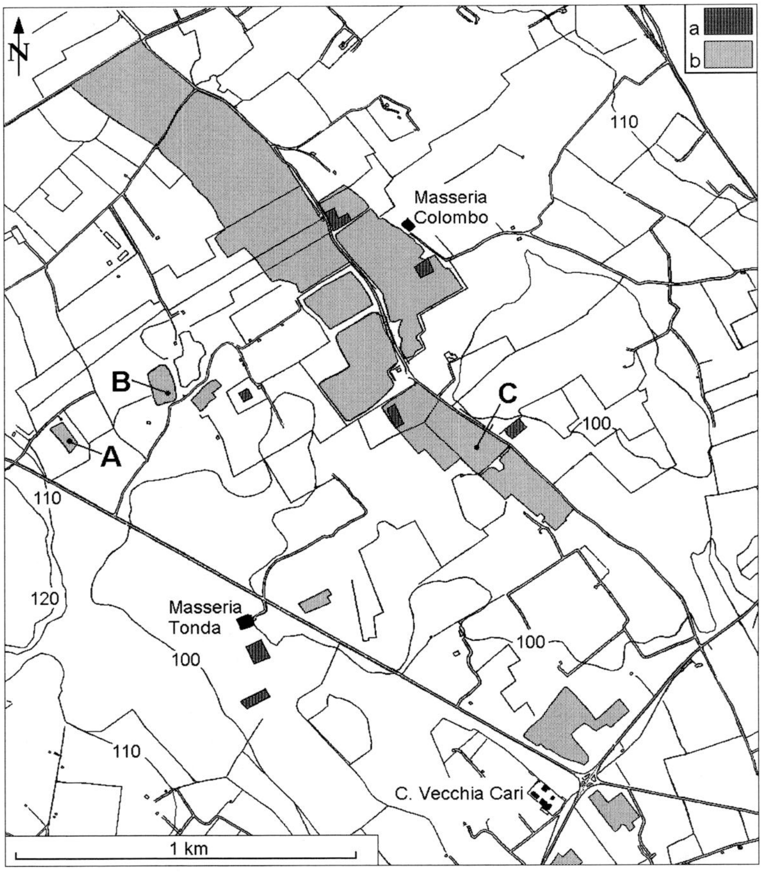

Fig. 6. Quarrying activity in the study area: a, in 1972 (from aerial photographic interpretation); and b, present situation. Location of the three sites (A, B and C) of illegal heaping of drums.

Table 1. *Chemical analysis on pollutants from the drums located in the quarries (analysis from ALcontrol Geochem kindly provided by Mr N. Gray). The water samples come from wells in the shallow aquifer located within the circled area cited in the text; the oil comes from drums. The soil was sampled below a drum at the quarry A*

Sample type	Oils	Polyaromatic hydrocarbons	Polychlorinated biphenyls	Phenols	Date of sampling	Laboratory of analysis
Water	0.12 ppm	Not analysed	Not analysed	Not analysed	October 2000	USL Lecce 2
Water	0.12 ppm	Not analysed	Not analysed	Not analysed	October 2000	USL Lecce 2
Oil	Not analysed	314.3 mg kg^{-1}	<1 μg l^{-1}	55 761 μg kg^{-1}	January 2001	ALcontrol Geochem
Oil	Not analysed	Not analysed	<1 μg l^{-1}	59 832 μg kg^{-1}	January 2001	ALcontrol Geochem
Soil	Not analysed	22.9 mg kg^{-1}	666 mg kg^{-1}	17 μg kg^{-1}	January 2001	ALcontrol Geochem
Water	Not analysed	Not analysed	Not present	Not analysed	February 2001	Multilab
Water	Not analysed	Not analysed	Not present	Not analysed	February 2001	Multilab
Water	Not analysed	Not analysed	Not present	Not analysed	February 2001	Multilab
Water	Not analysed	Not analysed	Not present	Not analysed	February 2001	Multilab

the method of LeGrand is currently used to support regional planning (Ibe *et al.* 2001), while that proposed by Aller *et al.* (1985) (DRASTIC) produces vulnerability maps, which can be useful for land management at moderate and small scales (Vias *et al.* 2005). Other methodologies (e.g. GOD and SINTACS) are applied to environmental planning, at small–medium scale, to prevent pollution. Because the models differ slightly in their formulation and type, it has to be expected that their use will provide the basis for making comparative analyses, checking for convergence of results and, hence, ensuring higher level of confidence in the predicted vulnerability.

The models involve parameter rating and point count systems, which are based on the evaluation of the factors related to groundwater contamination, in relation to their capacity to enhance or attenuate the contaminant diffusion in the subsoil systems. The methods differ from one another owing to inclusion of additional factors such as topographic slope change, hydraulic and chemical–physical characteristics of the soil, type of aquifer, etc. (Table 2). In each method scores are ascribed to every factor, within a predetermined range. The sum may be weighted to assign different emphasis to a particular factor, based on the local situation. The sum of the scores gives the overall vulnerability or possibility of pollution. In karst, the assessment of groundwater contamination is complex, as some of the factors are very difficult to estimate: for example, the lack of correspondence between topographic divides at the surface and the real limits of the hydrogeological basin (Bonacci 2004; Gunn 2007). The type and distribution of surface and subsurface karst features have to be carefully considered, as well as the presence of residual deposits; the latter are very common in karst areas, and may play a positive role in the sorption of pollutant (Foster & Hirata 1988).

EPIK is a method devoted to karst environments (Doerfliger *et al.* 1999), the acronym standing for Epikarst (E), Protective cover (P), Infiltration conditions (I) and Karst network development (K). This multi-attribute weighted-rating method assesses semi-quantitatively the groundwater sensitivity of karst terrain. The four attributes include considerations about the special features of karst that are not fully considered in the previously mentioned methods: for example, speleological knowledge, that is all the information deriving from caving activity, is taken into account. In the EPIK method, the higher the score, the greater the protection of the area, i.e. the less vulnerable the area is.

The use of models to simulate the behaviour and movement of pollutant in the unsaturated and saturated zone for groundwater management is a controversial issue within the scientific community as well

Fig. 7. Quarry A (location shown in Fig. 6) was originally filled with residual deposits. Removal of terra rossa, together with the quarrying activities, resulted in the present situation, which shows a depth of 21 m.

as among the water authorities. Yet, many of those responsible for groundwater quality protection consider models to be the most pragmatic approach to a complex problem. Therefore, models appear to have a guiding role in policymaking, as well as in developing environmental regulations and establishment of remedial action for regulatory agencies responsible for groundwater (Davidson *et al.* 1988).

In the following, we first apply the aforementioned four methods to the case study area, and discuss their main outcomes, as well as their limits of application in karst. Then, the EPIK method is applied in order to show its greater sensitivity and applicability to groundwater pollution evaluation in karst environments.

Preliminary evaluation of the impact

Site vulnerability as the first level of analysis

A first assessment of the impact of quarrying on the vulnerability of the aquifers at Burgesi has been performed using the DRASTIC (Aller *et al.* 1985) and LeGrand (LeGrand 1964, 1983) methods. The database used as input in both methods is mainly derived from geological mapping and hydrostratigraphical survey, and, subordinately, from interpretations of the available hydrogeological maps.

As aforementioned, our attention at Burgesi focused on the top calcarenite unit, which has been intersected by quarrying activity, and this has been largely extracted in the past, reducing its thickness by several metres at many sites in the study area. This unit was examined in detail as regards its petrography and hydrogeological features, as the unsaturated zone above the shallow aquifer is located within it. Three representative samples of the unit were collected in the Burgesi graben, by selecting the following varieties: surficial, weathered, material consisting of reddish bioclastic grainstone (B3), and yellowish bioclastic grainstones with low (B1) and medium packing density (B2). The latter were sampled within one of the quarries, at 4 and 12 m below the land surface, respectively. The varieties are composed mainly of low-magnesium calcite ($CaCO_3 > 97\%$) with an insoluble residue that consists of clay minerals with gibbsite and goethite, especially in sample B3. At the macroscopic scale, the calcarenites appear weathered with abundant vugs, even though they are locally well cemented with a carbonate crust. For each of the three varieties of

Fig. 8. Solid wastes, including drums of industrial oils, dumped in the doline at site A (photograph taken in October 2000).

calcarenite, thin sections were prepared following the bedding in the rock mass. Fabric observations and photographs of the thin sections were undertaken with a Leitz Diaplan polarizing microscope and 35 mm camera (Fig. 11). The fabric is typical of loosely packed biocalcarenites (biosparites), moderately sorted, with a self-supporting framework of skeletal remains of marine organisms and algal oncoids. The micritic matrix is absent or recrystallized and hardly distinguishable from the microcrystalline cement. The carbonate skeletal grains consist of fragments and rarely whole shells, and include benthic and planktonic foraminifera, bryozoa, lamellibranchs, gastropods, calcareous algae and serpulid worm tubes. Some of these were dissolved after deposition leaving empty casts with thin recrystallized micrite envelopes representing only the outer layers of these shells. The carbonate cement is not evenly distributed and mostly occurs at the contact between grains or on the grain walls in open pore spaces as a fine encrustation of calcite microcrystals. Late generation sparry calcite occurs in both interparticle pore spaces and skeletal moulds, and is characterized by greater sizes and a lighter colour; in sample B2, this cement constitutes a pore-filling mosaic characterized by calcite crystals that increases in size towards the centre of the voids. All the calcarenite types have a porosity in which the major contribution is made by vugs, especially in sample B1. Intergranular, intragranular and moldic porosity characterize all the varieties. Physical properties were determined by means of standard geotechnical laboratory tests (ASTM 1989) performed on three samples of each calcarenite type, and a specific gravity of 2.70 was assumed, based upon the chemical composition of the rocks (Table 3). Hydraulic conductivity (k) was determined on three samples of each variety of calcarenite using the test procedure outlined in Andriani & Walsh (2003). Water permeability tests were conducted in a purpose-built cell on cylindrical rock samples (71×140 mm) using the falling head method. Testing temperatures ranged from 19 to 25 °C; the hydraulic conductivity (k) was expressed in ms^{-1} and then standardized at 20 °C. Measurements were performed for a range of hydraulic gradients between 0.5 and 15. The average values of the hydraulic conductivity standardized at 20 °C were 1.2×10^{-5}, 2.3×10^{-5} and 4.5×10^{-5} m s^{-1} for types B2, B3 and B1, respectively. Grain size seems to have no direct influence on the permeability, which was lower in the calcarenites with a higher degree of packing and cement contents, and a lower number of vugs.

Fig. 9. One of the many quarries (today abandoned) in the area, with solid wastes within.

Fig. 10. The drums of waste industrial oils that were stored in one of the quarries (site B in Fig. 6) in the year 2000, and which are still there below the plastic cover (photograph taken in June 2003).

In contrast to the porous Plio-Quaternary calcarenites, the Cretaceous bedrock has a secondary permeability derived from tectonic discontinuities that affect the rock mass and from dissolution. Based on permeability field tests carried out in boreholes, the typical values of hydraulic conductivity for these rocks are in the range 10^{-3}–10^{-4} m s^{-1} (Calò *et al.* 1992).

On the basis of the geological setting, the local stratigraphy and the laboratory tests, a preliminary evaluation of the impact from quarrying activities on both the shallow and deep aquifer at Burgesi has been attempted. As shown schematically in Table 4, the DRASTIC and LeGrand methods show a significant change in the vulnerability of the shallow aquifer in the quarrying areas. A similar outcome was obtained by application of the GOD (a method similar in many respects to the LeGrand model) and SINTACS (which is analogous to DRASTIC) methods. The DRASTIC and SINTACS methods can detect changes in the relative vulnerability of the deep aquifer, whereas the LeGrand method and GOD do not allow recognition

Table 2. *Summary of the methods to assess the first level of pollutant potential and list of the considered parameters*

Parameters	Methods			
	Legrand (1964, 1983)	DRASTIC (Aller *et al.* 1985)	GOD (Foster 1987)	SINTACS (Civita 1991)
Topographic slope variability	–	Yes	–	Yes
Characteristics of soil	Yes	Yes	–	Yes
Characteristics of unsatured zone	Yes	Yes	Yes	Yes
Groundwater recharge	–	Yes	–	Yes
Depth to water table	Yes	Yes	Yes	Yes
Type of aquifer	–	–	Yes	–
Characteristics of the aquifer	–	Yes	–	Yes

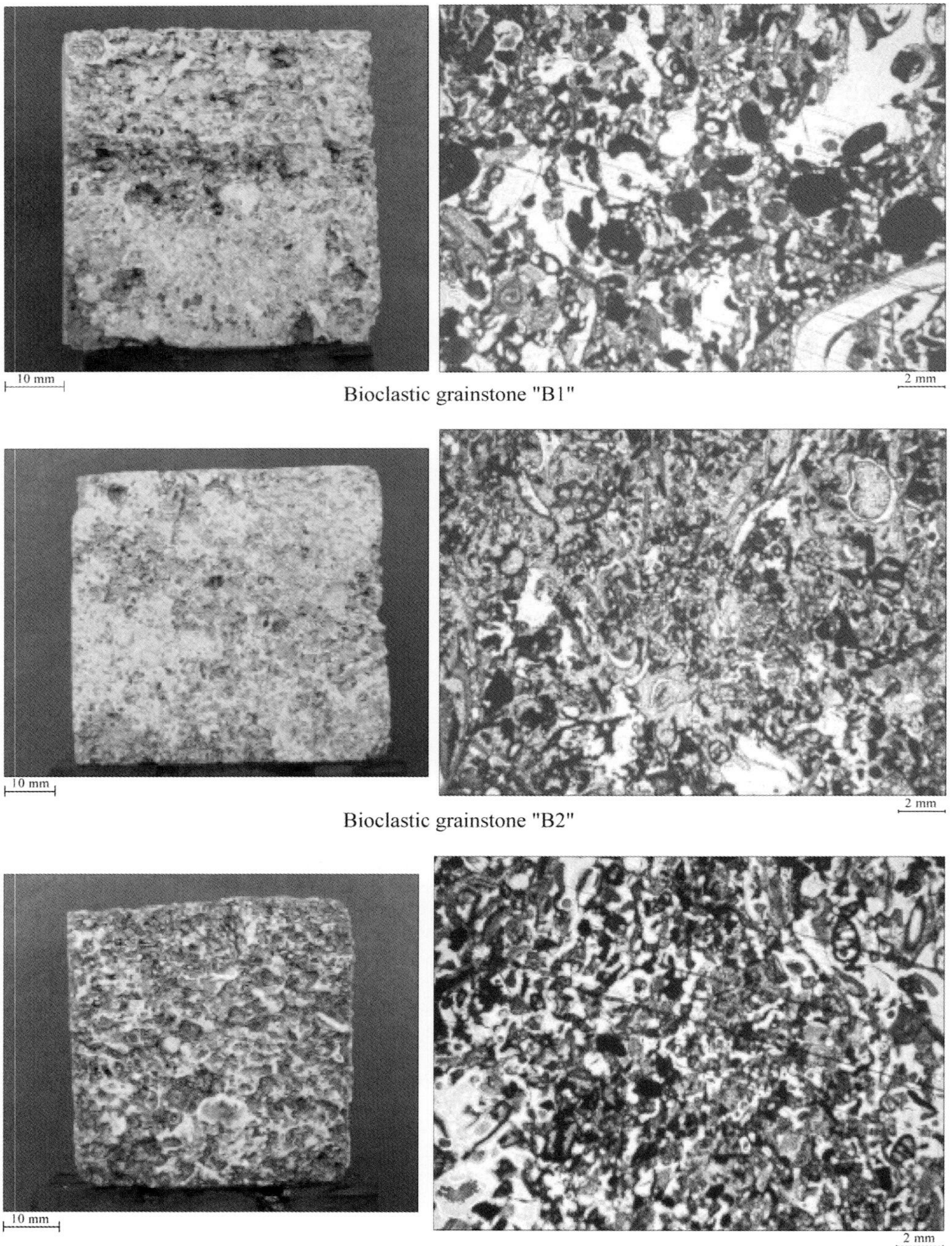

Fig. 11. Medium-grained bioclastic grainstones of the Burgesi area: on the left, images of samples used for geotechnical analysis; on the right, petrographic thin sections in plane-polarized light.

Table 3. *Average values of physical parameters of tested calcarenites*

	Samples		
	B1	B2	B3
Specific gravity, G_s	2.7	2.7	2.7
Dry unit weight, γ_d (kN m^{-3})	13.83	18.25	13.24
Saturated unit weight, γ_{sat} (kN m^{-3})	17.06	20.01	17.17
Porosity, n (%)	47.5	31.4	50.4
Water absorption, w_a (%)	22.7	10.7	30.6
Degree of saturation, Sr (%)	67.6	61.5	82.3

of variations in the parameters before and after quarrying.

At this first level of analysis, some further considerations must be provided about point centre A (Fig. 12). Before quarrying, the presence of a doline filled with several metres of residual terrain could constitute a barrier against leaching of pollutant. When using DRASTIC for the vulnerability assessment of the shallow aquifer before quarrying, the parameters ‘depth of water table’, ‘impact of vadose zone’ and ‘soil media’ had a very low value (1.5–3), and the resulting vulnerability was in the medium range. After removing the residual deposits, and quarrying the top calcarenites of the Plio-Quaternary sequence, the three aforementioned parameters reach the maximum values (9–10) and, as a consequence, the vulnerability of the shallow aquifer becomes very high. Similar results were again obtained at point centre A using the LeGrand method whose outcomes change from ‘very improbable possibility of pollution’ before quarrying to ‘imminent possibility of pollution’ in the present situation.

The deep aquifer (which represents the only supply of drinkable water for most of Apulia) does not seem to be significantly affected by quarrying, and the results shown in Table 4 and Figure 12 indicate its low vulnerability to pollution. However, it is very important to notice that the methods used for the first level of analysis do not take into account the possibility of fast movement of fluid from the shallow to the deep aquifer through the discontinuity systems and karst conduits in the rock mass. Therefore, a second level of analysis is an absolute necessity to provide appropriate tools of policy and management of the groundwater reservoirs.

Karst features and second level of analysis

The second level of analysis in evaluating the impact of anthropogenic activities at Burgesi is to consider the local karst features, to better define the vulnerability of the related aquifers and the possibility that pollutants might degrade the quality of groundwaters. Knowledge of the surface and subsurface features produced by karst processes is of crucial importance for this. However, this type of data, and particularly those dealing

Table 4. *Vulnerability (or possibility of pollution for LeGrand method) of the shallow aquifer (A) and the deep aquifer (B), before and after quarrying. The results of DRASTIC, GOD and SINTACS methods are normalized in five intervals of vulnerability (very low, low, medium, high and very high) with an equal range of values*

	Before quarrying	After quarrying
A		
DRASTIC	High vulnerability	Very high vulnerability
LeGrand	Possible but not likely	Probable
SINTACS	High vulnerability	Very high vulnerability
GOD	High vulnerability	Very high vulnerability
EPIK	High vulnerability	Very high vulnerability
B		
DRASTIC	Medium vulnerability	Medium–high vulnerability
LeGrand	Very improbable possibility of pollution	Very improbable possibility of pollution
SINTACS	Medium–high vulnerability	High–very high vulnerability
GOD	Low vulnerability	Low vulnerability
EPIK	High vulnerability	Very high vulnerability

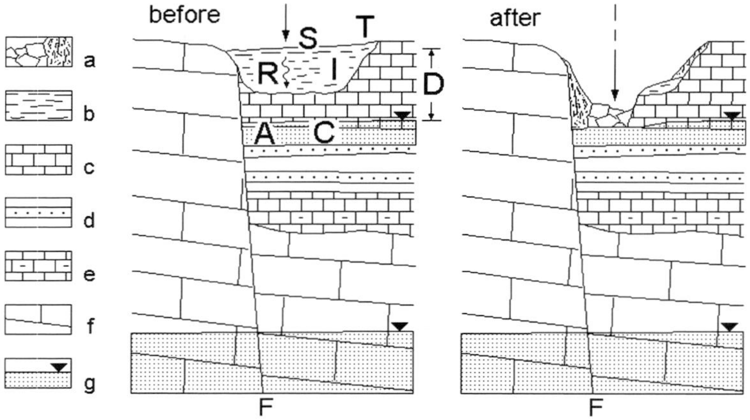

	agricultural weights	before	after
D	*5*	**3**	**10**
R	*4*	**9**	**9**
A	*3*	**10**	**10**
S	*5*	**1,5**	**10**
T	*3*	**9**	**9**
I	*4*	**3**	**9**
C	*2*	**8**	**8**
	total	**143,5**	**245**
vulnerability		medium	very high

Fig. 12. Example of vulnerability assessment for the superficial water table at the site A in Figure 6, performed by applying the DRASTIC method. Key: a, detritus; b, terre rosse and clay deposits; c, top calcarenites; d, argillaceous silts with sandy intercalations; e, bottom calcarenites; f, limestones; and g, water tables. F indicates faulting in the Cretaceous limestones, through which movement of water may occur from the surficial to the deep water table.

with speleology and development of underground cave systems, is not always available, which makes the analysis more difficult.

The surface karst landforms in the study area are mostly represented by dolines, partly filled by terre rosse and clay deposits (Fig. 2). The dolines have been formed by dissolution and/or by collapse; in this latter case, failure of the rock roof into an underlying cave in the calcarenites probably started the doline formation (Waltham *et al.* 2005). Unfortunately, at present it is not possible to discriminate between the two types of dolines, since these karst landforms have been strongly modified by anthropogenic activities and have lost their typical features. However, based on interpretation of pre-anthropogenic activity aerial photographs, we feel confident that most of the dolines formed by collapse. This hypothesis is supported by their well-defined circular shape, subvertical walls and small size (depths generally no greater than 10 m, maximum diameter of a few tens of metres). These elements are typical of collapse dolines, and differentiate them from solution dolines (Culshaw & Waltham 1987), which are generally much wider and show less abrupt walls.

In addition, the geology of the study area indicates that stratigraphy and tectonics probably played an important role in the development of dolines, some of which appear to be controlled by the presence of faults, while others could have been started at the contact between different lithologies (Gams 1994). Whatever the origin of the dolines, both fault dolines and collapse dolines can be considered as possible point centres for pollutants owing to the cleaning of residual deposits and quarrying activities performed at the sites.

As for the subsurface features of karst, conduit systems can be hypothesized, with subhorizontal galleries developed at the boundary between the argillaceous silts and the top calcarenites. Karst systems connecting the surficial runoff with the groundwater circulation allow a fast aquifer recharge along the mainly vertical network of shafts related to the fault system delimiting the SW boundary of the graben. This area is characterized by diffuse fracturing in the Cretaceous rock mass and, even though with less discontinuities, in the calcarenites. Besides, it has to be noted that the Cretaceous limestone bedrock is highly fractured and karstified, whereas the saturated zone is characterized by preferential flow levels related to the palaeogeographic history of Salento and especially to eustatic fluctuations of the sea and the karst base level (Tadolini & Tulipano 1981). Water flows mainly along these subhorizontal levels that constitute, at the same time, preferential pathways for pollutant migration. Vertical drainage of fluids is favoured by the normal faults that bound horst and graben. The local hydrogeological setting is thus characterized by possible drainage from the surficial to the deep aquifer through the faults bounding the flanks of the graben (Fig. 2), in a manner similar to that pointed out in other sectors of the Salento peninsula by Tulipano & Fidelibus (1989) and Delle Rose *et al.* (2000, 2003).

Taking into account the above delineated karst features, and especially the possibility of connection between the two aquifers in the area, the karst

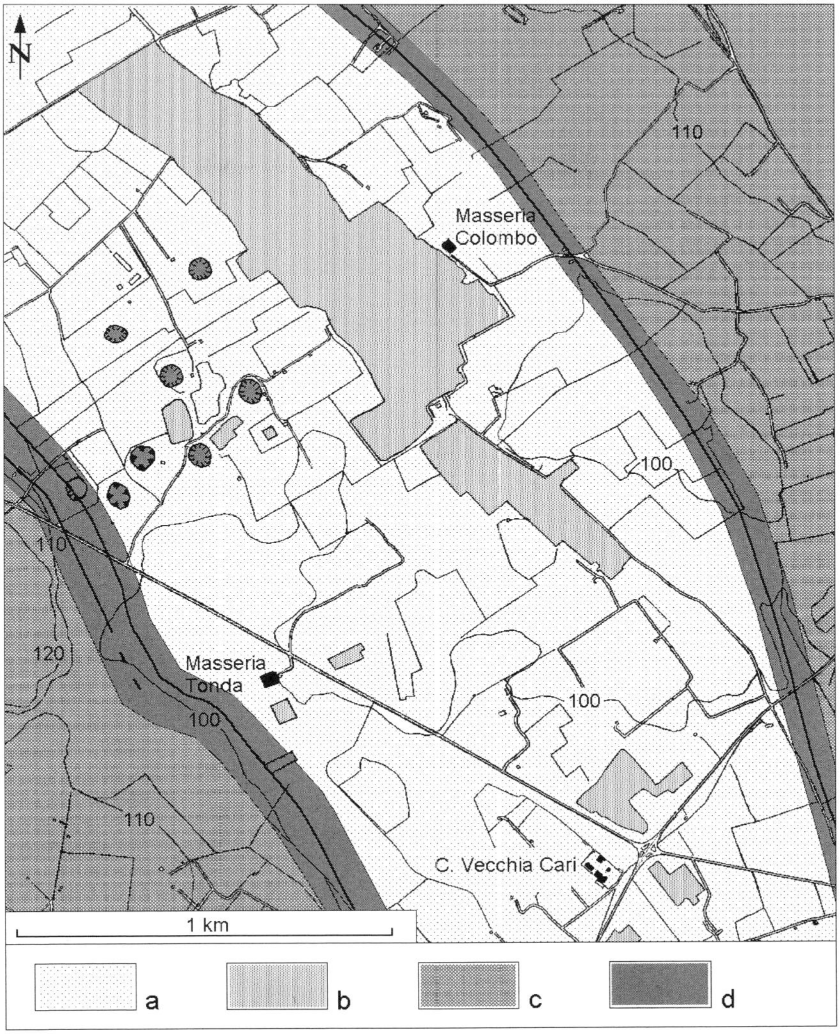

Fig. 13. Vulnerability zonation at the second level of analysis, with indication of the point centres (fault zones, collapse and solution dolines, quarries, etc.). Explanation of vulnerability levels: a, low; b, moderate; c, high; and d, very high.

aquifers appear to be much more vulnerable than in the first level of analysis, with probable severe consequences for the quality of groundwater. This is, again, particularly well evidenced by the quarry at site A, where, before quarrying, the residual deposits guaranteed a protection of groundwater from pollution, forming a filter able to reduce the leaching of pollutants toward the aquifers, and to absorb several substances such as trace metals and complex organic, reducing their concentration as they moved underground. Even though these deposits show a marked variability in both the horizontal and vertical sense, which makes their filtering action extremely variable from site to site, residual soils are generally considered good barriers against pollutant infiltration (Foster & Hirata 1988). Once the residual deposits have been removed, the occurrence of a fractured and karstified bedrock in probable connection with the shallow aquifer, and the presence of pollutants in the dolines and quarries of the area, result in a high to very high level of vulnerability.

At the whole scale of the Burgesi graben, the final zonation of vulnerability to groundwater pollution (Fig. 13) shows the highest vulnerable areas as corresponding to the contact zones between carbonate bedrock and top calcarenites, and at all the sites where the original karst landscape has been impacted. Quarries and dolines are considered to be possible point centres of pollution, from where immission of pollutants underground may occur.

Application of the EPIK method

The EPIK method was designed for karst environments, and takes into account the special features of these areas that are often underestimated, or not considered at all, in methods designed for non-karstic environments. The first attribute of EPIK, 'Epikarst', was characterized starting from the geomorphological and surface karst features at Burgesi. Owing to the presence of several dolines and of medium to high fracturing, the latter concentrated at the contact between the Cretaceous bedrock and the Plio-Quaternary sequence, the attribute Epikarst in the study area was considered as 'highly developed', that means class E_1.

The second attribute, 'Protective cover', includes any deposit covering the aquifer, and considers the related stratigraphy and thickness. In our specific case study, the presence of very thin soils caused this attribute to be defined as P_1 (absence of protected cover). Water residence time is related to thickness of the soil (Williams 1983), so the thinner the soil, the greater the vulnerability (Doerfliger *et al.* 1999).

'Infiltration conditions', the third attribute, concerns the type of recharge to the karst aquifer. In the EPIK method, owing to the different vulnerability related to diffuse and concentrate recharge, four infiltration classes have been identified (Doerfliger *et al.* 1999): concentrated; diffuse; an intermediate class where runoff may be important; and the rest of the catchment. The evaluation of the surface runoff can be difficult as it is dependent on the slope and its vegetation. Definition of this attribute at Burgesi was particularly difficult owing to the presence of features leading to concentrated recharge (swallow holes and dolines), which are combined, at the same time, with runoff on slopes of less than 10% in the cultivated areas and 25% in those devoted to pastures. In order to be conservative, class I_2 was assigned.

The fourth attribute, 'Karst network development', represents the main difference between EPIK and the other methods as it takes into account the presence of solution openings greater than 10 mm in diameter, that is the effective minimum aperture for turbulent flow (Bögli 1980). As pointed out by the authors of the EPIK method, lack of detailed maps of the karst network in many areas has led to use of a single value of the attribute per catchment (Doerfliger *et al.* 1999). At Burgesi, we observed the presence of small conduits with decimetric openings. These were the only karst conduits at the surface and in the quarry rock walls; however, the possibility of local greater size conduits (e.g. in the main dolines, below the residual cover) cannot be excluded. These features led to the karst network being defined as poorly developed and thus assigned class K_2.

After definition of the four attributes, the EPIK method allows calculation of the protection factor F_p by a basic formula where the attributes are weighted and summed. At Burgesi, the protection factor F_p ranges from a maximum value of 15 to 10 at some of the more vulnerable sites such as the original doline (quarry A in Fig. 6). It is well below the value under which the vulnerability is classified as very high (Protection area S_1, with F_p lower or equal to 19). Low F_p values correspond to high vulnerability. Thus, overall the Burgesi area presents a very high vulnerability, which become still higher at the site of quarry A.

Discussion

In this paper we have attempted to evaluate the impact of quarrying on the vulnerability of a shallow and a deep aquifer in a karst area of Apulia. The evaluation has been performed by means of widely used methods for groundwater vulnerability assessment (DRASTIC, SINTACS, LeGrand and GOD), and the EPIK method, which takes into account the specific features of karst environments.

Regarding the first group of methods, at a first level of analysis the similar approaches, and the very close parameters considered in the rating systems, led to similar results: a significant increase in the vulnerability of the shallow aquifer (both spatially, and at a point centre where a more detailed analysis was performed), whereas no substantial variation of the parameters was registered as regards the deep aquifer before and after quarrying. However, the first level of analysis did not take into account the possibility of fast movement of fluid from the shallow to the deep aquifer through the discontinuity systems and the karst conduits in the rock mass.

A second level of analysis (not originally included in the four methods) was thus carried out, in order to include in the analysis further data representative of the particular situation of karst, both at the surface and underground. This second level considered the local karst features to better define the vulnerability of the related aquifers, and the possibility that pollutants might degrade the quality of groundwaters. This might occur through drainage from the shallow to the deep aquifer through the connecting faults. The presence of underground karst systems could further increase the vulnerability of both the aquifers, since these might drain any pollutant from the ground surface to the deep aquifer in a matter of a few hours. Even though the approach was qualitative, inclusion of karst data in the second level of analysis brought about a better definition of the problem, and indicated the high possibility of pollutant movement to the deep aquifer, which provides the only supply of potable water for most of the region.

The EPIK method gave due weight to the specific features of karst, such as the epikarst, protection cover, infiltration conditions and karst network development. Application of the method resulted in very high vulnerability assessment, with the highest values at the sites where the original karst landscape was heavily modified by human activities. Even though this evaluation is still semi-quantitative, the outcome expresses the great fragility of karst environment and the serious consequences that anthropogenic activities may produce.

Monitoring and protection of karst water resources is now an absolute priority in Apulia (Delle Rose *et al.* 2003), and there is a strong need for the development of policies that will prevent further contamination of this valuable and limited natural resource. Among the possible strategies aimed at managing the contamination of carbonate aquifers, the only acceptable long-term approach, given the lack of a complete comprehension of the local hydrogeological setting, appears to be control of the sources of pollution (Rajagopal & Tobin 1989; Smith 1993; Williams 1993; El Naqa 2004). In this sense, the recent activities at Burgesi, with a project to restore the natural environment in the area, represent a positive step towards recovery of the site. Given the recent history of degradation at the Burgesi graben, however, monitoring of the area has to be continuously performed.

The complex and special characteristics of karst aquifers have been recently highlighted by Daly *et al.* (2002) and by Bakalowicz (2005), who showed how the study methods used in classical hydrogeology are generally invalid and unsuccessful in karst. Thus, there is a need for further studies about the assessment of vulnerability of karst aquifers. A thorough knowledge of the surface and subsurface karst and the local hydrogeology, integrated by data from caving explorations and from tracing experiments (Gunn 2007), appears to be crucial in order to provide the scientific basis that local authorities can draw upon for any decision regarding land planning and exploitation of the natural resources in karst environments.

The Province of Lecce kindly provided some data regarding geological investigations. Mr N. Gray made available to us the results of the chemical analysis on pollutants from the drums in the quarry. C. Fidelibus (Technical University of Zurich, Switzerland) and L. Tulipano (University 'La Sapienza', Rome, Italy) provided useful comments on a preliminary version of the article. R. Civilla performed the inventory of wells.

References

ALLER, L., BENNET, T., LEHR, J. H. & PETTY, J. R. 1985. *Drastic: A Standardized System for Evaluating Groundwater Pollution Potential Using Hydrogeologic Setting*. US Environmental Protection Agency, Washington, D.C., **600/2-87-035.**

ANDRIANI, G. F. & WALSH, N. 2003. Fabric, porosity and water permeability of calcarenites from Apulia (SE Italy) used as building and ornamental stone. *Bulletin of Engineering Geology and the Environment*, **62**, 77–84.

ASTM. 1989. *ASTM C97-83 Standard Test Methods for Absorption and Bulk Specific Gravity of Natural Building Stones*. American Society for Testing Materials, Philadelphia, PA, 1–2.

BAKALOWICZ, M. 2005. Karst groundwater: a challenge for new resources. *Hydrogeology Journal*, **13**, 148–160.

BÖGLI, A. 1980. *Karst Hydrology and Physical Speleology*. Springer-Verlag, Berlin.

BONACCI, O. 2004. Hazards caused by natural and anthropogenic changes of catchment area in karst. *Natural Hazards and Earth System Sciences*, **4**, 655–661.

BONDESAN, A. & MENEGHEL, M. 1990. Impact by limestone exploitation in western Lessini Mountains (north-eastern Italy). *Studia Carsologica*, **2**, 7–18.

CALÒ, G., GNONI, R. & STANI, M. 1992. *Caratteri idrogeologici delle falde superficiali della Penisola Salentina e valutazione della vulnerabilità degli acquiferi*. Amministrazione Provinciale di Lecce, **31**.

CULSHAW, M. G. & WALTHAM, A. C. 1987. Natural and artificial cavities as ground engineering hazards. *Quarterly Journal of Engineering Geology*, **20**, 139–150.

D'ALESSANDRO, A. & MASSARI, F. 1997. Pliocene and Pleistocene depositional environments in the Pesculuse area (Salento, Italy). *Rivista Italiana di Paleontologia e Stratigrafia*, **103**, 221–258.

DALY, D., DASSARGIES, A. *ET AL.* 2002. Main concepts of the 'European approach' for karst-groundwater vulnerability assessment and mapping. *Hydrological Journal*, **10**, 340–345.

DAVIDSON, J. J., RAO, P. S. & RAO, C. 1988. Using model to solve ground water quality problems. *In*: *US NRC, Hazardous Waste Site Management: Water Quality Issues*. Water Science and Technology Board.

DELL'ANNA, L., FIORE, S. & LAVIANO, R. 1985. The mineralogical, chemical and grain-size features of some clay deposits from Terra d'Otranto (Puglia, Southern Italy). *Geologia Applicata e Idrogeologia*, **20**, 111–123.

DELLE ROSE, M. 2001. Geological constraints on the location of industrial waste landfills in Salento karst areas (southern Italy). *In*: *Proceeding of the 6th International Conference on Groundwater Pollution, Rodi (Greece)*, 57–68. WIT Press, Southampton.

DELLE ROSE, M., FEDERICO, A. & FIDELIBUS, C. 2000. A computer simulation of groundwater salinization risk in Salento peninsula (Italy). *In*: *Proceedings of the 1st Conference on Risk Analysis, Bologna (Italy)*, 465–475. WIT Press, Southampton.

DELLE ROSE, M., FIDELIBUS, C., INTERNÒ, G. & PARISE, M. 2003. The experience of southern Apulia (Italy) coastal karst aquifer: indication from the management. *In*: *Proceedings of the 1st Conference on Sustainable Planning and Development, Skiathos (Greece)*, 453–561. WIT Press, Southampton.

DOERFLIGER, N., JEANNIN, P. Y. & ZWAHLEN, F. 1999. Water vulnerability assessment in karst environments: a new method of defining protection areas using a multi-attribute approach and GIS tools (EPIK method). *Environmental Geology*, **39**, 165–176.

EL NAQA, A. 2004. Aquifer vulnerability assessment using the Drastic model at Russeifa landfill, northeast Jordan. *Environmental Geology*, **47**, 51–62.

EKMECKI, M. & GUNAY, G. 1997. Role of public awareness in groundwater protection. *Environmental Geology*, **30**, 81–87.

ESCOLERO, O. A., MARIN, L. E., STEINICH, B., PACHECO, A. J., CABRERA, S. A. & ALCOCER, J. 2002. Development of a protection strategy of karst limestone aquifers: the Merida Yucatan, Mexico case study. *Water Resources Management*, **16**, 351–367.

FORD, D. C. & WILLIAMS, P. W. 1989. *Karst Geomorphology and Hydrology*. Unwin Hyman, London.

FOSTER, S. S. D. 1987. Fundamental concepts in aquifer vulnerability, pollution risk and protection strategy. *In*: *Proceedings of the Conference on 'Vulnerability of Soil and Groundwater to Pollution*,' Volume 38, 69–86. TNO Committee on Hydrological Research, Delft, The Netherlands.

FOSTER, S. S. D. & HIRATA, R. 1988. *Groundwater Pollution Risk Assessment. A Methodology Using Available Data*. Pan American Center for Sanitary Engineering and Environmental Sciences, Lima.

GAMS, I. 1994. Type of contact karst. *Geografia Fisica e Dinamica Quaternaria*, **17**, 37–46.

GAMS, I., NICOD, J., SAURO, U., JULIAN, E. & ANTHONY, U. 1993. Environmental change and human impacts on the Mediterranean karsts of France, Italy and the Dinaric region. *In*: WILLIAMS, P. W. (ed.) *Karst Terrains: Environmental Changes and Human Impact*. Catena, Supplement 25. Catena, Cremlingen-Destedt, 59–98.

GILLIESON, D. 1996. *Caves*. Blackwell, Oxford.

GUNN, J. 1993. The geomorphological impacts of limestone quarrying. *In*: WILLIAMS, P. W. (ed.) *Karst Terrains: Environmental Changes and Human Impact*. Catena, Supplement 25, 187–197.

GUNN, J. 2004. Quarrying of limestone. *In*: GUNN, J. (ed.) *Encyclopedia of Cave and Karst Science*. Routledge, London, 609–611.

GUNN, J. 2007. Contributory area definition for groundwater source protection and hazard mitigation in carbonate aquifers. *In*: PARISE, M. & GUNN, J. (eds) *Natural and Anthropogenic Hazards in Karst Areas: Recognition, Analysis and Mitigation*. Geological Society, London, Special Publications, **279**, 97–109.

IBE, K. M., NWANKWOR, G. I. & ONYEKURU, S. O. 2001. Assessment of ground water vulnerability and its application to the development of protection strategy for the water supply aquifer in Owerri, southeastern Nigeria. *Environmental Monitoring and Assessment*, **67**, 323–360.

KACAROGLU, F. 1999. Review of groundwater pollution and protection in karst areas. *Water, Air, and Soil Pollution*, **113**, 337–356.

LAMOREAUX, P. E., POWELL, W. J. & LEGRAND, H. E. 1997. Environmental and legal aspects of karst areas. *Environmental Geology*, **29**, 23–36.

LANGER, W. H. 2001. *Potential Environmental Impacts of Quarrying Stone in Karst–A Literature Review*. USGS Open-file Report, **0F-01-0484**.

LEGRAND, H. E. 1964. System for evaluating the contamination potential of some waste sites. *American Water Works Association Journal*, **56**, 959–974.

LEGRAND, H. E. 1983. *Standardized System for Evaluating Waste Disposal Sites, A Manual to Accompany Description and Rating Charts*. National Water Well Association, Worthington.

MARTINIS, B. 1970. *Note illustrative della carta geologica d'Italia – Foglio 223 S. Maria di Leuca*. Servizio Geologico d'Italia, Roma, **64**.

MEMON, B. A. & PROHIC, E. 1989. Movement of contaminants in karstified carbonate rocks. *Environmental Geology & Water Sciences*, **13**, 3–13.

NICOD, J. 1972. *Pays et paysages du calcaire*. Presses Universitaires de France, Paris.

NICOD, J. & SALOMON, J. N. 1999. Impacts of agricultural transformation on the principal karstic regions of France. *International Journal of Speleology*, **26B** (1/4), 15–31.

PARISE, M. & PASCALI, V. 2003. Surface and subsurface environmental degradation in the karst of Apulia (southern Italy). *Environmental Geology*, **44**, 247–256.

RAJAGOPAL, R. & TOBIN, G. 1989. Expert opinion and ground-water quality protection: the case for nitrate in drinking water. *Groundwater*, **27**, 835–847.

RICCHETTI, G., CIARANFI, N., LUPERTO SINNI, E., MONGELLI, F. & PIERI, P. 1988. Geodinamica ed evoluzione sedimentaria e tettonica dell'Avampaese Apulo. *Memorie Società Geologica Italiana*, **41**, 57–82.

ROBBA, E. 1969. Il Plio-Pleistocene della zona di Taranto. *Rivista Italiana di Paleontologia e Stratigrafia*, **75**, 605–672.

SMITH, D. I. 1993. The nature of karst aquifers and their susceptibility to pollution. *In*: WILLIAMS, P. W. (ed.) *Karst Terrains: Environmental Changes and Human Impact*. Catena Supplement 25. Catena, Cremlingen-Destedt, 41–58.

TADOLINI, T. & TULIPANO, L. 1981. The evolution of fresh-water/salt-water equilibrium in connection with withdrawals from the coastal carbonate and karstic aquifer of the Salentine peninsula. *Geologisches Jahrbuch*, **29**, 69–85.

TOZZI, M. 1993. Assetto tettonico dell'Avampaese Apulo meridionale (Murge meridionali – Salento) sulla base dei dati strutturali. *Geologica Romana*, **29**, 95–111.

TULIPANO, L. & FIDELIBUS, M. D. 1989. Temperature of groundwaters in coastal aquifers: some aspects concerning salt-water intrusion. >*In*: *Proceedings of the Salt-water Intrusion Meeting, Ghent (Belgium)*, 308–316.

VIAS, J. M., ANDREO, B., PERLES, M. J. & CARRASCO, F. 2005. A comparative study of four schemes for groundwater vulnerability mapping in a diffuse flow carbonate aquifer under Mediterranean climatic conditions. *Environmental Geology*, **47**, 586–595.

WALTHAM, T., BELL, F. & CULSHAW, M. 2005. *Sinkholes and Subsidence: Karst and Cavernous Rocks in Engineering and Construction*. Springer, Berlin.

WHITE, W. B. 1988. *Geomorphology and Hydrology of Karst Terrains*. Oxford University Press, New York.

WILLIAMS, P. W. 1993. Environmental change and human impact on karst terrains: an introduction *In*: WILLIAMS, P. W. (ed.) *Karst Terrains: Environmental Changes and Human Impact*. Catena Supplement 25. Catena, Cremlingen-Destedt, 1–19.

Natural and anthropogenic hazards in the karst of Jamaica

M. J. DAY

Department of Geography, University of Wisconsin-Milwaukee, P.O. Box 413, Milwaukee, Wisconsin 53201, USA (e-mail: mickday@uwm.edu)

Abstract: About two thirds of Jamaica is karst landscape, and karstic hazards affect much of the country and about half of the population, mostly in rural areas. The karst includes extensive areas of dolines and dry valleys, together with poljes and classical tropical tower and cockpit karst. With population and urbanization increases, and as infrastructure is developed, karstic hazards are becoming more prevalent and risks are increasing. One major natural hazard is seasonal drought, which disrupts water supplies, particularly in rural areas where groundwater resources are poorly developed and residents depend on rainwater and springs. Conversely, seasonal flooding, particularly that associated with tropical storms, causes property damage and human death, injury and displacement. Ground surface subsidence and collapse threatens developing infrastructure, dwellings and livestock, but the potential for catastrophic karstic failure of industrial facilities such as dams and retention ponds, including the storage facilities associated with bauxite mining and processing, appears to be relatively limited. Slope failure also occurs, but is not often recognized as a hazard and has not been studied in detail. Human impacts include quarrying, bauxite mining, groundwater abstraction, urbanization, agricultural development and tourism. Groundwater contamination is a serious anthropogenic hazard, particularly associated with the bauxite industry. Less than 10% of the karst area is within protected areas.

The setting

Jamaica is located south of the Cayman Trough, which represents the northern boundary of the Caribbean tectonic plate. Carbonate rocks have been forming throughout the western two thirds of the island since the Eocene, when deposition of the relatively impure Yellow Limestones began within subsiding troughs and block and belt structures defined by two major fault systems trending east–west and NW–SE (Draper 1987; Robinson 1994). Further subsidence was followed by deposition of the deeper-water White Limestones, karstification of which began with renewed uplift and folding in the Miocene. Subsequent denudation exposed the underlying Yellow Limestones and older non-carbonate rocks, producing a series of inliers from which surface drainage converged upon the White Limestones, leading to the development of a mixed autogenic–allogenic karst hydrological system, particularly on the northern side of the island. The most significant karst is developed on the Clarendon Block, the Newmarket–Montpelier Belt and the North Coast Belt (Draper 1987; Robinson 1994) (Fig. 1).

It is the Eocene–Miocene-aged White Limestone Group that is the most widespread and most extensively karstified, essentially because of its high degree of purity, commonly in excess of 99%, and because of the development of effective secondary permeability along the lines of structural weakness resulting from block faulting. The White Limestones are dominantly biomicrites and biosparites, although the Troy Formation is dolomitized in part. The overlying Upper Miocene Montpelier Formation is dominantly micritic and relatively impure, and is often described as a chalk. Karst development is most pronounced on the harder, more crystalline members (Day 1982).

Carbonate deposition, uplift, faulting and karstification have continued throughout the Palaeogene and into the Neogene. The north coast in particular is characterized by a series of raised carbonate marine terraces, representing Quaternary uplift, on which is developed an extensive suite of littoral karren.

The karst is currently subject to a seasonally dry tropical humid climatic regime, classified as Aw in the Koppen system, which promotes karstification and accentuates the inherent variability in the karst hydrological regime. The climate is dominated by the Northeast Trade Winds, temperatures range from 19 to 32 °C, and annual rainfall totals range from less than 1000 mm in drier leeward areas to over 3000 mm at higher elevations to the windward. A marked dry season, with considerable risk of drought, extends from December to April, and a wet season, with flooding risks, from May to October. The late summer hurricane season, typically June–September, may bring torrential

From: PARISE, M. & GUNN, J. (eds) *Natural and Anthropogenic Hazards in Karst Areas: Recognition, Analysis and Mitigation*. Geological Society, London, Special Publications, **279**, 173–184. DOI: 10.1144/SP279.14
0305-8719/07/$15.00

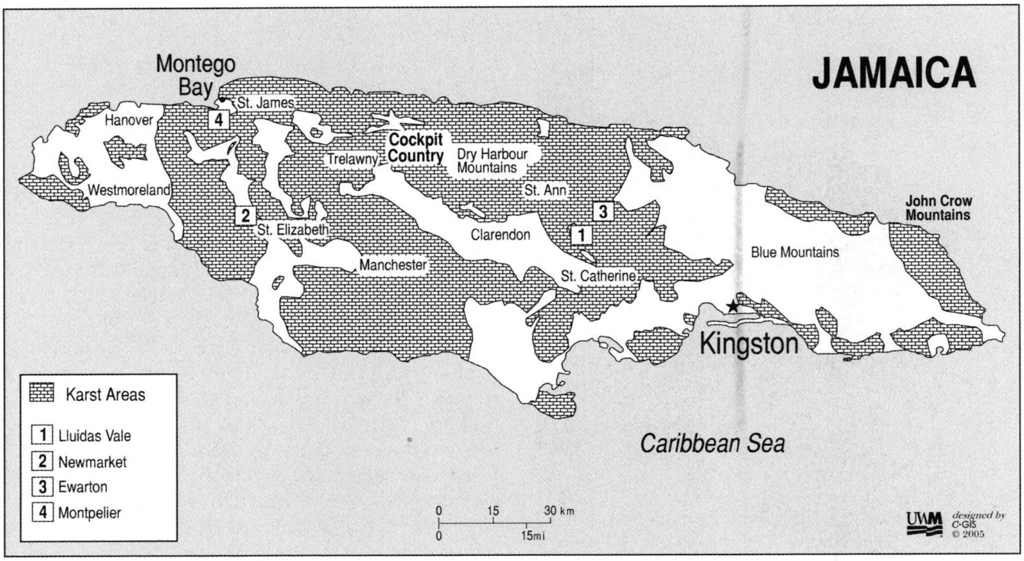

Fig. 1. The Jamaican karstlands.

rainfall accompanied by flooding and damage (O'Hara 1987, 1990).

The karst landforms include a wide range of holokarst and fluviokarst features, including dolines, dry valleys, cockpits, towers, poljes, stream sinks, caves and springs (Sweeting 1958; Versey 1972; Gardner *et al.* 1987; Troester *et al.* 1987). The extent and importance of the doline karst is generally underestimated, but it is significant particularly in some Yellow Limestone areas in the SW on the Manchester Plateau and in the North Coast belt north of the Duanvale fault system, for example in the vicinity of Browns Town (Day 1976) (Figs 1 & 2). Dry, ephemeral or seasonally intermittent river valleys are particularly well developed along the north coast (Day 1985) (Fig. 3). Cone, kegel or cockpit karst is the best known, most spectacular and widespread form, accounting for about 60% of

Fig. 2. Doline karst, St Ann.

Fig. 3. Dry valley, North Coast Belt.

the Jamaican karst, and the Cockpit Country, centred on Trelawny Parish, is the 'type-example' of the style (Day 1979; Day & Chenoweth 2004) (Figs 1 & 4). Many of the cockpits and other depressions contain bauxite deposits, probably derived from weathering of bentonitic clays of late Cenozoic volcanic origin (Comer 1974).

Poljes, structurally controlled extensive flat-floored depressions with steep sides (Bonacci 2004), are developed particularly peripheral to the main karst areas where allogenic drainage produces lateral planation and brings in alluvium. Perhaps the best-known examples are Lluidas Vale and the Queen of Spain's Valley (Draper & Fincham 1997). From these alleviated planar features rise more or less isolated and often steep-sided limestone hills or towers (Day & Tang 2004) (Fig. 5).

Drainage within karst depressions is predominantly centripetal, and may disappear underground through distinct stream sinks (Day 1976, 1979). Sinking streams are even more characteristic where allogenic drainage from the non-carbonate inliers encounters the limestone. This underground drainage then passes through a complex series of cave systems (Fincham 1997), finally emerging in peripheral springs, some of which drain from the karst as pocket valleys or steepheads.

Natural hazards

Beyond the overall roughness of much of the limestone terrain, which makes access and construction challenging in general, one major problem remains that of drought and water supply within the karst, particularly in rural areas where groundwater utilization is limited by finances and logistics and there is still considerable dependence on rainwater and springs. The perennial problem of the paucity of surface drainage assumes even greater seasonal dimensions throughout Jamaica

Fig. 4. The Cockpit Country, Trelawny.

Fig. 5. Polje and towers, Queen of Spain's Valley.

because rainfall itself is sparse and/or temporally unreliable. Short-term dry season drought is a recurrent problem in the Jamaican karst, but may also extend over longer periods and become severe, such as during 1975–1976 (Vickers 1979). Dry season drought has affected the karst more recently too, with serious problems in 1996–1998, 2000, 2004 and 2005 leading to crop and livestock losses, bush fires and emergency distribution of water supplies.

Although considerable progress has been made throughout the Jamaican karst in the provision of reliable urban water supplies via wells and pumps, much of the rural water supply still relies in large part upon collection of rainwater in tanks from roofs and gutters and in concrete catchments, locally termed *barbecues* (Fig. 6). These collection systems, although simple and environmentally friendly, have their limitations. Storage capacity is limited, and supply is unreliable, being characterized by periods of deficit or surplus. Storage may be further compromised by evaporation and leakage, or by accidental contamination, and the water may be deficient in minerals such as calcium that would otherwise be dissolved during percolation.

Springs remain the other important source of rural water supply, particularly around the periphery of the karstlands. Major perennial springs, such as those supplying Windsor, in Trelawny Parish, are the most reliable, but other communities also utilize seasonal and ephemeral springs when discharge is adequate. Drought conditions often require authorities or individuals to bring in water by trucks from non-karst areas or from remaining sources within the karst itself. Such drought conditions have a secondary impact in the form of increased probability of bush fires, resulting from either natural or anthropogenic causes.

Conversely, and perversely, flooding poses a greater hazard than drought and has more serious short-term consequences, including human death,

Fig. 6. Barbecue used for rainwater catchment.

injury and displacement, and damage to homes and other structures. Flooding of the karst landscape is an integral component of the natural karst hydrology, and is to be expected on a seasonal or intermittent basis, although it does not occur everywhere throughout the karst landscape and it remains problematic to predict. Flooding in the karst has been documented throughout the history of its study, dating back to the 19th century (Sawkins 1869). Twentieth century studies regularly noted temporary flooding of cockpits and other karst depressions, with upwelling via estavelles – basal sinks that later acted as drains for the floodwaters (Zans 1951; Sweeting 1958; Versey 1959; Day 1976). At the time this caused little disruption in sparsely settled areas, but this hazard is now accentuated by increasing population and settlement within the karst (Barker & Miller 1995; Miller 1998) (Fig. 7).

Flooding within the karst occurs through a number of distinct, but often complementary, mechanisms and affects only certain parts of the karst landscape. First, heavy and/or prolonged rainfall, such as that associated with tropical storms, may produce significant overland flow when surface and epikarstic infiltration capacities are exceeded. This is rare on exposed limestone surfaces where flowpaths are generally short and drainage sinks rapidly into open fissures, but it does occur occasionally (Day 1979). Such surface flow is more common where the surface is mantled by bauxite or other regolith, and is often indicated by rills or gullies. Surface flooding of this nature poses potentially greater hazards, particularly where natural drainage paths are provided by elongated, compound depressions, known in Jamaica as *glades* (Day 1976; Chenoweth & Day 2001).

Second, flooding of the karst may occur where rivers carrying increased allogenic drainage from adjacent non-karst terrain exceed the intake ability of the karst drainage system and overflow their channels. This is particularly likely in border poljes, such as Lluidas Vale, where such occurrences are sufficiently frequent as to be predictable on a broadly seasonal basis. Poljes, whose flat surfaces are essentially a product of alluviation during such overbank floods, may also flood more locally and unpredictably via dispersed estavelles.

Third, local flooding within the karst may result from filling and overtopping of the epikarstic reservoir, which lies close to the surface within the upper portion of the vadose zone. Such flooding is unpredictable and uncommon, but is usually localized and temporary.

Fourth, flooding on a broader scale may result from elevation of groundwater levels and upwelling of groundwater via estavelles. This again occurs on

Fig. 7. Flooding, St Catherine, 1974.

a more-or-less predictable seasonal basis in larger, low-lying cockpits, such as Bamboo Bottom and Guthrie Bottom on the northern periphery of the Cockpit Country. Here it causes little damage, although local residents report losing cattle to the floods in the past. Elsewhere, however, it is this regional groundwater upwelling that represents the most serious flooding hazard, particularly in more densely populated, low-lying areas (Fig. 7). Increased groundwater levels and volumes may also lead to increased spring output, which may in turn lead to stream flooding on the downstream sides of the karst.

Finally, normally dry valley systems within the karst may become activated after prolonged or heavy rainfall, through a combination of groundwater elevation, epikarstic overflow, increased spring output and/or allogenic inputs. This has occurred in the north coast valley system (Day 1985), where it poses serious hazards to coastal tourist infrastructure such as hotels and roads, as well as to local residents. Dry valleys on the southern side of the island, such as Sixteen Mile Gully, also carry ephemeral drainage.

In recent decades, serious floods in the Jamaican karst occurred in 1963, 1974, 1977, 1979, 1986, 1988, 1993, 1995, 1998, 2001 and 2002. The 1977 floods claimed 11 victims, and damage was estimated at Ja$6m (US$2.8m). Eyre (1983) and O'Hara (1990) documented the severe flooding in the Newmarket–Hopewell area of St Elizabeth, St James, Hanover and Westmoreland parishes in June 1979, which killed over 40 people and caused damage estimated at Ja$70m (US$33m). Predisposing geomorphic conditions included the following:

- the Newmarket and other regional basins are large, fault-bounded grabens in echelon, floored by alluvium or terra rossa clays;
- at shallow depth, the Yellow Limestone deflects water laterally into the grabens;
- within the White Limestones there is locally low sinkhole density and restricted water transmissivity (Donaldson & Walters 1981);
- Sinkholes were blocked by sediment.

Rainfall on 12 June 1979 totalled up to 865 mm, following 3 weeks during which precipitation was almost three times that normally recorded (Blake 1981). Intense surface runoff began extremely rapidly and 'substantial rivers literally burst from within the limestone, often with a tremendous noise, and were observed discharging huge volumes of water within a matter of minutes' (Eyre 1983, 13). A river, 10 m wide and 2 m deep, appeared almost immediately at Welcome, flooding the glade. The depression containing the Shafston Blue Hole, a permanent vauclusian spring, was inundated and overflowed, with water cascading 86 m into a second depression, which filled with 55 000 m^3 of water. This in turn was overtopped, with water descending another 22 m to fill a third depression with 113 000 m^3 of water. This itself overflowed, devastating the village of Brighton. Elsewhere the flood waters eroded a channel up to 6 m deep and 4 m wide over a distance exceeding 1 km. The resultant floods formed 90 individual lakes, covering more than 10 km^2, some of which took longer than a year to recede.

Since 1987 at least 20 flood events in the Jamaican karst have been reported. Most recently, the karst island-wide was flooded during heavy rains during May–June 2002, and that in Manchester Parish was further disrupted by flooding associated with tropical storms Isidore and Lili in September 2002, causing displacement of more than 40 families.

Early ideas about the development of the Jamaican cockpit karst stressed the role of surface collapse, but more recent studies suggest that this is of only secondary significance, with epikarstic dissolution being the primary formative mechanism. This notwithstanding, occasional surface collapse and subsidence does occur within the karst, although it has yet to cause more than minor localized damage and inconvenience. Subsurface dissolution, cavity formation, surface subsidence and collapse represent major hazards in many karstlands (Waltham *et al.* 2005), but Jamaica has not yet suffered their effects as catastrophically as elsewhere. Day (1976) noted the presence of subsidence/collapse features within dolines north of Browns Town, St Ann, and demonstrated subsequently that, although the overall collapse pit distribution approximates that predicted by a random set of processes, certain sites within depressions are especially prone to collapse. In particular, 55% of collapses occur between 20 and 50 m of doline bases and 57% occur within internal topographic lows. By contrast, only 25% of collapses occur within centripetal drainage channels and only 22% occur along depression long axes (Day 1984).

Cumulative data on the location of both pre-existing subsidence and collapse features and contemporary (1974–1999) collapse/subsidence events within the Jamaican karst suggest that the topographic distribution is as follows: depression bases 26%, lower slopes 51%, upper slopes 15%, corridors 6%, saddles 3% and hilltops and ridges <1% (Day 2003). However, only about 10–15% of all topographic sites investigated show any evidence of collapse or subsidence, suggesting that the 100-year collapse probabilities are as follows: depression bases 3.6–5.4%, lower slopes 7.2–10.8%, upper slopes 2.0–3.0%, corridors 0.8–1.2%, saddles 0.4–0.6%, and hilltops and ridges 0% (Day 2003).

Despite these relatively low probabilities, both ground surface collapse and subsidence represent an increasing threat to developing infrastructure, such as highways and public service facilities, plus a minor hazard to rural dwellings and livestock (Day 1984). Minor cavitation problems have recently been encountered during construction of the new Northern Coastal Highway.

There also exists the potential for karst-related deterioration or catastrophic karstic failure of industrial facilities such as dams and retention ponds, including the storage facilities associated with bauxite mining and processing. Detailed site-specific subsurface studies provide the most acceptable way to detect imminent cavity collapse and sinkhole formation, but probabilistic studies allied with site assessment are valuable tools where such investigations are not warranted or not possible for other reasons. In the context of the proposed decommissioning, dewatering and revegetation of the red mud bauxite waste disposal pond at Mount Rosser, Ewarton, Jamaica (Fig. 8), the risk of main and auxiliary dam instability and/or failure as a result of the development and collapse of underlying karst cavities (sinkholes) was assessed employing a probabilistic model developed from previous studies of collapses in differing topographic positions within the Jamaican karst (Day 2003). This approach indicated that the probability of auxiliary dam failure within the expected life of the dams is between 0.5 and 3.0%, with a site average estimate of about 1.0%, and that the probability of main dam failure under the same scenario is between 1.0 and 3.0%, with a site average estimate of about 2.0%. A rapid geomorphological assessment (RGA) and site inspection revealed no evidence of factors that might exacerbate the potential for failure, and the overall assessment was that the Mount Rosser dams are unlikely to fail as a result of the formation or expansion of voids in the carbonate bedrock.

Slope failure also poses a minor to moderate hazard to buildings, roads and other structures, although one that is rarely recognized. Although karst slopes are often considered to be stable, localized failure does occur, particularly along weathered planes of weakness, such as joints, and through shifting of talus. Karst bedrock failure may occur as slab failure, rock collapse or rockfall, the precise mechanism depending essentially on the configuration of the planes of weakness and the internal structure of the rock itself (Day 1978). Slab failure is uncommon, but does occur in the more crystalline formations where near vertical joints are undercut, for example by fluvial erosion of tower bases in poljes such as Lluidas Vale. Many of the White Limestone formations are rubbly, and here rock collapse and individual

Fig. 8. Red mud storage pond, Mount Rosser, Ewarton.

rockfall are the principle failure mechanisms. Talus is a significant component of Jamaican cockpit karst terrain (Chenoweth & Day 2001; Day & Chenoweth 2004), and is potentially unstable, particularly when disturbed. This represents a potential threat not only to individuals traversing talus slopes, but also to construction projects.

Anthropogenic hazards

As throughout the Caribbean and Central America, human impact on the Jamaican karst has been long term and is increasing, particularly via surface bauxite mining, quarrying, increased groundwater abstraction, urbanization, agricultural development and tourism (Day 1993). Human impacts on the karst may be schematized by reference to a broad impact–process–consequence model involving human activity, geomorphic process change and landscape response, and problems for human occupancy and activity. This can be illustrated by reference to the Jamaican karst hydrology. Through time, human activities have increasingly disturbed this system through forest clearance, soil utilization, groundwater abstraction and modification of surface drainage courses. This disturbance has resulted in process changes such as increased runoff, decreased infiltration, increased surface sediment transport and decreased spring discharge. In turn, this has impacted the human population through increased flood susceptibility, desiccation of springs, accelerated soil erosion and surface subsidence (Day 1993).

Forest clearance and agriculture have had a profound effect on Jamaica's karst landscape. As of 1985, about 60% of the Jamaican karst was utilized for some form of agriculture (Day 1993), and these activities are increasing rapidly (Barker & Miller 1995; Miller 1998). Sediment eroded from agricultural lands frequently has blocked sinkholes and other karst drainage features, for example in the Cave River basin (Molina & McDonald 1987). Flooding in the karst may be exacerbated by human activities, such as infilling of sinks and depressions. Extensive flooding of a large depression at Muster Ground, St Ann Parish during 1973 and 1974 was caused by the infilling of the natural outlet with local quarry debris (Day 1976, 1978). Flooding in the Cave River basin, caused by eroded soil blocking sinkholes, has caused losses of lives and property, necessitating a flood prevention and mitigation project involving broad watershed management (Molina & McDonald 1987). Seasonal drought results in serious localized agricultural losses and increased fire hazard.

Bauxite mining and limestone quarrying both have had pronounced impacts on the Jamaica karstlands. Limestone quarrying for roadbed and general construction has had an impact both locally and nationally, but the most significant quarrying has been for cement production. Nationally, the annual production of limestone in the 1988 was about 5×10^6 tonnes, and cement production was nearly 350 000 tonnes (Day 1993). In 2003, limestone and cement production amounted to 3.6×10^6 tonnes and 276 000 tonnes, respectively (United States Geological Survey 2003). Lime production in 1988 was about 73 000 tonnes (Day 1993), rising to 276 000 tonnes by 2003 (United States Geological Survey 2003). There is also a limited amount of marble and gypsum mining, and small-scale mining of bat guano from caves still takes place, although less than in previous centuries.

Commercial bauxite production began in Jamaica in 1952, and prior to 1980 almost one third of the country was owned by five foreign mining companies (Sachak 1983). Jamaica is the world's third largest bauxite producer, and some 100 000 ha of northern Jamaican karst, particularly in the Dry Harbour Mountains, have been exploited for bauxite and alumina by surface mining. Annual production in 1988 was about 8×10^6 tonnes of dry bauxite and about 2×10^6 tonnes of alumina (Day 1993). By 2003, bauxite and alumina production amounted to 13.4×10^6 and 3.8×10^6 tonnes, respectively (United States Geological Survey 2003). Beyond the physical devastation of the surface karst landscape by bauxite mining (Fig. 9), the operations have caused deforestation, ecological damage and the displacement of thousands of local residents, particularly from St Ann Parish. The alumina production facilities have been linked to local increases in respiratory illness, and to degradation of coral reefs.

Groundwater and surface water contamination is also a serious hazard, particularly where industrial effluents and urban runoff enter underground drainage systems via point recharge. Groundwater contamination has been documented throughout the Jamaican karst where urban and industrial effluents enter the underground flow system (White 1979). One particular problem has been the contamination of surface waters and groundwater by 'red mud', the caustic waste from alumina production (Bell 1986). In some cases this has been severe, resulting in groundwater sodium concentrations of up to 1725 mg l^{-1} (Wedderburn 1977).

The Mount Rosser site, north of Ewarton in St Catherine Parish, was formerly plagued by such problems, although it is currently being decommissioned (Fig. 8). The present footprint of the tailings pond is about 35 ha and the red mud itself totals about 11×10^6 m^3, with a consistency averaging about 50% solids. The mud is classified as a high plasticity clay, and its pH is about 10.0. The

Fig. 9. Bauxite mining, Dry Harbour Mountains.

Mount Rosser site straddles the drainage divide between the White River system to the north and the Linstead system to the south. The pond floor was not lined, and regional groundwater studies have documented restricted sodium ion-rich contaminant plumes within both systems, probably representing turbulent conduit flow within the limestone aquifer (Wedderburn 1977). Contaminant levels have declined consistently from maxima of about 60 mg l^{-1} sodium since cessation of red mud disposal. Underground losses of red mud and liquor from the pond have been reported, notably in 1959, 1968 and 1973 (Wedderburn 1977); these appear to have been associated with sinkhole collapse in topographic lows that were inundated during red mud accumulation. Some mud leakage, particularly along the south abutment of an auxiliary dam occurred during operation, but this now appears to have ceased. Groundwater contamination has also occurred previously via direct infiltration of sodium-rich pond water into the limestones above the mud level, which accounted for more than 90% of seepage loss during pond operation. Downward seepage through the mud currently is estimated at 5.5×10^{-7} cm s^{-1}.

Other, less recognized and less documented anthropogenic hazards also characterize the Jamaican karst. Sweeting (1972, p. 327) commented that 'Lack of transport is probably the second most important problem in the karstlands after the lack of water', and this remains relevant today. Although substantial improvements in the national road network have been made since the 1970s, roads within the karst remain relatively few and many of those are in poor condition. Traffic accident rates in Jamaica are generally high (Government of Jamaica 2001), and roads through the karst are no exception.

Portions of the Jamaican karst are also utilized for cultivation of marijuana, which is widely used by residents of the karst, and growers are often protective of their crops, sometimes guarding them with weapons, including firearms. Police forays against growers and transporters may occasionally increase the risk of violence.

Conclusions

A wide range of natural hazards exists within the Jamaican karst, potentially affecting 65% of the land area and 50% of the human population. Although much of the karst area is currently rural, urbanization and other development is increasing rapidly, with the increasing probability of karst hazards having greater impacts both socially and economically. Seasonal drought and the provision of reliable water supply remain problematic and are becoming more so as domestic, agricultural and industrial demands increase. Periodic flooding has a long history of causing property damage and human death, injury and displacement, and new development in the karst needs to be acutely aware of vulnerable locations and situations. Ground surface subsidence and collapse also increasingly threatens developing infrastructure, but the potential for catastrophic karstic failure can be reduced considerably by appropriate site investigation and engineering practices (Waltham *et al.* 2005). Slope failure is not often recognized as a hazard in the karst and warrants further investigation.

Human impacts on the karst presently include quarrying, bauxite mining, groundwater abstraction and contamination, urbanization, agricultural development and mass tourism. All of these are increasing in scope and impact, and there is a real need for a comprehensive assessment of these activities, their consequences and their future sustainability. Failing this, anthropogenic hazards can only continue to increase exponentially.

The karstlands are also increasingly involved in ecotourism activities, but less than 10% of the total Jamaican karst area is afforded recognition in protected areas (Kueny & Day 1998). Cave and karst conservation throughout Jamaica is generally disorganized (Day & Koenig 2002), with only a handful of caves and karst sites afforded any meaningful protection. The commercially operated Green Grotto (Runaway Bay Caves) on the north coast is the only cave to have undergone an environmental impact study. Windsor Great Cave, in Trelawny Parish, while exploited for guano and tourism for the past 70 years, is starting to receive some conservation because the previous owner donated the land title to WWF-UK with stringent terms of reference, and because of the presence of the nearby Windsor Research Center.

Some karst is protected in the John Crow and Blue Mountains National Park and, since 1950, much of the Cockpit Country has been designated as a forest reserve, although there has been little enforcement of conservation directives. Between 1981 and 1987, the Cockpit Country lost 14% of its world-class forest cover (Eyre 1990). The immediate vicinity has a population of some 10 000 people, and is exploited for bauxite mining and agriculture. In the Forest Reserve, illegal logging, farming, hunting and trapping for the pet trade are particular problems (Barker & Miller 1995; Chenoweth *et al.* 2001). More recently, the Cockpit Country has been proposed as a UN World Heritage Site, and there are currently plans to inscribe it as a national park (Eyre 1995; Chenoweth *et al.* 2001; Day 2004).

To date, stakeholder reaction to natural and anthropogenic hazards within the karst has been mixed. International attitudes to hazard mitigation and landscape conservation, particularly those of NGOs, have been generally positive. Likewise, national NGOs, such as the Jamaica Conservation Development Trust, have been supportive of hazard awareness and reduction. Local residents are most directly affected, but their attitudes are ambivalent. Some people recognize the hazards and seek to avoid or minimize them either individually or through broader societal action, but others are unaware or unable to do so, and others may actually benefit in various ways from the hazardous conditions and activities. Similarly, business and industrial concerns may benefit economically from activities which cause hazards, but they also incur losses when hazards disrupt facilities or productivity, so they too need to be informed and aware. Jamaican society as a whole stands to benefit from sustainable and rational development within the karst. However, the risks need to be understood and assessed as part of a comprehensive management strategy. Government agencies have sent mixed messages, some supportive of hazard reduction and karst conservation, others aligned with developments that underestimate hazards or increase them. Ultimately, the Jamaican karstlands require a comprehensive management plan, the formulation and implementation of which is becoming ever more urgent.

This research has been supported in part by grants from the Natural Environment Research Council (UK) and from the Center for Latin American and Caribbean Studies at the University of Wisconsin-Milwaukee.

References

BARKER, D. & MILLER, D. J. 1995. Farming on the fringe: small-scale agriculture on the edge of Cockpit Country. *In*: BARKER, D. & MCGREGOR, D. F. M. (eds) *Environment and Development in the Caribbean: Geographical Perpectives*. The Press, University of the West Indies, Kingston, 271–292.

BELL, J. 1986. Caustic waste menaces Jamaica. *New Scientist*, **3**, 33–37.

BLAKE, J. T. 1981. The meteorology of the June 12 disaster. *Journal of the Geological Society of Jamaica*, **20**, 3–13.

BONACCI, O. 2004. Poljes. *In*: GUNN, J. (ed.) *Encyclopedia of Caves and Karst*. Fitzroy Dearborn, New York, 599–600.

CHENOWETH, M. S. & DAY, M. J. 2001. Developing a GIS for the Jamaican Cockpit Country. *In*: BECK, B. F. & HERRING, J. G. (eds) *Geotechnical and Environmental Applications of Karst Geology and Hydrology*. Balkema, Rotterdam, 67–72.

CHENOWETH, M. S., DAY, M. J., KOENIG, S., KUENY, J. & SCHWARTZ, M. 2001. Conservation issues in the Cockpit Country, Jamaica. *Proceedings 13th International Congress of Speleology*, **2**, 237–241.

COMER, J. B. 1974. The genesis of Jamaican bauxite. *Economic Geology*, **69**, 1251–1264.

DAY, M. J. 1976. The morphology and hydrology of some Jamaican karst depressions. *Earth Surface Processes*, **1**, 111–129.

DAY, M. J. 1978. Engineering hazards in tropical karst terrain. *Applied Geography Conferences*, **1**, 288–298.

DAY, M. J. 1979. The hydrology of polygonal karst depressions in northern Jamaica. *Zeitschrift fur Geomorphologie N.F., Supplementbande*, **32**, 25–34.

DAY, M. J. 1982. The influence of some material properties on the development of tropical karst terrain. *Transactions of the British Cave Research Association*, **9**, 27–37.

DAY, M. J. 1984. Predicting the location of surface collapse within karst depressions: A Jamaican example. *In*: BECK, B. F. (ed.) *Sinkholes: Their Geology, Engineering and Environmental Impact*. Balkema, Rotterdam, 147–151.

DAY, M. J. 1985. Limestone valley systems in north-central Jamaica. *Caribbean Geography*, **2**, 16–33.

DAY, M. J. 1993. Human Impacts on Caribbean and Central American Karst. *In*: WILLIAMS, P. W. (ed.) *Karst Terrains: Environmental Changes and Human Impact*. Catena Supplement 25. Catena, Cremlingen-Destedt, 109–125.

DAY, M. J. 2003. An assessment of karstic collapse hazards at Mount Rosser, Ewarton, Jamaica. *In*: BECK, B. F. (ed.) *Sinkholes and the Engineering and Environmental Impacts of Karst*. American Society of Civil Engineers, Geotechnical Special Publications, **122**, 40–49.

DAY, M. J. 2004. Stakeholder reaction to the proposed establishment of the Cockpit Country National Park, Jamaica. *In*: BATELAAN, O., DUSAR, M., MASSCHELEIN, J., TRAN TAN, Van, VU THANH TAM & NGUYEN XUAN, KHIEN (eds) *Proceedings of the International Transdisciplinary Conference on Development and Conservation of Karst Regions*. RIGMR, Hanoi, 34–39.

DAY, M. J. & CHENOWETH, M. S. 2004. Cockpit Country Cone Karst, Jamaica. *In*: GUNN, J. (ed.) *Encyclopedia of Caves and Karst Science*. Fitzroy Dearborn, New York, 233–235.

DAY, M. J. & KOENIG, S. 2002. Cave monitoring priorities in Central America and the Caribbean. *Acta Carsologica*, **30**, 123–134.

DAY, M. J. & TANG, T. 2004. Tower Karst. *In*: GUNN, J. (ed.) *Encyclopedia of Caves and Karst Science*. Fitzroy Dearborn, New York, 734–736.

DONALDSON, L. & WALTERS, W. O. 1981. Hydrologic aspects of the June 12, 1979 flood rains on the Newmarket sub-basin. *Journal of the Geological Society of Jamaica*, **20**, 61–77.

DRAPER, G. 1987. A revised tectonic model for the evolution of Jamaica. *In*: AHMAD, R. (ed.) *Proceedings of a Workshop on the Status of Jamaican Geology. Journal of the Geological Society of Jamaica*, **10**, Special Issue, 120–150.

DRAPER, G. & FINCHAM, A. G. 1997. The geology of Jamaican caves. *In*: FINCHAM, A. G. (ed.) *Jamaica Underground: The Caves, Sinkholes and Underground Rivers of the Island*, 2nd edn. The Press, University of the West Indies, Kingston, 29–33.

EYRE, L. A. 1983. Flood hazard and tropical karst: The Newmarket syndrome. *In*: *Proceedings, National Council for Geographic Education Tropical Karst Landscapes Symposium, Ocho Rios, Jamaica*, 11–17. Department of Education, University of the West Indies, Kingston, Jamaica.

EYRE, L. A. 1990. The tropical national parks of Latin America and the Caribbean: Present problems and future potential. *Conference of Latin Americanist Geographers Yearbook*, **16**, 15–33.

EYRE, L. A. 1995. The Cockpit Country: A World Heritage Site? *In*: BARKER, D. & MCGREGOR, D. F. M. (eds) *Environment and Development in the Caribbean: Geographical Perspectives*. The Press, University of the West Indies, Kingston, 259–270.

FINCHAM, A. G. 1997. *Jamaica Underground: The Caves, Sinkholes and Underground Rivers of the Island*, 2nd edn. The Press, University of the West Indies, Kingston.

GARDNER, T. W., BACK, W. ET AL. 1987. Central America and the Caribbean. *In*: GRAF, W. L. (ed.) *Geomorphic Systems of North America*. Geological Society of America, Boulder, CO, 343–402.

GOVERNMENT OF JAMAICA. 2001. World wide web address: http://www.mtw.gov.jm/General%20Information/Policy%20Documents/rd%20safety.pol.pdf

KUENY, J. A. & DAY, M. J. 1998. An assessment of protected karst landscapes in the Caribbean. *Caribbean Geography*, **9**, 87–100.

MILLER, D. J. 1998. Invasion of the Cockpits: Patterns of encroachment into the wet limestone forest of Cockpit Country, Jamaica. *In*: MCGREGOR, D. F. M., BARKER, D. & EVANS, S. L. (eds) *Resource Sustainability and Caribbean Development*. The Press, University of the West Indies, Kingston, 373–389.

MOLINA, M. & MCDONALD, F. 1987. Sinkhole management and flooding in Jamaica. *In*: BECK, B. F. & WILSON, W. L. (eds) *Karst Hydrology: Engineering and Environmental Applications*. Balkema, Rotterdam, 293–298.

O'HARA, M. 1987. Physical hazard and risk assessment in Jamaica, West Indies. *In*: CULSHAW, M. G., BELL, F. G., CRIPPS, J. C. & O'HARA, M. (eds) *Planning and Engineering Geology*. Geological Society, London, Engineering Geology Special Publications, **4**, 311–322.

O'HARA, M. 1990. Flood hydrology of western Jamaica: A study in a karstic limestone environment. *Singapore Journal of Tropical Geography*, **11**, 100–116.

ROBINSON, E. 1994. Jamaica. *In*: DONOVAN, S. K. & JACKSON, T. A. (eds) *Caribbean Geology: An Introduction*. University of the West Indies Publishers Association, Kingston, 111–127.

SACHAK, N. 1983. Impact of bauxite in Jamaica. *Caribbean Geography*, **1**, 67–69.

SAWKINS, J. G. 1869. *Reports on the Geology of Jamaica*. HMSO, London.

SWEETING, M. M. 1958. The karstlands of Jamaica. *The Geographical Journal*, **124**, 184–199.

SWEETING, M. M. 1972. *Karst Landforms*. Macmillan, London.

TROESTER, J. W., BACK, W. & MORA, S. C. 1987. Karst of the Caribbean. *In*: GRAF, W. L. (ed.) *Geomorphic Systems of North America*. Geological Society of America, Boulder, CO, 347–357.

UNITED STATES GEOLOGICAL SURVEY. 2003. *Minerals Yearbook 2003*. World wide web address: http://minerals.usgs.gov/minerals/pubs/country/2003/caribbeanmyb03.pdf

VERSEY, H. R. 1959. The hydrological character of the White Limestone Formation of Jamaica. *In*: *Transactions of the Second Caribbean Geological Congress, Mayaguez, Puerto Rico*, 59–68.

VERSEY, H. R. 1972. Karst in Jamaica. *In*: HERAK, M. & SPRINGFIELD, V. T. (eds) *Karst: Important Karst Regions of the Northern Hemisphere*. Elsevier, Amsterdam, 445–466.

VICKERS, D. O. 1979. The rainfall of Jamaica. *Journal of the Geological Society of Jamaica*, **18**, 5–9.

WALTHAM, T., BELL, F. & CULSHAW, M. 2005. *Sinkholes and Subsidence: Karst and Cavernous Rocks in Engineering and Construction*. Springer, Berlin.

WEDDERBURN, L. A. 1977. Ground water pollution of a limestone aquifer by caustic waste. *In*: DILAMARTER, R. R. & CSALLANY, S. C. (eds) *Hydrologic Problems in Karst Regions*. Western Kentucky University, Bowling Green, KY, 388–404.

WHITE, M. 1979. Underground water resources of Jamaica. *Journal of the Geological Society of Jamaica*, **18**, 54–66.

ZANS, V. A. 1951. On the karst hydrology of Jamaica. *International Association de Hydrologie Scientifique*, **2**, 267–279.

Biotic versus abiotic calcite formation on prehistoric cave paintings: the Arcy-sur-Cure 'Grande Grotte' (Yonne, France) case

E. CHALMIN[1,2], F. D'ORLYÉ[1], L. ZINGER[3], L. CHARLET[2], R. A. GEREMIA[3], G. ORIAL[4], M. MENU[1], D. BAFFIER[5] & I. REICHE[1]

[1]*UMR 171 CNRS, Laboratoire du Centre de Recherche et de Restauration des Musées de France, Palais du Louvre, Paris, France (e-mail: ina.reiche@culture.gouv.fr)*

[2]*UMR 9995 CNRS/UJF, Laboratoire de Géophysique Interne et Tectonophysique, Domaine Universitaire, Grenoble, France (e-mail: chalmin@esrf.fr)*

[3]*UMR 5553 CNRS/UJF, Laboratoire d'Ecologie Alpine, Domaine Universitaire, Grenoble, France*

[4]*Laboratoire de Recherche des Monuments Historiques, Champs-sur-Marne, France*

[5]*DRAC Rhône-Alpes, 6, quai Saint Vincent, Lyon, France*

Abstract: The 'Grande Grotte' cave at Arcy-sur-Cure (Yonne, France) with its prehistoric paintings shows important calcite concretions. Two types of calcite have been observed on the wall: translucent yellowish layers and opaque white or grey layers that completely obstruct the paintings. Other calcite types are present in the lakes of the cave (floating calcite rafts at the surface of the lake and soft calcite at the bottom of the lake).

The morphology of the different calcites was observed at different scales by optical microscopy with normal and polarized light, scanning electron microscopy (SEM) and transmission electron microscopy (TEM). The elemental composition was measured by using particle-induced X-ray emission (micro-PIXE) and the structure by X-ray diffraction (XRD), infrared (FT-IR) and Raman spectroscopy. The bacterial diversity and its role in calcite formation were assessed by culture and 16S-SSCP in order to distinguish and to assess various abiotic and biotic formation mechanisms.

The investigation of calcite characteristics enables conclusions on the formation mechanism and on a biotic or abiotic origin of the calcites. The change of calcite types on the walls reveals changes of the environmental cave parameters. In addition, interactions of calcites with the prehistoric paint layer could be evaluated.

Palaeolithic rock art present in SW Europe is one of the most ancient arts currently known thanks to its preservation in limestone caves; however, this art is threatened by numerous alteration phenomena. Among them are the development of plants and moss, the washing out the paintings by water flow on the wall, anthropogenic degradation after the opening of the cave to visitors, degradation by micro-organism (mushrooms and bacteria) as in the cave of Lascaux (Allemand 2003) and the formation of calcium carbonate ($CaCO_3$) that may hide the paintings. The formation of $CaCO_3$ has been particularly observed in the cave called 'Grande Grotte' in the karstic network of Arcy-sur-Cure (24 500–28 000 BP, Yonne, France, Fig. 1) where rock art was discovered in 1990 (Baffier & Girard 1992).

In many cases, the calcium carbonate layer is translucent and has enabled good conservation of prehistoric paintings. However, in Arcy-sur-Cure and other caves, opaque calcium carbonate layers, formed during the Holocene period, hamper the visibility of the paintings and can endanger their conservation. The thickness of this white external layer ranges from a few hundred micrometres to a few millimetres. The restoration of part of the cave paintings has been carried out by removing the opaque white layer by mechanical abrasion while keeping the translucent layer, in order to show and to study the paintings (Girard *et al.* 2001). New figures have been brought to light during the restoration (Baffier & Girard 1998).

The formation conditions of different calcium carbonate types in this cave are not well understood. Indeed, the calcium carbonate formation depends on many different parameters such as climatic changes, anthropogenic factors and change of the vegetative cover of the outside of the cave (Spötl

From: PARISE, M. & GUNN, J. (eds) *Natural and Anthropogenic Hazards in Karst Areas: Recognition, Analysis and Mitigation*. Geological Society, London, Special Publications, **279**, 185–197. DOI: 10.1144/SP279.15
0305-8719/07/$15.00

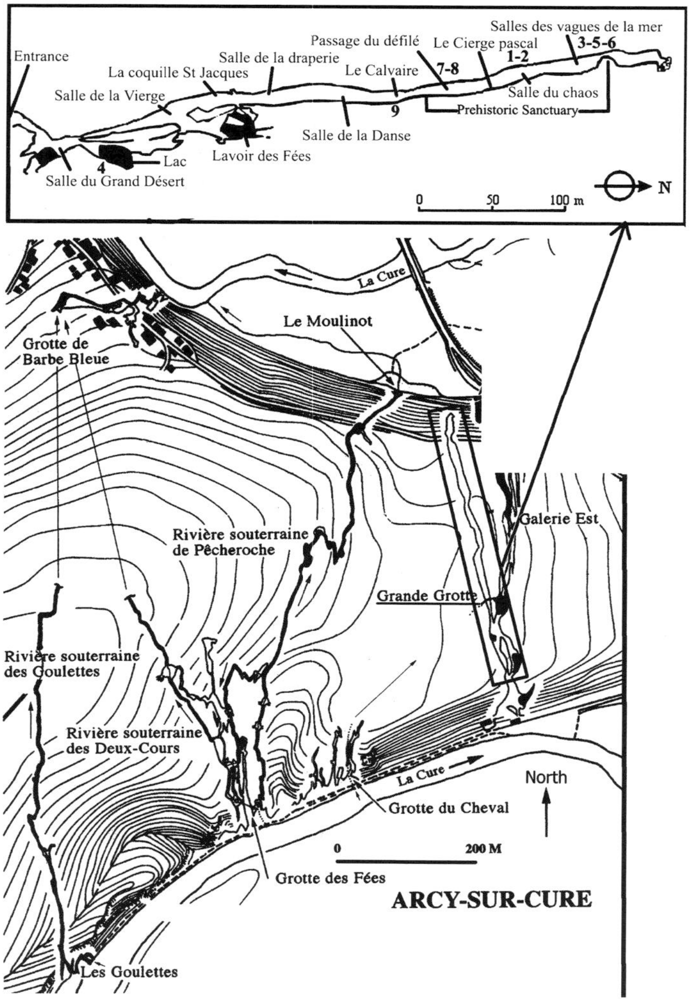

Fig. 1. Localization of samples coming from the 'Grande Grotte' cave at Arcy-sur-Cure.

et al. 2005). Calcium carbonate exists in three polymorphic forms: calcite, aragonite and vaterite (a metastable form). Moreover, the formation of calcium carbonate minerals could arise by biotic (biomineralization) or purely abiotic processes (dissolution–recrystallization). Concerning the biogenic calcium carbonate formation, a variety of bacteria have been shown to be involved, namely *Pseudomonas fluorescence* (Appanna *et al.* 1997), *Bacillus megatherium* (Adolphe *et al.* 1994), *cyanobacteria* (Schultze-Lam *et al.* 1992; Dittrich *et al.* 2003; Dittrich & Obst 2004), *Bacillus pasteurii* (Fujita *et al.* 2004) and *Myxococcus xanthus* (Ben Chekroun *et al.* 2004). Biomineralization processes are divided into two fundamentally different groups based upon their degree of biological control: 'biologically induced' mineralization and 'biologically controlled' mineralization (Weiner & Addadi 1997). Castanier *et al.* (1999) proposed that in karstic environment, the formation of calcite can be induced by some bacteria (*Isosphaera* and *Vitreoscilla*). They locally modify the physico-chemical conditions (e.g. the bicarbonate concentration) and so favour nucleation and crystal growth. The abiotic formation of calcium carbonate depends on several conditions (Ca concentration and alkalinity, pCO_2, pH, temperature, etc.), which in turn affect both nucleation and crystal growth of calcium carbonate. In karstic environment, speleothems are formed in caves by CO_2

degassing out of a Ca- and HCO_3^--rich water (White 2003).

Calcium carbonate samples from the wall, from lake sediment and from film (floating at the surface of the lake like a raft) in the 'large cave' at Arcy-sur-Cure were studied using a multidisciplinary approach including microscopic, spectroscopic, microbiological and molecular biology methods in order to assess the origin and the possible biogeochemical formation mechanisms of the various calcium carbonates present at the surface of Palaeolithic rock art.

Experimental methods

Sample description

Four types of calcium carbonate present inside the cave (translucent, opaque, floating and soft, NC#1–NC#7) were sampled in sterile conditions in most cases. The localization and the description of these calcium carbonates are presented in Table 1.

The position of the translucent and opaque calcium carbonates on the wall or on stalactites (NC#5) is depicted in Figure 2. This figure shows from inside outwards a columnar calcite, a red paint layer (3), a translucent layer (2) and a 'milky' or opaque layer (1).

The opaque powdered sample from the wall (NC#2) is compared with the other forms of calcium carbonate present inside the cave, floating calcium carbonate at the surface pool (NC#7) and soft calcium carbonate sediment from the lake (NC#4). Pool water (W#8) and water dropping from the ceiling (W#9) were also sampled to evaluate their bacterial activity and to compare their chemical composition with that of the floating calcium carbonate (NC#7).

Optical and electron microscopic methods

Thin sections of stalactite samples were prepared to facilitate their observation by optical microscope under polarized light (OMPL). Some other fragments of the stalactite were embedded in an epoxy resin to prepare polished cross-sections for optical microscopy (OM) and scanning electron microscopy (SEM) observation. For SEM, the samples were coated with carbon.

First, stalactite cross-sections were observed under the OM and the prepared thin sections with the OMPL. Powdered samples and cross-sections were also observed by SEM to analyse the crystallinity of $CaCO_3$ and the various layers of $CaCO_3$ as well as the paint layers. The observations were carried out on a Philips XL 30 CP instrument (operating voltage 20 kV).

Some sample fractions showing a small crystallinity were also characterized by transmission electron microscopy (TEM). TEM observations were carried out on the Jeol 3000 FX instrument (operating voltage 300 kV). Samples were first ground in an agate mortar. Then, the powder was dispersed ultrasonically in ethanol. Finally, a droplet of that mixture was deposited onto a carbon-coated gold grid.

Structural characterization methods

The powdered samples were studied without conditioning by Raman micro-spectroscopy and by X-ray diffraction (XRD). However, Fourier transform infrared spectroscopy (FTIR) requires specific sample preparation. For FTIR studies, KBr pellets of 2 mm in diameter were prepared by mixing sample powder with anhydrous KBr in proportion 1:100 in mass. FTIR was carried out in transmission mode on KBr pellets with a Perkin Elmer Spectrum 2000 spectrometer in a region ranging from 400 to 4000 cm^{-1} with a resolution of 4 cm^{-1}.

The XRD measurements were carried out on a Siemens Krystalloflex 810 diffractometer (experimental conditions: $\theta - 2\theta$, 5°–80°/0.02°/12 s/step, Co K_α radiation). The sample powders were separated by passing through a sieve under 50 µm. Identification of diffraction patterns was made by comparison with the JCPDS database.

Table 1. *Name and localization of calcite samples from the 'Grande Grotte' cave at Arcy-sur-Cure*

No.	Description	Localization	Sterile?
NC#1	Powdered opaque white calcium carbonate	'Chaos' Room	Yes
NC#2	Powdered opaque grey calcium carbonate	'Chaos' Room	Yes
NC#3	Stalactite	'Salle des Vagues' ceiling	No
NC#4	Soft calcium carbonate	'Lac' sediment	Yes
NC#5	Stalactite with paint layer	'Salle des Vagues' ceiling	No
NC#6	Translucent calcium carbonate sampled on the stalactite NC#5	'Salle des Vagues' ceiling	No
NC#7	Floating calcium carbonate	Pool in the 'Passage du défilé'	Yes
W#8	Water	Pool in the 'Passage du défilé'	Yes
W#9	Streaming water	'Calvaire' stalactites	No

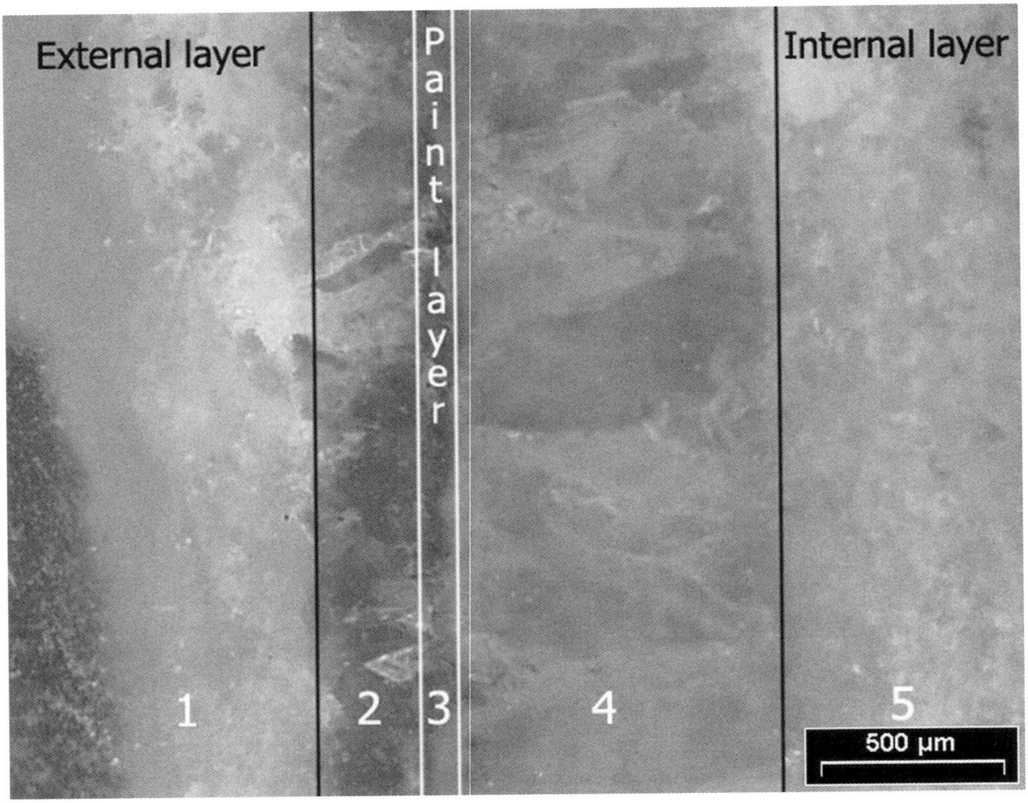

Fig. 2. Polished cross-section of the stalactite (NC#5) observed by optical microscope presenting a stratigraphy of layers of various calcite types: opaque and white calcite in layers 1 (external) and 5 (internal), translucent calcite in layers 2 and 4. Layer 3 represents a paint layer with aluminosilicates and iron oxides.

Raman spectra were obtained with a LABRAM (ISA-Jobin Yvon) spectrometer using a He–Ne laser at 632.8 nm. The experimental conditions are a laser power of 9.7 mW and an objective 80 ULWD with lateral resolution of 1.2 μm^2. Energy resolution of spectra is of 2 cm^{-1}.

Elemental composition

Cation analyses on water samples were carried out by inductively coupled plasma atomic emission spectrometry (ICP-AES), Perkin-Elmer, Optima 3300DV. The natural water was conserved with some drops of nitric acid.

Element concentrations on powder and on stalactite cross-sections were measured for major, minor and trace elements using particle-induced X-ray emission (micro-PIXE) at the external microbeam of the 2 MV tandem accelerator AGLAE facility installed at the C2RMF. The experimental conditions were 3 MeV protons in a He atmosphere, a beam diameter of 30 μm and 2 Si(Li) detectors: a low-energy detector for matrix elements and a high-energy detector for trace element analysis equipped with a 50 μm Al filter. Concentration profiles with a step size of 40 μm were also measured on the stalactite cross-sections under the same conditions (Calligaro *et al.* 2004). Spectrum analysis was performed with the Gupix package (Maxwell *et al.* 1988) assuming all elements contained in the spectrum exist as oxides except Cl and Ca. The latter was considered as carbonate. The sum of all oxides and carbonate is equal to 100%. For trace element analysis, the Ca content is used to normalize the high- and low-energy spectrum. The error of the determined concentrations lies in the range of 5–10%.

Bacterial counting, isolation and identification

Cultivable bacteria were isolated from two different types of samples: floating calcium carbonate

(NC#7) and water coming from a pool inside the cave (W#8). Each sample was duplicated. Solid samples were crushed in a mortar with a pestle and suspended in Tween 80 surfactant (0.5%). Aliquots (1 ml) of the suspension were plated on PCA (Plate Count Agar: peptone 5.20 g, yeast extract 2.5 g, D-glucose 1 g, agar 14 g, per litre of distilled water) and incubated at 30 °C for 3 days. CFU (Colony Forming Units) were scored and calculated per milligramme of sample. Individual colonies were then selected by the morphology, isolated by repeated streaking on PCA (three times). The API Tests (Biomérieux, Marcy l'Etoile, France) were used to identify each strain. To establish the bacterial strains phylotype, the complete 16S DNA was amplified using the 16Sc and 16Sd primers (Amann *et al.* 1995). The polymerase chain reaction (PCR, 3 min at 94 °C followed by 30 cycles of min 94 °C, 2 min 50 °C and 2 min at 72 °C, and 10 min at 72 °C) was carried out on a PE 2400 thermocycler (Perkin-Elmer, Cortaboeuf, France). The PCR products were purified with the QiaQuick PCR Purification Kit (Qiagen, Cortaboeuf, France), according to the manufacturer's instructions. Sequencing was performed, on the 5′ end of both strands, using the ABI PRISM Dye Terminator Cycle Sequencing ready Reaction Kit (Perkin Elmer) in a 20 μl volume containing 35 ng of purified DNA and 2 μM of primer. Sequencing reactions were performed for 25 cycles of 30 s at 96 °C, 30 s at 50 °C and 4 min at 60 °C. The excess dye terminators were removed by spin-column purification. Sequencing reaction products were separated by electrophoresis for 3 h on a ABI PRISM 377 DNA sequencer (Perkin-Elmer) in a 5% Long Ranger gel (FMC). The sequencing reactions were corrected using SeqScape 2. The phylotype was deduced using BLAST (Altschul *et al.* 1990) and the Ribosomal Database Project (Cole *et al.* 2005).

Molecular microbiological analysis

DNA from cave samples were extracted from floating calcium carbonate (NC#7) or pool water (W#8) using the Power Soil DNA (Carlsbad, USA) extraction kit according to manufacturer's recommendations. Three independent extractions were performed per sample. For Single Strand Conformational Polymorphism (SSCP), PCR was performed using the universal primers W49 and W104 (FAM) (Duthoit *et al.* 2003). The PCR mixture (25 μl) contained 50 μM of each dNTP, template DNA (0.5–5 ng), Isis® polymerase 0.5 U, 0.4 μM of each primer in the buffer supplied. The amplification parameters were 2 min at 94 °C, 30 cycles of 15 s at 94 °C, 15 s at 56 °C, 15 s at 72 °C, ending and a final step of 7 min at 72 °C with 10 min at 72 °C. The PCR products were separated by electrophoresis in a 1.5% agarose gel to check the presence of the amplified band of 200 bp. The diluted PCR product was mixed with Formamide, NaOH and molecular size markers (GeneScan 400 Rox, Perkin-Elmer), denatured at 94 °C during 2 min, chilled on ice and submitted to capillary electrophoresis for 30 min on 5% of CAP gel (Perkin-Elmer) in an automated sequencer ABI 377 (Perkin-Elmer). The collected electrophoresis patterns were first corrected with GeneScan Analysis® 3.7 (Perkin-Elmer). The readable range was established between 50 and 400 bp.

Results

Structural characterization

The XRD patterns obtained for all powder samples (NC#1, NC#2, NC#4, NC#6 and NC#7) indicate the presence of very pure and well-crystallized calcite (Fig. 3). No other polymorphic calcium carbonate could be detected and no distinction between the various calcites – except some changes in intensity ratio owing to preferential orientation in particular for floating calcite (NC#7) – could be made.

The FTIR spectra of the powdered samples depict the large carbonate group band at 1420 cm^{-1} corresponding to the C=O stretching, and two fine bands at 800 and 700 cm^{-1} corresponding to OCO group vibrations. A shift to 1431 cm^{-1} of the C=O stretching band is observed for the white and grey opaque calcite. This phenomenon seems to be linked to the size effect. The FTIR did not reveal the presence of organic matter contained in the calcite samples.

The Raman spectra confirm that all powder samples (NC#1, NC#2, NC#4, NC#6 and NC#7) are characterized as very pure and well-crystallized calcite, except for the soft calcite that presents a small quantity of sulphates (line at 1018 cm^{-1}) at some calcite surfaces. The great purity of calcite appears as an isotopic peak at 1065 cm^{-1} (Fig. 4a), corresponding to the symmetric elongation mode of the carbonate group (ν_1). The sample NC#7 of floating calcite shows a textural orientation, as already observed by XRD, evidenced by a change in peak intensity as a function of the analysed face of the crystal.

The Raman spectrum of the white layer of the cross-section (NC#5) presents an enlargement of the base line in comparison with the translucent calcite, but no specific peak ratio (Fig. 4). In contrast, the translucent layer shows a change in peak intensity linked to a specific crystal orientation (most intense peak at 712 cm^{-1}). So, the variation of peak intensity is in relation with the presence

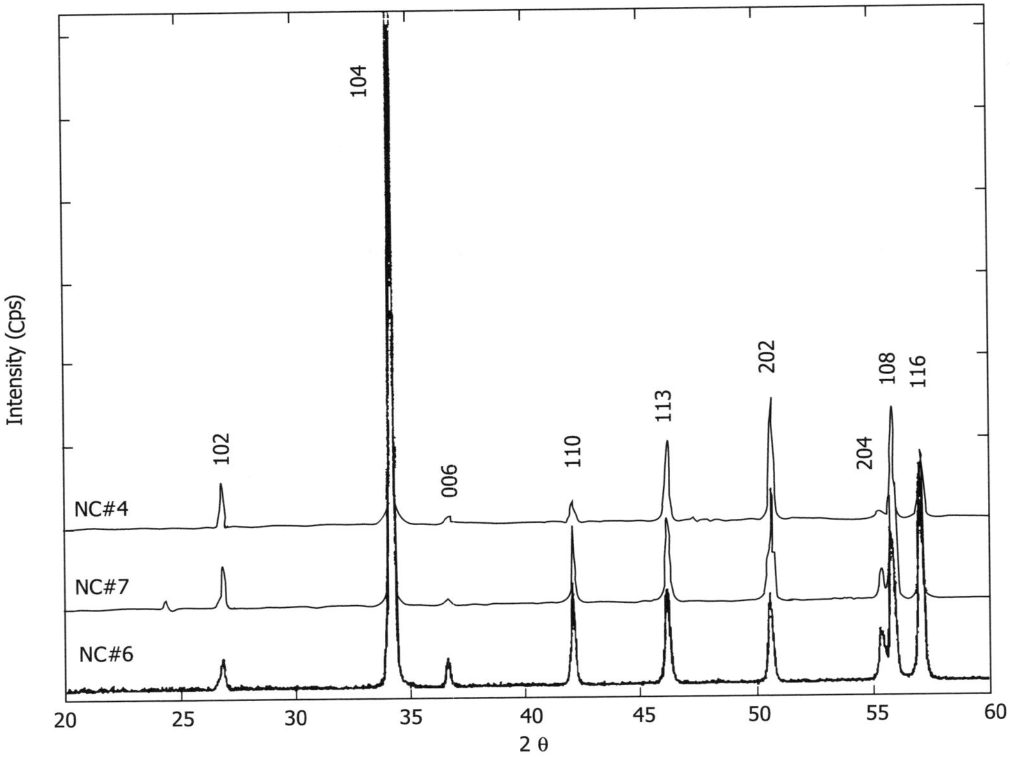

Fig. 3. DRX patterns of the translucent calcite (NC#6), the floating calcite (NC#7) and the soft calcite (NC#4). 101, indexation of calcite reflections.

or absence of a preferential calcite crystal orientation. The Raman spectrum of the red layer in the cross-sections of the stalactite sample NC#5 (Fig. 2) also demonstrates the presence of iron oxides (hematite) (Fig. 4).

Morphological characterization of the different types of calcite

OMPL observation of the stalactite thin sections reveals an important difference in grain size among the different calcite layers. The translucent calcite is formed of very large crystals (>500 μm) and is orientated around the stalactite central axis for the first layer ('Soda straw') and perpendicular to this axis for the external layer (NC#3). It seems that the crystal growth of translucent calcite (sparitic calcite) is not perturbed by the presence of the 20 μm-thick iron oxide layer as evidence on NC#5 (Fig. 5).

In contrast, the intercalated white or 'milky' calcite is generally microcrystalline (micritic) with some nanometric crystals present, as can be seen on the TEM micrograph (Fig. 6a). The grey calcite presents the same nano-morphology.

The calcite crystals formed in contact with water (in sediment or at the lake water–atmosphere interface) are very different. They have rhombohedral shape and a size ranging from 10 to 20 μm as observed by SEM (Fig. 6b).

Chemical analysis

The quantity of Ca in the calcite samples is less than 40 wt% (nominal value). These samples contain various other cations such as Mg, Zn, Na, Sr, Mn and Fe, which could be absorbed onto or included in the calcite structure. Globally, the total amount of foreign cations does not exceed 1300 ppm, except for the sediment calcite (NC#4) with more than 1 wt% foreign cations, mainly containing Mn (260 ppm) and Na (8500 ppm).

Figure 7 shows the Al, Si and Fe content determined by micro-PIXE on powder calcite (NC#1, NC#2, NC#4, NC#6 and NC#7). We can distinguish two types of calcite samples as a function of the amount of Si, Al and Fe: on one side, the wall

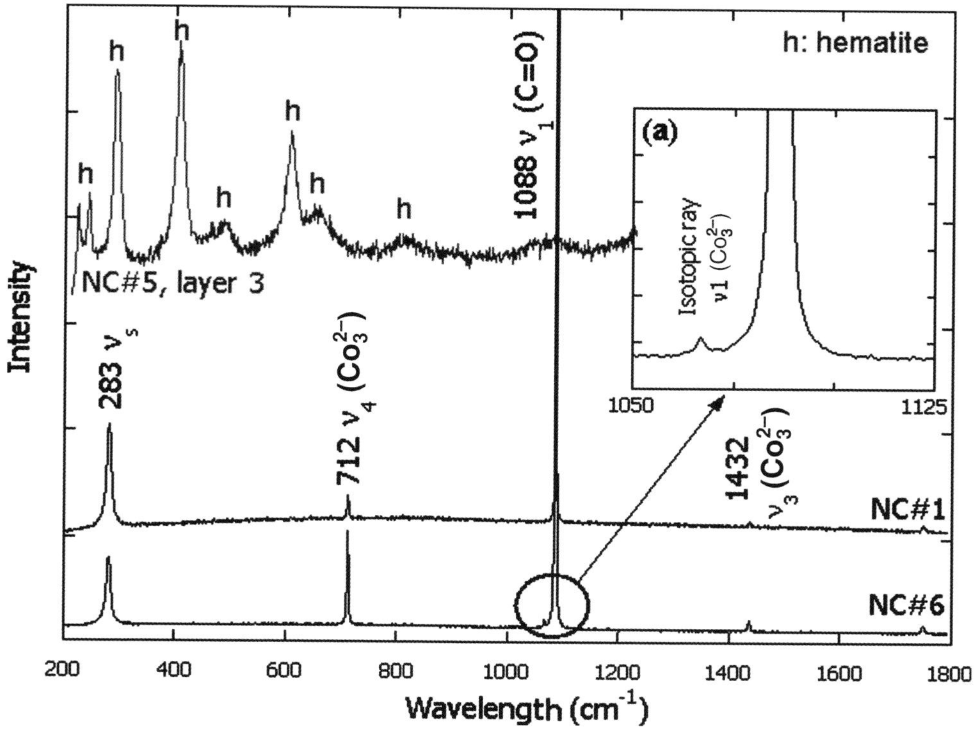

Fig. 4. Raman spectra of translucent (NC#6), opaque (NC#1) calcite and the red painting layer 3 (NC#5). (**a**) Inset of an isotopic peak, ν_1 (CO_3^{2-}) at 1065 cm^{-1}.

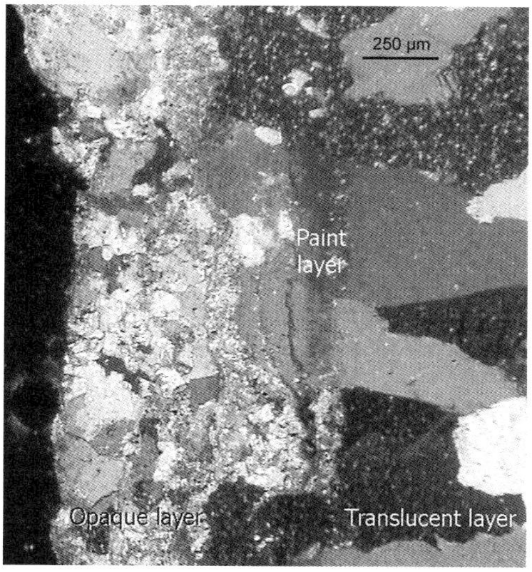

Fig. 5. Thin section of the NC#5 (stalactite) observed by OMPL showing the external opaque calcite layer (small crystals of micritic calcite), the thin paint layer (20 μm thick) and the internal translucent calcite layer (large crystals of sparitic calcite).

calcite (NC#1, NC#2 and NC#6), very pure containing small amounts of impurities; and on the other side, the calcite in contact with lake water (NC#4 and NC#7), characterized by the presence of large amounts of aluminosilicates (over 1500 ppm of Al and of Si).

Cave water samples show a large concentration of Ca (about 2.5×10^{-3} mol l^{-1} for the lake and for the streaming water) that corresponds to favourable conditions for calcite precipitation (ICP-AES analysis). This water is also enriched in Si (about 3200 ppm) and in chloride (about 0.51×10^{-3} mol l^{-1} for the 'Lavoir des Fées', 0.28×10^{-3} mol l^{-1} for the 'Lac' and 0.20×10^{-3} mol l^{-1} for the streaming water), sulphate (less than 0.21×10^{-3} mol l^{-1}) and nitrate (about 0.53×10^{-3} mol l^{-1} for the 'Lavoir des Fées' and 1.68×10^{-3} mol l^{-1} for the 'Lac'). The streaming water is more enriched in nitrate (almost 0.11 mol l^{-1}) suggesting pollution by fertilizers owing to extensive cereal cultivation on the plateau above the cave.

The micro-PIXE concentration profiles of Ca, Al, Si, Fe, Na, Mg, Cl, K, Ti and Sr obtained on the stalactite cross-section NC#5 (Fig. 8)

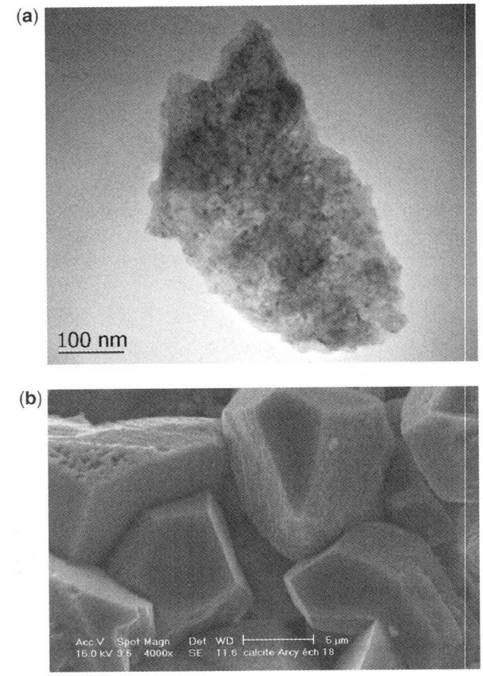

Fig. 6. Various morphologies of calcite: (**a**) TEM micrograph showing the polycrystallinity of the opaque calcite crystal (NC#1); and (**b**) SEM micrograph showing large rhombohedral crystals of floating calcite (NC#7).

demonstrate the presence of an iron-enriched aluminosilicate layer (layer 3) corresponding to the red layer. We can assume that this paint layer is composed of a mixture of clay and red iron oxides. Only slight chemical changes between different calcite layers are measured, except directly in the paint layer and in the outer layers enriched in Mg and Na.

Microbiological studies

The strategy to study the bacterial content of the samples consisted of a bacterial culture and analysis of the DNA isolated from the sample. On one hand, culture of the pool water and floating calcite revealed the presence of a significant amount of bacteria (more than 3×10^3 CFU mg^{-1}) and resulted in the isolation of many bacterial species (such as *Pseudomonas fluorescence*, *Bacillus* sp. identified by the API gallery). The DNA coding for the 16SRNA was amplified by PCR and sequenced. Comparison of the sequences with a database allowed us to identify the closest phylogenic strains (Table 2). On the other hand, the molecular approach (SSCP) gives a hint on the phylogenetic structure of bacterial populations. The DNA isolated from the cave samples served as template for the amplification of a low conserved region of the 16S DNA. The mixture was resolved by capillary electrophoresis. The peak corresponding to the isolated bacteria was also determined in calibration experiments (Table 2). As judged by

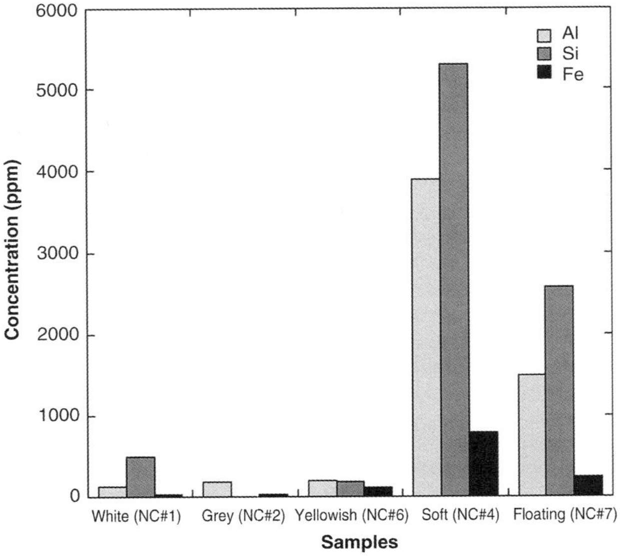

Fig. 7. Elemental composition of various calcites (NC#1, NC#2, NC#6, NC#4 and NC#7) in Al, Si and Fe determined by 3 MeV micro-PIXE.

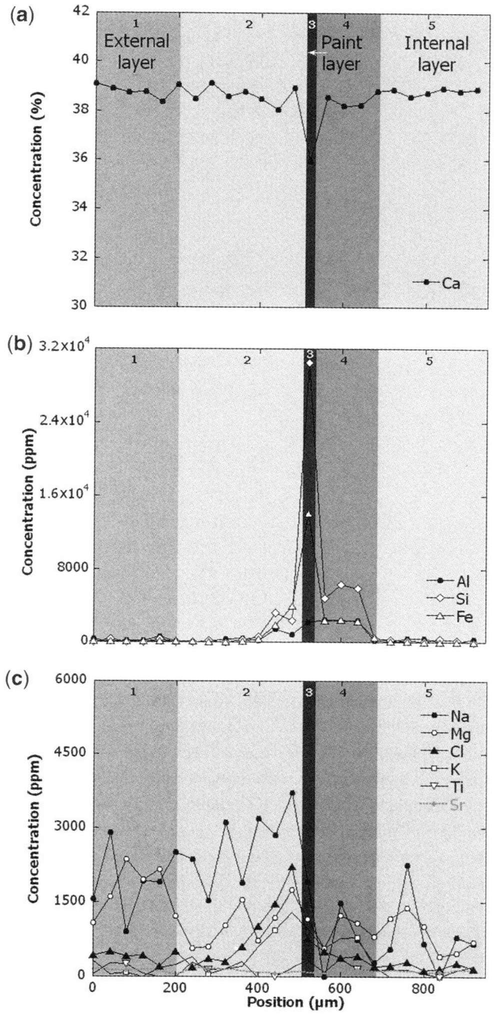

Fig. 8. Profiles of concentration obtained using micro-PIXE on the stalactite cross-section (NC#5). Experimental conditions: beam of 3 MeV protons with 40 µm diameter, 40 µm step.

the number of peaks and the fluorescence intensity (FI, Fig. 9), the bacterial diversity of floating calcite depends on the sample, most probably owing to spatial heterogeneity. This heterogeneity is striking when studying the bacterial diversity of different water samples. It is worth noting that the DNA isolated from the non-sterile sample NC#2 renders only minor peaks. This result indicates that the background of DNA owing to the sampling procedure is very low, supporting our studies. DNA co-migrating with *R. erythropolis* and *Pseudomonas* sp. 16SDNA was present in most of the

Table 2. *Bacterial strains isolated from floating calcite (NC#7) after amplification of 16S DNA by PCR and separation by SSCP*

Closest sequence (% identity)	t SSCP (pb)
Rhodococcus erythropolis (100)	191.67
Variovorax sp. (99)	214.99
Paenibacillus chitinolyticus (98)	218.53
Pseudomonas sp. (98)	219.45

samples at significant levels (Fig. 9), indicating that these isolated species are not contaminant. A peak related to *Variovorax* sp. was present only in floating calcite, but not in the wall calcite. Concerning the *Paenibacillus* sp., it was only found in floating calcite of a particular pool. Cultivated bacteria represent only a minor part of the bacterial species present in the samples. Indeed, bacteria corresponding to non-shadowed peaks in Figure 9 were not isolated. In this regard, it is interesting to note that bacteria corresponding to peaks I, II and III may be related to each type of calcites. While peak I and II are present in water and grey calcite; peak III is only present in white calcite. Thus, the later bacteria may interfere in the crystallization process of the calcite.

Discussion

Characteristics of calcite types in the cave and indications of the formation mechanisms

Seven natural calcite samples of the 'Grande Grotte' cave at Arcy-sur-Cure were studied using elemental, structural, morphological and biological analyses. A better understanding of the formation mechanism is now possible thanks to a multidisciplinary approach. Powder samples and stalactite cross-sections were investigated. The characterization of these calcite samples and the comparison with other calcite grown at the water–air interface on lakes present within the cave enable us to differentiate various types of calcites. The prehistoric wall paintings (24 500–28 000 BP) are coated by two of these distinct calcite types. We could distinguish translucent yellowish, opaque white and opaque grey calcite (also called 'milky' calcite). The different calcites found on the paintings reveal changes of the environmental cave parameters over time.

The translucent calcite is characterized by large crystals (over 150 µm, up to 1 mm in diameter) containing very few impurities. Its structure is highly textured (with specific orientation). This morphology and structure can be linked to a

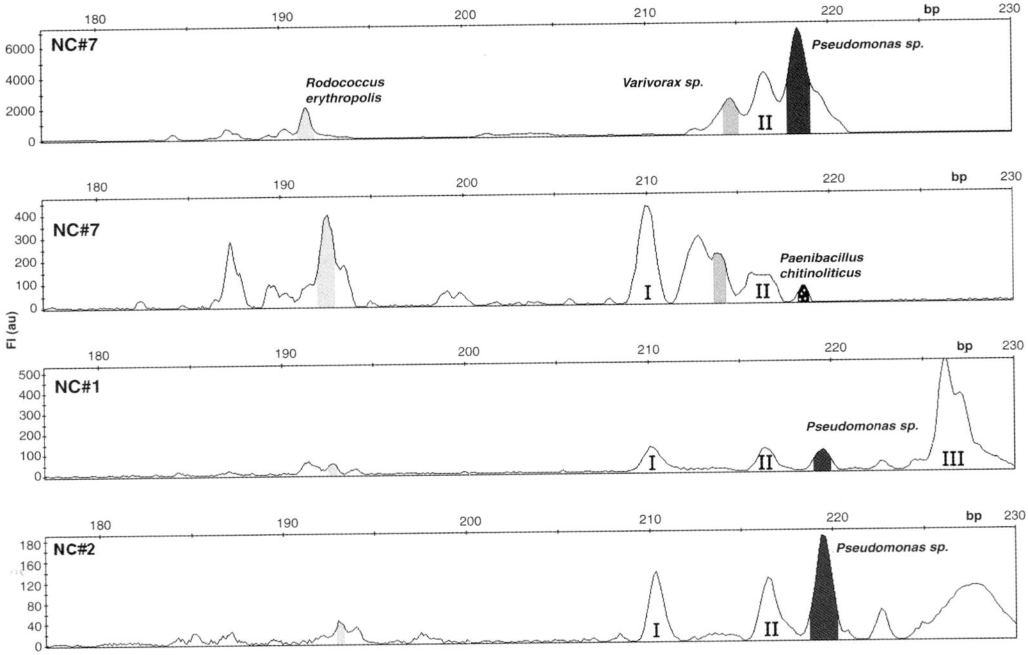

Fig. 9. Evaluation of bacterial diversity by SSCP. Peaks representing isolated bacteria are shaded. Grey-shaded peaks correspond to the migration of 16S DNA of the isolated strains.

formation by a slow crystallization process of CO_2 degassing from the water percolating over the cave walls in a poorly concentrated Ca medium. Indeed, the precipitation rate is strongly related to the level of supersaturation (Zhong & Mucci 1993). Constant relative humidity, pCO_2 and impurity water content favour the formation of the large oriented crystals found in the translucent layer (Spötl *et al.* 2005). Further aerological and environmental parameters also have to be taken into account during the crystal growth (Y. Perrette pers. comm.) but the biotic origin of the calcite seems to be negligible.

The white opaque calcite layer present at the external surface of the wall layers contains small randomly orientated crystal sizes (<10 μm). This 'milky' calcite has specific Raman and XRD spectra. The enlargement of the peak base for both white and grey 'milky' calcite can be linked to the small crystal size. The grey and white calcites have the same morphological and structural characteristics as those obtained by a rapid crystallization process in an oversaturated medium or in the presence of an inhibitor compound such as phosphate compound (Lin *et al.* 2002). However, extraction of bacterial DNA shows distinct DNA signatures between the grey and white 'milky' calcites. We have found that the *Pseudomonas* sp. is present in both the white and grey calcite, while a significant amount of *R. erythropolis* is only present in white calcite. Although *Rhodococus* strains were not reported to be involved in calcite formation, they may play an indirect role in calcite formation. As phylotypes of peak III are only present in white calcite, they may also play an important role in calcite formation. Further studies including DNA cloning and sequencing are necessary to identify these phylotypes and their implication on the calcite formation. Finally, strains present in the lake water and in the floating calcite seem to have grown recently in the karst. Thus, the chemical and nutrient modifications in the percolating water may influence both the bacterial species and the growth rate. All these factors can contribute indirectly to the formation of calcite.

Thanks to these various characteristics, the following hypotheses concerning the formation mechanism of the 'milky' calcite can be established.

- If one considers only a purely abiotic precipitation mechanism, the variation of texture among layers of calcite may be linked to hydrological conditions and variations in CO_2 pressure. The formation of the opaque calcite could be due to a short cycle of moistening–drying out of the wall, which involves more nucleation sites and fast precipitation of calcite during the moistening disrupted by the drying out (White 2003). This could be a result of forest clearance that occurred sometimes before the 17th century). Once the fields are under cultivation (as vineyards), the soil water retention capacity decreases and overflow increases, and the faults can dry out between two water saturated seasons.

- However, the presence of the bacterial DNA is a potential indicator of the biotic origin. Bacteria contained in the calcite can have two distinct effects on the formation of calcite: on one hand, they can favour the nucleation of calcite; and, on the other, they can inhibit calcite crystal growth (Schultze-Lam *et al.* 1992).

Soft and floating calcites in contact with water were studied for comparison with wall calcite. Their distribution in crystal size is more homogeneous. We can suppose that the soft and floating calcite crystal growth is less influenced by seasonal parameters (saturation conditions, pCO_2, T, pH, humidity, etc.) than wall calcite crystal growth. From a chemical point of view, the presence of impurities (aluminoscilicates enriched in iron, sulphate, phosphates, etc.) is very specific for the sediment and floating calcite and, can be explained by the composition of water which is also rich in Si, sulphate and nitrate. We suppose that during crystal growth cations can be included inside the structure of calcite, or can be absorbed at the surface. Cations present as trace element in water (such as Fe^{2+}, Mn^{2+}, Zn^{2+}) are coprecipitated within calcite.

Impact of calcite growth on the prehistoric paint layer

The study of the stalactite section enables the presence of distinct calcite layers to be shown over one paint layer, composed by ochre (aluminosilicate enriched in iron) and considered by archaeologists as a trace of a prehistoric painting. The presence of the paint layer seems not to disturb the crystal growth of the underlying translucent calcite layer as evidenced on a thin section in the optical microscope with polarized light. The traces of painting seem instead to be protected by the translucent calcite. The presence of Al and Si in the neighbouring calcite layers suggests an interaction between the pigment and calcite. However, no real transport of colouring elements contained in the paint layer such as iron is evidenced. The paint layer is also a boundary in chemical composition between inside and outside calcite layers. Mg and Na are more concentrated in the outer layer, and Treble *et al.* (2003) has shown that Mg incorporation is sensitive to groundwater residence time controlling prior calcite precipitation and Na is sensitive to growth rate. So these chemical variations may already indicate the impact of anthropogenic activities such as agriculture development on the plateau above the cave.

Transition between the two kinds of calcite

The two types of wall calcite, translucent and opaque, have distinct specific crystal sizes and shapes. The stratigraphic study demonstrated a change in texture, size and orientation of calcite crystals. The transition between layers is clear and explained by a variation in crystallization mode, induced by a modification of growth parameters. This modification is linked to local climatic changes, influencing directly the kinetics of calcite precipitation (White 2003). According to Spötl *et al.* (2005), seasonally changing air flow results in pCO_2 variation and in different seasonal supersaturation conditions. The thickness of the water streaming film often influences crystal size (Frisia *et al.* 2000). So it could be important to better define the hydrological and atmospheric conditions in the cave and to follow their evolution with seasons, in order to fully explain the changes in crystal size between the translucent and the 'milky' calcite layers.

Another parameter important to consider that could modify the cave environment is the ground floor of the vegetative cover. Indeed, a direct link between outside activities with speleothem growth patterns has been reported by Perrette *et al.* (2000). Use of EPR spectroscopy and laser-induced fluorescence showed that the development of agriculture induce the incorporation of small amounts of organic complexes into the calcite. Concerning Arcy-sur-Cure, some information has been obtained from the Arcy-sur-Cure cave archives on recent land-use change: until the 17th century and up until the beginning of the 20th century, wine covered the plateau, the ground then lay fallow and then until the 1970s cereals have been cultivated. However, since we cannot date the moment at which calcite shifted from translucent to 'milky' pattern, and as we do not know when forest clearance first occurred, we cannot show a direct relationship between the changes of wall calcite layers and the variation of culture.

Conclusion

Various calcites samples of the 'Grande Grotte' cave at Arcy-sur-Cure have been studied by a multimethod approach including microscopic, spectroscopic and biological analysis to understand the formation of different calcite layers on the top of the Palaeolithic paintings (Magdalenian period). Different kinds of translucent, opaque, soft and floating calcites could be distinguished in the cave. The translucent calcite layer protecting the paintings is characterized by large crystals and could have formed predominantly by a slow inorganic process. The formation of an opaque or 'milky' layer obstructing the paintings could have been initiated by a rapid abiotic process, but could also have been biotic. Indeed, the micro-biological studies evidence the presence of two bacteria, *Pseudomonas* and *Bacillus* sp., known for their calcifying properties. Nevertheless, their presence is not sufficient to prove the biologically induced calcite formation. The change in crystal size and texture between these two calcite types can be explained by a variation in the calcite precipitation kinetics explained by changes of land use, of hydrology or other environmental parameters. It is necessary to date the beginning of the formation of the 'milky' layer and to link it with the date of the forest clearance in order to demonstrate more accurately the link between external parameter changes and changes in calcite crystallinity. The determination and identification of organic matter in each calcite layer by EPR and laser-induced fluorescence analysis is in progress. The floating and sediment calcites that have morphological, structural and chemical similarities can arise predominantly from abiotic process, favoured by the water lake parameters.

A part of the Palaeolithic paintings has been restored by removing the opaque layer. The translucent calcite stabilizes and protects the precious paintings, acting like a varnish. Future work will focus on a gentle way of preserving the cleaned paintings and to prevent the opaque layer continuing to grow, for example by controlling the environmental conditions, in order to conserve the paintings for future generations.

We wish to thank M. Le Comte de La Varende (owner of the cave at Arcy-sur-Cure) who permitted the study and gave authorization for visiting and sampling in the cave. We are grateful to S. Joiret and the Laboratoire Interfaces et Systèmes Electrochimiques (LISE, UPR 15, CNRS) for the help during the Raman study, and to the Laboratoire de Thermodynamique et Physico-Chimie Métallurgique (LTPCM, CNRS/INPG, Grenoble, France) for the TEM access. We would like to thank M. Girard and E. Guillamet as well as the CORA association (J.-C. Liger, D. Molez and G. Souchet) for their help and their support on site. The thin sections of stalactite were kindly prepared by A. Leclaire (C2RMF). We also thank J. Gury (LECA) for the constructive discussion concerning the micro-organism activity. This project is supported by the National PRNC programme 2004–2006 financed by the French Ministry of Culture and Communication.

References

ADOLPHE, J. P., PARADAS, J. & SOLEILHAVOUP, F. 1994. An example of carbonate biosynthesis in karst. *In*: SASOWSKY, I. D. & PALMER, M. V. (eds) *Breakthroughs in Karst Geomicrobiology and Redox Geochemistry, Conference Proceedings. Colorado Spring, USA*. Karst Waters Institute, Leesburg, VA.

ALLEMAND, L. 2003. Qui sauvera Lascaux? *La Recherche*, **363**, 26–33.

ALTSCHUL, S. F., GISH, W., MILLER, W., MYERS, E. W. & LIPMAN, D. J. 1990. Basic local alignment search tool. *Journal of Molecular Biology*, **215**, 403–410.

AMANN, R. I., LUDWIG, W. & SCHLEIFER, K. H. 1995. Phylogenetic identification and in situ detection of individual microbial cells without cultivation. *Microbiological Review*, **59**, 143–169.

APPANNA, V. D., ANDERSON, S. L. & SKAKOON, T. 1997. Biogenesis of calcite: A biochemical model. *Microbiological Research*, **152**, 341–343.

BAFFIER, D. & GIRARD, M. 1992. Découverte de peintures paléolithiques à la Grande Grotte d'Arcy-sur-Cure (Yonne), France. *International Newsletter on Rock Art*, **2**, 2–3.

BAFFIER, D. & GIRARD, M. 1998. *Les cavernes d'Arcy-sur-Cure*. La maison des roches, *Terres Préhistoriques. Paris.*

BAFFIER, D., GUILLAMET, E., CHILLIDA, J., GIRARD, M., HARDY, M. & BRUNET, J. 1998. La Grande Grotte d'Arcy-sur-Cure. De nouvelles découvertes par amincissement de la calcite. *International Newsletter on Rock Art*, **21**, 28–29.

BEN CHEKROUN, K., RODRIGUEZ-NAVARRO, C., GONZALEZ-MUNOZ, M. T., ARIAS, J. M., CULTRONE, G. & RODRIGUEZ-GALLEGO, M. 2004. Precipitation and growth morphology of calcium carbonate induced by mixococcus Xanthus: implications for recognition of bacterial carbonates. *Journal of Sedimentary Research*, **74**, 868–876.

CALLIGARO, T., DRAN, J.-C., SALOMON, J. & WALTER, P. 2004. Review of accelerator gadgets for art and archaeology. *Nuclear Instruments and Methods in Physics B*, **226**, 29–37.

CASTANIER, S., LE MÉTAYER-LEVREL, G. & PERTHUISOT, J.-P. 1999. Ca-Carbonates precipitation in limestone genesis – the microgeobiologist point of view. *Sedimentary Geology*, **126**, 9–23.

COLE, J. R., CHAI, B. *ET AL*. 2005. The Ribosomal Database Project (RDP-II): sequences and tools for high-throughput rRNA analysis. *Nucleic Acids Research*, **33**, D294–D296 doi: 10.1093/nar/gki038.

DITTRICH, M. & OBST, M. 2004. Are picoplankton responsible for calcite precipitation in lakes? *Ambio*, **33**, 559–564.

DITTRICH, M., MÜLLER, B., MAVROCORDATOS, D. & WEHRLI, B. 2003. Induced calcite precipitation by cyanobacterium Synechococcus. *Acta Hydrochimica et Hydrobiologica*, **31**, 162–169.

DUTHOIT, F., GODON, J. J. & MONTEL, M. C. 2003. Bacterial community dynamics during production of registered designation of origin Salers cheese as evaluated by 16S rRNA gene single-strand conformation polymorphism analysis. *Applied and Environmental Microbiology*, **69**, 3840–3848.

FRISIA, S., BORSATO, A., FAIRCHILD, I. J. & MCDERMOTT, F. 2000. Calcite fabrics, growth mechanisms, and environments of formation in speleothems from the italian alps and Southwestern Ireland. *Journal of Sedimentary Research*, **70**, 1183–1196.

FUJITA, Y., REDDEN, G. D., INGRAM, J. C., CORTEZ, M. M., FERRIS, F. G. & SMITH, R. W. 2004. Strontium incorporation into calcite generated by bacterial ureolysis. *Geochimica et Cosmochimica Acta*, **68**, 3261–3270.

GIRARD, M., BAFFIER, D., BRUNET, J. & GUILLAMET, E. 2001. L'intervention directe sur les parois: un apport à la connaissance des tracés préhistoriques. Le cas de la Grande Grotte d'Arcy-sur-Cure (Yonne). *In*: *L'art avant l'histoire, la conservation de l'art préhistorique*. SFIIC, Paris, 197–207.

LIN, R.-Y., ZHANG, J.-Y. & ZHANG, P.-X. 2002. Nucleation and growth kinetics in synthesizing nanometer calcite. *Journal of Crystal Growth*, **245**, 309–320.

MAXWELL, J. A., CAMPBELL, J. L. & TEESDALE, W. J. 1988. The Guelph PIXE Software Package. *Nuclear Instruments and Methods in Physics Research B*, **43**, 218–230.

PERRETTE, Y., DELANNOY, J. J., BOLVIN, H., CORDONNIER, M., DESTOMBES, J. L., ZHILINSKAYA, E. A. & ABOUKAIS, A. 2000. Comparative study of a stalagmite sample by statigraphy, laser induced fluorescence spectroscopy, EPR spectrometry and reflectance imaging. *Chemical Geology*, **162**, 221–243.

SCHULTZE-LAM, S., HARAUZ, G. & BEVERIDGE, T. J. 1992. Participation of cyanobacterial S layer in fine-grain mineral formation. *Journal of Bacteriology*, **174**, 7971–7981.

SPÖTL, C., FAIRCHILD, I. J. & TOOTH, A. F. 2005. Cave air control geochemistry, Obir Caves (Austria): Implications for speleothem deposition in dynamically ventilated caves. *Geochimica et Cosmochimica Acta*, **69**, 2451–2468.

TREBLE, P., SHELLEY, J. M. G. & CHAPPELL, J. 2003. Comparison of high resolution sub-annual records of trace elements in a modern (1911–1992) speleothem with instrumental climate data from southwest Australia. *Earth and Planetary Science Letters*, **216**, 141–153.

WEINER, S. & ADDADI, L. 1997. Design strategies in mineralized biological materials. *Journal of Materials Chemistry*, **7**, 689–702.

WHITE, W. B. 2003. Paleoclimate records from speleothems in limestone caves. *In*: SASOWSKY, I. D. & MYLROIE, J. (eds) *Studies of Cave Sediment. Physical and Chemical Records of Paleoclimate*. Kluwer Academic/Plenum, New York, 135–175.

ZHONG, S. & MUCCI, A. 1993. Calcite precipitation in seawater using a constant addition technique: a new overall reaction kinetic expression. *Geochimica et Cosmochimica Acta*, **57**, 1409–1417.

Index

Page numbers in *italic* denote figures. Page numbers in **bold** denote tables.